T0189889

AGRICULTURAL DRONES

A Peaceful Pursuit

AGRICULTURAL DRONES

A Peaceful Pursuit

K. R. Krishna

Apple Academic Press Inc.
3333 Mistwell Crescent
Oakville, ON L6L 0A2
Canada

Apple Academic Press Inc.
9 Spinnaker Way
Waretown, NJ 08758
USA

© 2018 by Apple Academic Press, Inc.

First issued in paperback 2021

Exclusive worldwide distribution by CRC Press, a member of Taylor & Francis Group
No claim to original U.S. Government works

ISBN 13: 978-1-77-463642-8 (pbk)
ISBN 13: 978-1-77-188595-9 (hbk)

All rights reserved. No part of this work may be reprinted or reproduced or utilized in any form or by any electric, mechanical or other means, now known or hereafter invented, including photocopying and recording, or in any information storage or retrieval system, without permission in writing from the publisher or its distributor, except in the case of brief excerpts or quotations for use in reviews or critical articles.

This book contains information obtained from authentic and highly regarded sources. Reprinted material is quoted with permission and sources are indicated. Copyright for individual articles remains with the authors as indicated. A wide variety of references are listed. Reasonable efforts have been made to publish reliable data and information, but the authors, editors, and the publisher cannot assume responsibility for the validity of all materials or the consequences of their use. The authors, editors, and the publisher have attempted to trace the copyright holders of all material reproduced in this publication and apologize to copyright holders if permission to publish in this form has not been obtained. If any copyright material has not been acknowledged, please write and let us know so we may rectify in any future reprint.

Trademark Notice: Registered trademark of products or corporate names are used only for explanation and identification without intent to infringe.

Library and Archives Canada Cataloguing in Publication

Krishna, K. R. (Kowligi R.), author
Agricultural drones : a peaceful pursuit / K.R. Krishna, PhD.

Includes bibliographical references and index.
Issued in print and electronic formats.
ISBN 978-1-77188-595-9 (hardcover).—ISBN 978-1-315-19552-0 (PDF)

1. Aeronautics in agriculture. 2. Drone aircraft. 3. Agriculture--Remote sensing. I. Title.

| S494.5.A3K75 2017 | 631.3 | C2017-907439-3 | C2017-907440-7 |

Library of Congress Cataloging-in-Publication Data

Names: Krishna, K. R. (Kowligi R.), author.
Title: Agricultural drones : a peaceful pursuit / author: K. R. Krishna.
Description: Waretown, NJ : Apple Academic Press, 2017. | Includes bibliographical references and index.
Identifiers: LCCN 2017055544 (print) | LCCN 2017056541 (ebook) | ISBN 9781315195520 (ebook) | ISBN 9781771885959 (hardcover : alk. paper)
Subjects: LCSH: Aeronautics in agriculture. | Drone aircraft. | Agricultural innovations.
Classification: LCC S494.5.A3 (ebook) | LCC S494.5.A3 K75 2017 (print) | DDC 623.74/60243381--dc23
LC record available at https://lccn.loc.gov/2017055544

Apple Academic Press also publishes its books in a variety of electronic formats. Some content that appears in print may not be available in electronic format. For information about Apple Academic Press products, visit our website at **www.appleacademicpress.com** and the CRC Press website at **www.crcpress.com**

ABOUT THE AUTHOR

Dr. K. R Krishna has authored several books on International Agriculture, encompassing topics in agroecosystems, field crops, soil fertility and crop management, precision farming and soil microbiology. More recent titles deal with topics such agricultural robotics, drones and satellite guidance to improve soil fertility and crop productivity. This volume deals in detail about agricultural drones. Agricultural drones are set to reduce farm drudgery and revolutionize the global food generating systems.

He is a member of several professional organizations such as International Society for Precision Agriculture, the American Society of Agronomy, the Soil Science Society of America, the Ecological Society of America, the Indian Society of Agronomy, the Soil Science Society of India, etc.

CONTENTS

PREFACE

Agricultural drones are set to revolutionize global food generating systems. Agricultural drones are already flocking and hovering over farms situated in a few agrarian zones. Their usage is still rudimentary in many other regions, but drones are destined to engulf almost every cropping belt. They are set to offer a very wide range of services to farmers and reduce drudgery. Drones make crop production more efficient and economically advantageous. Drone's greatest advantage is in providing accurate data to farmers, which is actually picked from vantage locations above the crop. It was never possible for the past several millennia. The imagery and digital data that drone's sensors offer is simply not feasible using human scouts, particularly at that rapidity, accuracy and cost. Drones offer data about status of crop and perform tasks such as spraying liquid fertilizers and pesticides at relatively rapid pace, and they offer greater accuracy compared to farm workers. They effectively replace usual human skilled scout's and farm workers' drudgery in fields.

This book is about agricultural drones that are destined to reduce human drudgery to its lowest limits and yet offer better crop productivity, in any of the agrarian zones. These 'agricultural drones' are contraptions for 'peaceful pursuits'. They offer automation of farms, so that in future, fewer farmers will manage larger farms. They inspect large farms of over 10,000 ha in a matter few hours, which is beyond human capabilities. They are less costly, versatile and offer a wide range of services related to farm imagery, crop status, irrigation needs and pest attacks in farms. All these are achieved using sensors (cameras) that pick digital data and are attached electronically with computer stations, iPads and push buttons.

Drones are among the most recent gadgets to invade the agrarian regions of the world. They seem to spread into all the different agro-ecoregions of the world and dominate during accomplishment of a variety of agronomic procedures. Along with ground robots (e.g., GPS-guided autonomous planters, sprayers, weeders and combines) and regular satellite guidance, they could offer total automation of farm production procedures. Drones are not expected to clog the skyline above the farms/crops.

They are required to fly past crops at low altitudes perhaps a couple of times at each critical stage of the crop, to obtain ortho-images and/or spray chemicals. Drones are expected to amalgamate well into farmers' chores and get counted as yet another gadget such as tractors, sprayers or combine harvesters.

Agricultural drones are peaceful to the hilt. Drones do not disturb soil neither its biotic factors nor the physicochemical properties. They do not disturb the standing crop, except when copters fly close to crop canopy inducing leaves to flutter a bit. Drones do not touch the crop. They quietly analyse and collect data staying at a distance from the crops, using sensors. Drones operate above the crop in the atmosphere. At the same time, soil type or crop has least influence into their functioning or efficiency in terms of gathering digital information or spraying herbicides and other chemicals. Drones could be deployed in any agrarian zone, be it wet tropics, arid and dryland stretches, hilly mountain farming terrains, or flat/undulated prairies with cereal stretches. They are expected to help farm managers with unmatched accuracy and efficiency compared to other methods presently in vogue. On a different front, we are making much noise about a stray drifting drone that may be potentially used to spy neighbours. Regulations for usage of drones, in general, and those used in farms are being finalized in United States of America and other nations.

Drones are being employed to detect droughts, nutrient dearth in field, disease and pest attacks on variety of crops grown in agrarian regions. Drones are expected to throng the agricultural fields worldwide, rather very soon. 'Agricultural drones' and their operators have a great role to play in the protection of crops, crop belts and food security. Drones could become a worldwide phenomenon sooner than we anticipate. We have to note that global crop protection using drones to scout for diseases/pests periodically, as a routine, is a clear possibility. When this is followed by pesticide spray at variable-rates, it is directly related to food security and nourishment of billions of humans. Consistent with the theme of this book, 'Agricultural Drones' bestow peace on earth through better grain harvests worldwide. They scout cropping belts rapidly and work towards better distribution of fertilizers, pesticides and water in crop fields. Global crop production is expected to become much easier through the use of drones. Ultimately, drones could be a boon to human kind by allowing us to achieve higher grain harvests. Drones could minimize human drudgery in fields markedly.

Agricultural drones are at the threshold of spreading into every nook and corner of North American farming zones. Drones have already caught the imagination and secured a role in fertile farming belts found in Europe. They say, in major farming nations of Europe, such as Germany, France, Hungary and Poland, the agricultural sector is already 'ploughed-up' to receive drones in large number. Drones are expected to initiate a techno-logical revolution through automation of European farms. Drones have the potential to flourish in Asian farming belts as well. Here, we may have to be shrewd enough to consider drones and drone technology as an aspect not too closely connected with precision farming. The tendency to asso-ciate drone technology with precision farming, in general, has to be really weighed out well and done carefully. Drone technology applied indepen-dently has its ability to impart excellent advantages to farmers and agri-cultural researchers.

This book has ten chapters. An introductory chapter offers historical information about drones, their development and use in military, civilian surveillance, transportation and natural resource monitoring. It also lists and briefly describes various types of drones and their specifications. Further, a jist of our current knowledge about drones are provided. Chap-ters 2–9 offer greater details and discussions that concentrate on various aspects of natural resources and agricultural crop production. Agricultural drones are employed to obtain aerial imagery and accomplish a range of agronomic procedures in the crop fields. Drones are also used to detect drought, floods, soil erosion and crop stand. Chapters 3–7 deal with above aspects in detail and offer most recent information. Chapter 8 deals with one of the most important aspects of farming, namely yield forecast, using aerial imagery. Drones are used regularly to derive digital images of fields, crops, their growth, grain formation and maturity. Crop maladies, if any, are also imaged by drones. Digital images of crop growth and grain formation for several seasons are collated, layered and studied in conjunction with soil type, its fertility status, pest/disease incidence and drought incidence. Finally, computers with appropriate software are employed to decide agronomic procedures, plus, develop a grain yield forecast for the current season. Chapter 9 deals with economic aspects of agricultural drones. The cost of a drone unit, its operation, servicing and eventual fiscal advantages to farmers are also discussed. Farmers employing drones in their farms and other commercial settings have to observe certain rules and regula-tions. Firstly, they have to register the drone (vehicle) with appropriate governmental agency. Some of these topics are also stated in Chapter 9. Chapter 10 offers a summary of aspects discussed in the entire volume. It also lists some unique points about the way, drones have moved from exclusively military zones to agricultural crop fields and natural vegetation monitoring.

ACKNOWLEDGEMENTS

During the course of preparation of this volume on agricultural drones, several of my colleagues, friends and staff from agricultural research institutions, universities, and agricultural drone companies have been kind to offer information, research papers, and permissions to utilize photographs and figures. Several others have been a source of inspiration as I compiled the book. I wish to express my thanks to the following people, CEOs, and experts from drone companies, professors, and researchers:

- *Mr. Adam Najberg*, DJI Inc., Nanshan District, Shenzen, China and Department of Intellectual Property, DJI Inc., Shenzen, China.
- *Drs. Antonio Liska and Domenica Liska*, Directors, Robota LLC, Lancaster, Texas, USA.
- *Prof. Guido Morgenthal*, Ascending Technologies, Technologien in Bauwesen, Germany.
- *Dr. Hans-Peter Thamm*, Geo-Technic, Neustr. 40, 53545 Linz am Rhein, Germany.
- *Mrs. Lia Reich*, Director Communications, Precision Hawk Inc., Noblesville, Indiana, USA.
- *Mrs. Lincy Prasanna*, Venvili Unmanned Systems Inc. Cedar Rapids, Iowa, USA.
- *Dr. Mac McKee, M.*, Director, Utah Water Research Laboratory and Professor, Civil and Environmental Engineering, Utah State University, 8200 Old Main Hill, Logan, Utah, USA.
- *Mr. Matthias Beldzik*, Ascending Technologies GmbH, Konrad-Zuse-Bogen, 482152 Krailling, Germany.
- *Mr. Matt Wade*, Marcomms Manager; SenseFly Ltd—a Parrot Company, Cheasseux-Lausanne, Switzerland.
- *Dr. Michael Dunn*, Anez Consulting LLC., Little Falls, Minnesota, USA.
- *Dr. Mitchell Fiene*, DMZ Aerial Inc., 1501 Parkview Court, Prairie du Sac, Wisconsin, WI 53578 USA.
- *Dr. Sean Wagner*, Spatial Land Analysis LLC, Bakersfield, California, USA.

- *Mr. Tom Nicholson*, AgEagle Aerial Systems Inc., Neodesha, Kansas, KS 66757, USA.

I also wish to thank Dr. Uma Krishna, Mr. Sharath Kowligi, and Mrs. Roopashree Kowligi, and I offer my best wishes to Ms. Tara Kowligi.

ABOUT THE BOOK

Agricultural Drones: A Peaceful Pursuit is a treatise that deals with the role of aerial robots in managing natural resources and agricultural farms. It is currently a sought-after aspect within the realm of agriculture. Agricultural drones are billed to revolutionize the way we conduct agronomic procedures and maintain natural vegetation on earth.

Agricultural drones are a recent phenomenon that have been introduced into agrarian regions. They are expected to spread into crop lands of different continents. These are small unmanned aerial vehicles that are operated using remote control or flown above the crop fields using pre-programmed flight paths. They are fitted with a range of cameras that pickup images of fields, soil status, crop growth, and grain formation. The cameras obtain images from a distance of 100–400 ft. above the crop, using sensors at visual, near infrared, or infrared spectral bandwidths. Agricultural drones provide accurate images that provide details about seed germination, seedling establishment, crop growth, and maturity status. These drones are used efficiently to collect data about leaf chlorophyll and plant nitrogen status in order to decide on fertilizer dosages. Drones fitted with thermal sensors help in detecting water status of crops/soil and in prescribing irrigation at variable rates. Agricultural drones help in keeping vigil over crop fields, particularly in regard to disease and pest attacks, if any. They are destined to become most efficient in conducting plant protection procedures, such as spraying and dusting. These drones are gaining in acceptance in agricultural experiment stations since they offer spectral data related to a crop's performance. They are highly useful during plant breeding and genetic evaluation of elite genotypes. Drones fly very rapidly past the crop fields and obtain digital data, rather too slowly, like human scouts.

Agricultural drone companies (start-ups) are now a growing trend in North America and Europe. There are several models produced for farmers to use and reap better harvests at lowered cost. Agricultural drones replace human farm workers, and reduce use of fertilizers and plant protection chemicals. Agricultural drones are expected to improve economic advantages during crop production.

This book, *Agricultural Drones: A Peaceful Pursuit* has ten chapters. An introductory chapter provides historical data, details about various models of drones, most recent and popular agricultural drones in usage, and a glimpse about drones in farming. The other chapters deal specifically with topics such as drones in soil fertility, in production agronomy, in irrigation, in weed control, in pest and disease control, in grain yield forecasting, and about economic gains due to drones. The last chapter provides a summary.

Agricultural drones are really a recent topic. There are no detailed treatises on this topic. Hence, this book will be timely and useful to professors, agricultural extension officers, and students. Farmers and farm consultancy agencies will find the book useful to becoming conversant with recent developments about drones. This book is also informative to the general public. This book should serve as an excellent textbook for students in agriculture, engineering, geography, etc.

LIST OF ABBREVIATIONS

3D	three-dimensional
ADC	analog-to-digital converter
CEC	cation exchange capacity
CIMMYT	International Maize and Wheat Centre
CSM	crop surface models
CWSI	crop water stress index
DSLR	digital single-lens reflex
DSM	digital surface model
EC	electrical conductivity
FAA	Federal Aviation Agency
GAI	green area index
GIS	geographic information system
GM	genetically modified
GNDVI	green normalized difference vegetative index
GPS	Global Positioning System
HALE	high-altitude long endurance
HT	herbicide tolerant
IC	internal combustion
IR	infrared
LAI	leaf area index
MALE	medium-altitude long endurance
NASA	National Aeronautics and Space Agency
NBI	nitrogen balance index
NDVI	normalized difference vegetative index
NIR	near infrared
SAVI	soil adjusted vegetation index
SLR	single-lens reflex
SOM	soil organic matter
SPAD	soil plant analysis development
SSM	surface soil moisture
UAS	unmanned aerial systems
UAVs	unmanned aerial vehicles

USDA	United States Department of Agriculture
Vis	vegetative indices
VRAs	variable-rate applicators
VRT	variable-rate technology
VTOL	vertical take-off and landing

CHAPTER 1

DRONES IN AGRICULTURE

CONTENTS

1.1 INTRODUCTION

Agricultural drones are flocking and hovering over crop fields nowadays. They are expected to spread to every nook and corner of agrarian belts on earth. Drones offer aerial images from vantage points above a crop, 'a bird's eye view' but with great analytical detail and accuracy obtained using sophisticated visual, infrared (IR), near-infrared (NIR) and thermal sensors. They offer farmers with insights about their crops in such a detailed way that was never possible during the past several millennia. Drones allow us to study and compare soil types, assess influence of agronomic procedures and assess performance of several hundred crop genotypes with high accuracy. Drones collect and allow us to store a large posse of digital data about crops, soil and disease/pest attack that are required during precision farming. Drones offer economic advantages to farmers by reducing inputs and need for farm labour. They are quick in action, accurate, reduce farm drudgery to a large extent and make crop production efficient. Agricultural drones are a peaceful pursuit to the hilt, so much that they do not even touch or disturb the soil or the standing crop. Drones are totally non-destructive while analysing crops. At a stretch, drones stay in the atmosphere above the crop from few minutes to a maximum of couple of hours only, as required by farmers.

1.2   HISTORICAL ASPECTS OF DRONE TECHNOLOGY

First efficient use of drones was seen during military conflicts starting from early 20th century. For a long stretch of time, say a few decades, their usage was confined and stayed within the preserves of military engineering groups. Drones were initially developed to counter the enemy Zeppelins in World War I. It seems, earliest drone to be used in military warfare was developed in 1916. During the period between World Wars I and II, there were several modifications and improvements to drone technology (Nicole, 2015). It led to development of a range of models to suit different purposes related to military. Keane and Carr (2013) and Newcome (2004) have reviewed the historical aspects of drone technology. They suggest that efforts to develop drones for military use are at least 95–110 years old now. Actually, unmanned aerial vehicles (UAVs) have been around, for a duration much longer than most people imagine. Their development got initiated during World War I. They were known as aerial torpedoes or fling bombs. At present, drones are among the most dreaded military weapons in the West Asian conflict region.

Regarding data on development of drones in recent history, one of the lists suggests earliest use by British in the Mediterranean region. Drones were launched from an aircraft carrier named 'HMS Argus'. They were also used by Germans during the combat in 1944. In the same year, the United States of America used drones to bomb Japanese positions in Ballele islands (Arjomani, 2013).

Major advancements in drone technology that occurred in the United States of America during the 1950s and 1960s were represented by the use of *Ryan Firebees* series. During recent decades, UAVs engaged in warfare named Global Hawk, Predator and so forth are noteworthy. These are high-altitude drones of long flight endurance. They cover long distance in a day to seek the targets (Tetrault, 2014). Global Hawks are among the most efficient drones used by the United States of America in the Gulf War of 1991. It seems, until past decade, most of the drones used in surveillance and military zones were fixed-wing type. However, drones with rotary copters are also in vogue in many regions of the world. Reports suggest that 'RQ-*

Fire Scout' was among the earliest of drone copters to be used in missile launching and enemy scouting (Tetrault, 2014). It has been pointed out that characters such as excellent adaptability, safety and greater accuracy offered by copter drones are important during military actions. Such characters provide an edge over other types of war machinery. Report by Schwing (2007) suggests that drone technology went through a period of stagnation and lack of recognition until their consistent success in Vietnam. They have been utilized as most versatile and low-cost, effective military offence gadgets by Israelis since past two decades (Schwing, 2007). Drone technology for military and civilian usage was initiated in the United States of America in 1960. It was done under a program code named 'Red Wagon'. Reports suggest that drone usage in the U.S. military has ranged from apparent disinterest prior to 1980s to deployment for aggression, if need be, during surveillance and bombing enemy positions (Kennedy, 1998).

It seems even now, in 2015, the United States of America and Israel are the major users of drones for military purposes. Reports in 2013 suggested that over 52 nations are regularly using drones for variety of purposes (Gale Encyclopaedia of Espionage and Intelligence, 2014). They are used mainly to obtain high-resolution aerial imagery of enemy positions. They are equipped with computer decision systems and payload with bombs to destroy targets with great accuracy. Drones are efficient and quick in providing action and are highly economical as they cost less and their loss is not felt much. Fleets of drones could be manufactured fresh in a matter of days by ancillary industries that support military.

It is believed that there are several catalysts that have induced development and use of drones. In the general realm of public affairs and maintenance, it is the need for surveillance of events, towns, installations and natural resources that has given rise to constant demand for improvement in drone technology. Drones are essential in places that are dangerous and treacherous to human beings. For example, drones are required while handling and/or transporting dangerous chemicals. Most recent and supposedly most *prolific* reason is the agricultural uses of drones. Drones could potentially revolutionize crop production techniques. They are expected to reduce use of farm work force and human energy to, perhaps, the lowest levels that we can imagine (Gogarty and Robinson, 2012). Removing agricultural drudgery and making crop production more efficient is after all a priority concept since ages. Latest development in many of the developed nations is the formation of agencies that cater drone services. They cover a range of aspects, such as general aerial survey, aerial video monitoring and movie making, monitoring mines and industries, natural resource management, monitoring volcanoes and lava flow for example, in the Pacific North Western United States of America, aerial archaeology and local weather reports (Drone Services Hawaii, 2015). Such drone service facilities have sprung up in high numbers in the North American continent and Europe. We may note that unarmed drones have been effectively used to study the wild life, animal migration, and to guard the monuments within Egypt and other parts of North Africa (Gounden, 2013).

Drones were first utilized to study the weather pattern and follow thunderstorms/tornados by meteorologists of the United States of America during 1946. Since then, drones have been regularly used by the U. S. Meteorological Department to obtain weather data and gauge atmospheric processes. It is both interesting and useful to note that drones, particularly smaller versions that cost less, have been used to fly into clouds, cloud formations, storms and tornadoes. The aim is to collect data and study a range of parameters. This is actually done to analyse and accrue knowledge about factors that cause such weather patterns. For example, in the Central Plains of North America, Oklahoma, known as a tornado valley, drones have been sent in to detect and relay data about tornado's core, its periphery and impact on ground. Particularly, they intend to know effect of tornadoes on crops and farm infrastructure (Juozapavicius, 2013).

1.3   HISTORICAL ASPECTS OF AGRICULTURAL DRONES

Historically, agricultural world has been introduced with variety of implements, gadgets, automatic machines, new crop species, improved cultivars and methods to supply inputs such as soil nutrients and water or those that control diseases and pests. From ancient times, simplest and earliest of the implement to impinge the agrarian zones has been the wooden plough. It allowed farmers to grow crops on soil with better tilth. Plough hastened soil disturbance process and reduced weed infestation. Its use required a certain degree of drudgery but offered better crop stand and fairly systematic plant spacing. The spread of ploughing as an advanced technique must have taken a long stretch of time. This is attributable to constraints such as lack of interaction by humans situated in different regions/continents.

We should note that inability to travel frequently and easily was a major constraint to transmit new techniques during ancient period.

Later, during medieval period, a wide range of contraptions were devised to aid agrarian pursuit. Their spread was dependent on human migratory trends and ability to produce the new gadgets. During recent history, some of the most striking inventions to intrude the agrarian regions were the McCormick harvester in the mid-19th century, and later tractors with internal combustion (IC) engines that were energised by petrol. They reduced farmers' drudgery in fields. Sprayers and dusters during the early 20th century and combine harvesters in the mid-20th century were other machineries to impinge agrarian belts. These farm vehicles and gadgets revolutionised crop production procedures. As a consequence, large farms could be managed by farmers. Farmers could easily break limits stipulated by human physiological traits such as insufficient power and fatigue. Yet, human preoccupation with soil, water and crops supported well by farm drudgery were essential if farmers intended to operate large crop production units. Aspects, such as crop scouting, gathering accurate data about crop health, supplying fertilizers and water, spraying plant protection chemicals and so forth, all needed long stretches of farm labour. Further, availability of farm labour became a constraint in many regions.

As stated earlier, drones have been associated mostly with military pursuits such as reconnaissance and targeting enemy position with guided bombs. However, recent history about these aerial contraptions clearly shows that they have begun to play a peaceful role in global food generation. Approximately, since 2000–2005 A.D., agricultural drones have been evaluated and used to accomplish a range of different agronomic procedures in field and plantation locations. Actually, drones are among the latest gadgets to impinge the agrarian zones and their environments. They are relatively highly mobile, versatile and useful to farmers in several ways. They may also, in due course, affect several other aspects of agricultural world directly or indirectly.

In 1983, Yamaha Motor Company (Japan) launched the now popular RMAX copter drone. It seems it was actually designed and developed in response to a request by the Japanese Agricultural Department. They had requested Yamaha Company Ltd. to supply them with crop-dusting drone (Yamaha, 2014). First commercial use of Yamaha's R50 had begun by 1987. At present, RMAX is a popular agricultural drone in the farms of Far East. It is used to spray pesticides to rice crop.

Most recent reports suggest that use of field drones of small-to-medium size may gain in popularity, in the near future (Glen, 2015). They may be adopted to gather information about almost each and every plant that thrives in the field. They offer data good enough to be used during precision farming, that is, variable-rate techniques. A few types of such small drones are in testing stage, and a couple of them are still in the drawing board. However, there are also fantasies galore about the range of farm operations that could be accomplished by agricultural drones.

Agricultural drones made of low-cost wood and cameras attached to them have been in vogue, since past 5 years in Latin American nations such as Peru. Such low-cost drones are gaining in acceptance in smaller farms. They are relatively recent introductions into cropping zones. They provide excellent high-resolution imagery of crops such as potato and wheat (Cisneros, 2013).

Drone service agencies that cater to farmer's immediate needs are a very recent development within the realm of agriculture. Such commercial drone agencies began appearing in 2010. These drone agencies along with facility for satellite imagery are becoming invaluable, to large farms in Americas and Europe. They offer a range of services from aerial imagery of large cropping belts to a small management block in a farm. Drone companies, in particular, offer instantaneous images of crop growth and nutrient status, monitor diseases/pests and offer digital data to variable-rate applicators (Drone Services Hawaii, 2015; Homeland Surveillance and Electronics LLC, 2015).

Historically, drones are recent introductions into farming belts, say during 2007–2012. Dobberstein (2013) states that drones, having got a foothold in military, are now engulfing agrarian regions. They may soon spread out into farming belts all over the world. Drones are actually set to change the face of no-till systems. With their ability to rapidly scout the farms, they can provide clear images about weed infestation. Weeds are generally rampant during no-tillage farming systems. Drones could be conducting several of the major farm tasks such as three-dimensional (3D) imagery and scouting, spot spraying on pathogen-/pest-attacked patches, also spraying liquid fertilizers to supply in-season split dosages. Drones are getting popular because they also help farmers in decision-making.

There are many reports suggesting that drones are probably the most important next wave of agricultural technology. Drones may not entirely replace farm workers in some situations. For example, Ottos (2014) believes that drones do the primary imaging, rough sketching and show the farmers where to look for problems and where to apply amendments such as nutrients, herbicides, pesticides and so forth. At present (i.e. 2012–2015), industries producing drones for agricultural usage are being initiated and developed in large numbers.

History of drone introduction into farming stretches is no doubt very recent. Perhaps, they have become slightly conspicuous since past 5–10 years. This is despite absence of clear regulatory instruction for their commercial use in agriculture (Bowman, 2015). Drone usage is already in vogue on some crops. Soybean belt in the United States of America received drones in 2010 and since then they are getting common in the agriculture through their usage on soybean (United Soybean Board, 2014). Main purposes for which drones are employed on soybean crop are mapping the crop with high-resolution visual, NIR and IR cameras; crop scouting and monitoring work progress in soybean fields; assessing seedling emergence and crop stand so that replanting, if any, could be done efficiently. Drones are also used to obtain data on pest and disease, to fix crop-dusting schedules (United Soybean Board, 2014). The enthusiasm to introduce drones into agriculture has been increasing among farm vehicle producers and selling agents. Usually, a wide range of advantages are quoted for drones in farms, particularly, during precision farming. Polls and opinions among farm companies and farmers suggest that 75% believe that sky is the limit for drone-related advantages to farmers (Zemlicka, 2014). Drones have been called 'flying tractors' as in near future they are expected to throng farms, just like those other ground vehicles.

Drones are being tested in several different geographic locations, agrarian belts and on different crop species. The aim is to accomplish certain farm operations efficiently. For example, in Florida, drones are in use to monitor citrus groves for growth characters, weed infestation and to detect Huanglongbing disease since 2007 (Lee et al., 2008; Li et al., 2012; Garcia-Ruiz et al., 2013). Similarly, drones were introduced into the grape orchards of California in 2005 (Bailey, 2013).

1.3.1 AGRICULTURAL DRONES

It is now clear that, in addition to military, in due course, drones found their niche in different aspects of human endeavour. They were used to survey geologic sites, mines, natural reserves, public events and so forth. Drones were effectively used in highway patrolling and shadowing techniques. Drones found their way into agricultural regions only recently. Agriculture seems to support the most prolific use of drones to survey land, soil and to monitor the crop periodically. They offer excellent digital data for precision farming practices, so that farmers can apply fertilizers and chemicals at variable rates. Agriculture engulfs vast regions with wide variation in terrain, water resources, cropping system, disease/pest occurrence and economic returns. Drones that are apt for each situation could be expected to appear in different agricultural regions of the world. A look at the range of drone companies in North America, Europe and Far East clearly suggests that, drones are going to throng almost every corner of agrarian regions of the world. They are expected to help the farmers to reduce drudgery in fields, and obtain accurate information about crops in short time.

A drone could be defined as an aircraft (flat winged or copter) *sans* a human pilot. Drones were initially controlled using remotely situated ground stations with radio network connections. Incidentally, 'drone' is a terminology relatively primitive and more commonly employed in military. This term was used to identify totally or partially autonomous flying machines with ability for wide range of activities needed by military engineers. However, drones are not confined and exclusive to military usage anymore. Agricultural drones are gaining in popularity. During recent years, terms that are accepted as more accurate for autonomous aircrafts with computer-based decision-support systems and predetermined autonomous navigation are 'UAV' or 'unmanned aerial systems (UAS)'. UAVs and UAS come with iPads to control flight pattern and speed. They possess sensors with ability to fend obstacles. Drones that are offered with accessories for several different types of activities, such as spectral imagery, instantaneous transfer of images, computer decision systems, containers to hold agrochemicals and variable-rate nozzles, are called UAS. This explanation pertains to drones in agricultural farming. The Federal Aviation Agency of the United States of America has preferred to use the acronym 'UAVS', that is, 'unmanned aerial vehicle systems' (The UAV, 2015). The word 'system' denotes the entire range of ground stations with control equipment, range of sensors and computer programs to decode and evolve accurate imagery. It includes accessories for activities such as spraying pesticides, fertilizers, spreading seeds in replant regions and so forth.

1.3.2   TYPES OF DRONES AND THEIR CLASSIFICATION

Drones are classified and grouped using several criteria. On the basis of the purpose for which they are deployed and used in military, drones could be classified as 'target and decoy types'. They provide ground and aerial gunnery. Drones could be used exclusively for reconnaissance and battle intelligence. There are also combat drones providing the military with wide range of possibilities to attack enemy positions (e.g. dropping bombs).

Research and development drones are widely distributed all across the globe. They are used to watch large installations, industrial set-up, pipelines, roads, shipping lines and so forth. Civil and commercial drones are another set of drones. They are used in agriculture and several other activities related to commerce (AIRX$_3$ Visual Solutions, 2015; The UAV, 2015).

Several types of drones are produced by companies worldwide. Some are highly versatile and suitable for use in different situations. A few of them are specific to a particular function, say imagery of natural resources, vegetation and crops. A few are specific to pesticide application at variable rates, detection of disease and so forth. These are low-flying drones capable of close-up shots of crops, even leaves/canopy, so that occurrence of disease could be judged accurately. There are other drones such as 'RMAX' produced by Yamaha Motor Inc. or 'Venture Surveyor' by Volt Aerial Robotics Inc. that suit variety of situations. We may also note that it is now becoming common to develop and use drones with local material and expertise. They are targeted to overcome specific local problems related to crop production. Sophistications are added later as and when required (Zhao and Yang, 2011; Ministry of Agriculture, 2013). Drones could also be ordered to suit the specific role in agricultural crop production. Drone models are also leased to complete tasks such as pesticide spraying and so forth (Homeland Surveillance and Electronics LLC, 2015).

Drones are also classified on the basis of variety of traits. They could be either flat-wing types or copters. Flat-wing types need a stretch to gain height and become airborne, whereas copters have ability for vertical liftoff from any point in a crop field (see Krishna, 2016). However, recently, parachute-type drones have also been used effectively to map farms, soil types and demarcate 'management zones', during precision farming (Thamm, 2011; Thamm and Judex, 2006; Pudelko et al., 2008, 2012). A large balloon with helium or blimp can also serve as UAV. A modern blimp with its payload can be controlled using remote controller, but the balloon is generally unstable if the wind speed is beyond threshold (Yan et al., 2009).

Regardless of the type and purpose of the drones, some of the most important performance characters of drones are: (a) weight, (b) endurance and flight range, (c) maximum altitude, (d) wing load, (e) engine type and (f) power source (Arjomani, 2013). Therefore, drones have been frequently classified using the above characters as criteria.

Weight: First category of drones using their weight as a characteristic is 'Super heavy drones' that weigh over 2 t. They hold sufficiently large amounts of fuel. They fly at high altitudes serving military reconnaissance (e.g. Global Hawk). Next are 'heavy weight drones' that weigh between 200 and 2000 kg (e.g. Fire Scout). 'Medium-weight drones' weigh 50–200 kg. All of these drones are capable of holding more fuel and travel long distances with greater endurance. 'Light-weight drones' weigh 5–50 kg. They are useful in agricultural fields (e.g. RMAX, Yintong, Autocopter). 'Microdrones' are those weighing less than 5 kg. They are handy, quick to lift off and fly at relatively lower altitudes (e.g. Raven). These are more common in agricultural fields and are portable as they weigh less than 5 kg (e.g. Precision Hawk's Lancaster, eBee, Swinglet, CropCam).

Endurance: On the basis of endurance (flight period) and range of distance that drones can fly without fuel refill, they can be classified into long-, medium- and low-endurance types. Long-endurance UAVs can stay airborne for 24 h or more and travel 1500–20,000 km without refill, for example, Global Hawk. Medium-endurance drones fly for 5–24 h without brake. They transit long distances on the basis of speed. Low-endurance drones fly for an hour at a stretch and cover 100 km. Drones used in agricultural fields have very short endurance. It ranges from 30 min to 1 h. They cover over 50 km of predetermined path above the fields. It is generally believed that, as drones are powered by mechanical engines, their endurance in the air could be enhanced to help farmers to scout and survey fields for periods much longer than it is possible now (The UAV, 2015).

Altitude: Using altitude attained by drones during reconnaissance and imagery of ground conditions as a characteristic, we can classify the drones into low-, medium- and high-altitude drones. Low-altitude UAVs are most apt for close-up imagery of crop fields. Such drones fly just up to 100–1000 m above the crop canopy. Medium-altitude drones are also utilized in agriculture, particularly in obtaining NIR and thermal imagery of crops. They fly at about 1000–10,000 m above the crop height. High-altitude drones are commonly used in judging natural resources and for military reconnaissance. They fly at a height above 10,000 m above ground surface.

Wing Loading: Wing loading of a drone is yet another useful trait to classify them. Wing loading is actually a value derived by dividing the total weight of the drone by total wing area. The wing loading value for various agricultural drones or others may range from 5 to 250–300 kg·m⁻². Drones with wing loading of over 100 kg·m⁻² are classified as high wing loading types (e.g. Global Hawk). Those with 50–100 kg·m⁻² are classified as drones with medium wing loading (e.g. Fire Scout, X-45). Those with less than 50 kg·m⁻² are grouped as low wing loading ones. Most of the agricultural drones, particularly, the fixed-wing types are low wing loading types. A few copter drones have medium to high wing loading range.

Engine Type: Engines used in drones are usually run using petrol or electric batteries. Engine types commonly encountered are turbofans, two-stroke piston engines, turboprop, push and pull and electric with propeller. Agricultural drones, particularly, flat-winged small ones are energised mostly by electric batteries. Larger copter drones such as RMAX or Auto-copter are energised through petrol or diesel engines. Drones with electric batteries run only for short time. Their endurance is small, but they are highly useful in quick scouting and in obtaining close-up shots of crops.

Energy Source: On the basis of the energy source, we can group the drones into those possessing precharged electric batteries and a second group comprising of drones with IC engines dependent on petrol.

There are few other criteria that could be adopted to classify drones. According to Gogarty and Robinson (2012), modern drones could be classified as micro, small, medium and large. On the basis of the altitude and endurance considered together, they are referred as medium-altitude long endurance (MALE) or high-altitude long endurance (HALE) drones. Generally, drones for military surveillance, scouting and bombing are MALE or HALE types, because they have to travel long distances and high altitudes without detection, and return to barracks after completing the tasks. It seems drones such as Global Hawk and Predator could easily travel for 7500 miles from the launch site, identify the target, drop bombs and return. All these aspects are accomplished at the push of buttons and on the iPad or a computer screen (Gale Encyclopaedia of Espionage and intelligence, 2014). On the contrary, agricultural drones are usually micro or small, low-altitude or very low-altitude types. They have short endurance of around 30 min to 2 h. They are lightweight machines.

Agricultural drones could also be classified on the basis of range and quality of accessories. In other words, we consider cost of purchase, extent of complexity and tasks that drones can perform. There are also drones that may last for long period with ruggedized body or those with frames (platforms) made of light material. Hence, they require periodic corrections to platform.

On the basis of flight craft and takeoff, agricultural drones are grouped into vertical take-off multi-rotor drones. Flat-winged drones could be grouped as either short or long take-off drones. Next, on the basis of flight controllers used, agricultural drones could be semi-autonomous, when skilled farm technicians guide the drone's flight path, and autonomous when the flight path and imagery are predetermined using computer decision-support systems (see Krishna, 2016).

1.3.2.1   SENSORS ON DRONES

About 100 years ago, farming stretches in North America experienced a drastic change from animal-drawn implements to combustion-powered machines and tractors. The transition had impact on crops and their productivity. Thomasson (2015) suggests that a revolution or transition of similar proportion and impact has taken roots in agrarian belts. However, this time, it has been engineered by drones, sensors, Global Positioning System (GPS) and computer-aided decision systems. *Sensors are the centre piece of agricultural drone technology.* Most commonly used sensors on agricultural drones are the red, green and blue in the visual bandwidth, NIR and IR. Each and every improvement in sensors utilized in agricultural drones is always received eagerly by drone technologists and farmers. The resolution and accuracy with which they depict crop field and the happenings in greater detail decides how useful is the aerial imagery. Sensors should be simple to attach or detach. They should be small enough to fit the payload and space on the small drone. Current trends are to use small and swift drones to capture aerial images and collect digital data. Low-flying and hovering-type copters are preferable, if close-up shots with greater details of plant organs (leaf, twigs, pod etc.) are required. Whatever is the type of platform, sensors have to be really very sharp with high resolution. They should be able to get details of minutest items that farmers require, during monitoring of crops. Recently, researchers at Ecole Polytechnique Federal de Lausanne have developed 'super sharp sensors' in which three photodetectors forming a triangle are covered with a single lens. This optical device can focus on crops and relay images to microprocessers that map the images of fields. The optical device is very small—just 2 mm in length. It fits into any of the really small and micro-drones that hover above individual plants in a crop field (Floriano, 2015). Incidentally, this development mimics the compound eye of an insect and is destined to provide farmers with some excellent aerial imagery with details of their fields.

1.3.3 DEFINITIONS AND TERMINOLOGIES RELEVANT TO AGRICULTURAL DRONE TECHNOLOGY

Let us consider a few definitions, acronyms and important terminologies relevant to drones and their role in agriculture as described by Gogarty and Robinson (2012). They are as follows:

Unmanned Vehicle: It is any vehicle that is guided remotely without being driven/piloted by human beings. It includes several types of ground vehicles, robots and flying machines.

UAV: It refers to aerial robots or drones that fly on the predetermined path or using commands from a computer/command control system. Handheld remote controllers are most commonly used to control flight path of such UAVs.

The word 'drones' is the most commonly used and widely recognized synonym within the realms of military, agriculture and public usage. On the basis of the extent of autonomy of drones, they could be identified as semi-autonomous drones or fully autonomous drones.

Since many of the drone models are small and have short endurance within the realm of agriculture, farmers may use them in swarms, if possible. They can then complete the task of crop imagery at the quickest possible time (Gogarty and Robinson, 2012). This is comparable to using several tractors, simultaneously, to plough a large field, in time, to catch up with first rains. The entire farm of over 10,000 ha or more could be covered by 'drone swarms' in a matter of minutes. Drones could be directed to pick high-resolution pictures for definite purposes such as detection of disease, pests, leaf water status, leaf nutrient status and so forth. Therefore, 'drone swarms' may become more common in near future. Relay of digital information and inter-drone communication have to be sharp, accurate, quick and well thought out. Generally, the period of drone's flight is too small; therefore, coordination of flight path and imagery has to be accurate. In future, perhaps, 'drone swarms' could also be connected to ground robots so that farm operations could be done at one stretch with the help of digital information and directions from the drones. It seems easier said than done. Orchestration of ground vehicles and inter-vehicle communication with drones could be a complex task. It definitely requires further research input.

1.3.4   EXAMPLES OF AGRICULTURAL DRONES: THEIR NAMES, BRANDS, COMPANIES PRODUCING THEM AND THEIR USAGE DURING CROP PRODUCTION

At present, industries producing agricultural drones are well distributed in North America, Europe and Far East. In addition, there are innumerable computer agencies/companies that process the imagery derived from drones. Drone manufacturing units, in fact, are sprouting in big number because they reduce burden on hiring farm labour. In addition, they offer exceedingly accurate data about crop growth status. They can assess crop's nutritional status particularly plant-N, water status, weed infestation and pest/disease attack. Farmers need good knowledge about the drone models available. They should identify drones which suit their purpose best. For example, a lightweight, flat-winged drone (e.g. SenseFly's *eBee*; Precision Hawk's Lancaster; CropCam etc.) that passes over the crop field at a rapid pace is good enough to pick aerial imagery. However, if farmer wants to spray liquid fertilizer formulation or granules swiftly over a large field, then a rotor drone such as Yamaha's RMX with containers and variable-rate nozzles is best. Equally important is the knowledge about technical specifications of the drone model. We should note that brochures of certain companies that produce agricultural drones clearly compare and contrast various specifications, advantages and disadvantages. Further, they mention exact suitability of a particular drone model (see Zufferey, 2012). Usually, a wide range of accessories, particularly, those related to sensors are listed from which to pick. Following is a list of drones, their major function in farms and technical aspects. This list is only indicative. There are really a large number of agricultural drone models that are currently accessible to farmers.

1.3.4.1 FLAT-WINGED DRONES

1.3.4.1.1 Precision Hawk's Lancaster

This is a small fixed-wing drone used in agricultural surveillance and imagery services. It weighs about 1.5 kg. Physically, it is easy to handle this model as it spans just 4 ft in length and breadth. It holds sensors, mainly composed of visual, NIR, red-edge and thermal cameras. They can relay images swiftly to ground control computers that sew the images using appropriate computer software. They provide crucial information about crop genotypes and their growth status to the farmers. Precision Hawk's Lancaster can transmit images covering over 20 ac in a matter of few minutes (Reich, 2014). Such Lancasters (drones) can also offer useful data about water status of the crop and its variability, pests and disease incidence, if any, on the crop. It is produced by a company situated in Indianapolis, Indiana and Raleigh in North Carolina, United States of America. These are portable and cost about 3000–5000 US$ per piece. It comes with all accessories such as sensors, mapping and image processing software and facility to relay digital data and variable-rate applicators.

Technical Specifications: Precision Hawk carrying visual and multispectral cameras flies over crop fields and covers 500 ac in a 45-min flight period. There are options to replace broken/torn airframe and damaged cameras. The Lancaster platform has a 4×750 Hz Linus CPU with real-time embedded processors. It interfaces and interacts with commands via Wi-Fi, Ethernet, Bluetooth, USB and so forth. There are temperature processing units. The drone equipment is light in weight. It has a wingspan of 4 ft. Lancaster can take off from water or ground. Water kits could be attached when required. The package comes with live video streaming of crop fields via Internet. The resolution of cameras could be enhanced up to 6 mm per pixel. Cameras with lenses of longer focal length are fixed for high-resolution imagery. Precision Hawk usually carries a high-performance video/audio processor such as Texas Instruments' OMAP3730.

1.3.4.1.2   Swinglet Cam

It is a small flat-winged drone. It serves farmers with useful data about topology, soil type variations, crop growth and yield forecasts. It primarily offers high-resolution imagery of farmer's fields. It is a portable drone that fits into a small brief case, so that it can be carried to different locations and launched for use. Swinglet Cam costs about 7000 US$. It is produced by SenseFly Inc. situated at Cheseaux-Lausanne, in Switzerland.

The Swinglet package consists of drone and complete electronic system in ready-to-fly condition. It has a set of multispectral sensors and image storage memory card and cables. It also has radio modem for remote control and guidance during flight. It is powered by a set of lithium-polymer batteries and the package has a charger too. In addition, there are spare propellers, remote control spares, iPads and user manuals. Drone becomes ready for flight in a matter of few seconds, once removed out of the case. Swinglet Cam can be shifted from one spot to other without being dismantled or packed. Software user codes are also provided. For example, *eMotion 2* helps in controlling and setting up flight pattern. Next, *Postflight Terra LT* allows quick check on image overlap and calculates a rough ortho-mosaic in a matter of minutes, even while the drone is still above the crop field. The data from Swinglet Cam can be interfaced using *Postflight Terra* to rapid image processing centres located on the ground (SenseFly, 2013; Grassi, 2013). The data from computer decision-support systems could be later relayed, to variable-rate applicators on tractors or spray drones.

Technical Specifications: Swinglet Cam is a small drone with 80-cm wingspan. It weighs about 1.2 lb (600–700 g). It is powered by lithium batteries that keep the drone in flight at a stretch for 30–35 min, without recharging. Its endurance allows a flight distance of 36 km at a cruise speed of 10 m s^{-1}. It resists wind speeds of 25 km·h^{-1} during flight. Its radio contact equipment operates and keeps it linked for up to 1-km length. The cameras provide a 3-cm resolution on the pictures. In addition, Swinglet Cam has facility for data logging on board. It easily covers 6 km^2 area in one flight and offers accurate 3D images of crop field/terrain. The remote controllers and iPads allow farmers to simulate its flight path, prior to actual use. Swinglet drones can be used in swarms (SenseFly, 2013; Grassi, 2013). The package has software to correct flight path and avoid mid-air collision. Flight history can be stored in digital form and retrieved at any point of time.

1.3.4.1.3   eBee

It is a flat-winged, fully autonomous small drone. It is capable of obtaining high-resolution aerial imagery from above the crop fields. It offers digital data that can be converted instantaneously into 2D ortho-mosaics and 3D models. It has a ruggedized body and the entire drone fits into suitcase so that it can be carried to any location for launch. It is hand launched. At present, *eBee* is in use in the wheat production zones of Europe and in other continents. *eBee* is a recent model released for use in the year 2012. Its latest updated version was developed and released in 2015 (SenseFly Inc., 2015a, 2015b).

Technical Aspects: Its platform is flat winged and is powered by lithium batteries. *eBee* is also a very light drone weighing just 450 g. It is a very small drone and its payload is 0.15 kg. The endurance per flight is 45 min. The maximum speed while transiting above the crop fields is 90 kmph. It reaches altitudes higher than 1000 m. As the drone is small and light, it can only withstand wind speeds of 12–15 kmph in the atmosphere while in flight. *eBee's* flight path could be totally preprogrammed prior to launch. Its landing is smooth and predetermined. It carries a series of sensors such as cameras that operate at red, green and red-edge bandwidth. In addition, it has a thermal mapping facility. The digital data it relays can be used to prepare 2D ortho-mosaics and 3D models (SenseFly Inc., 2015a, 2015b).

1.3.4.1.4   Wave Sight

This is a multipurpose flat-winged drone used mainly to accomplish aerial scouting of farmland and crops, at different growth stages. These drones are launched using a catapult. Wave Sight has bays, where in, we can fix different cameras operating at different wavelength band. Cameras such as visual single-lens reflex, NIR, red edge and thermal IR could be fixed. The camera bay usually houses a 20 megapixel visual band camera, 20 megapixel NIR camera and an IR radiometric camera with ability to store GPS data. These cameras could be fixed with zoom lenses for close-up shots of crops during pest/disease detection. The sensors allow mapping of crops spread over 12 km^2 or 3000 ac per flight of 2 h. Wave Sight drones come with their own ground stations and remote controller, if the intention is to guide the path of the drone. Drones could be provided with predetermined flight instructions.

Technical Aspects: The wingspan of 'Wave Sight' is 7.5 ft, length is 4.5 ft, and it weighs about 20 lb. Wave Sight can gain a cruise speed of 73 kmph. It operates safely in the atmosphere withstanding wind speeds of 45 kmph. Flight endurance is 2 h (Volt Aerial Robotics, 2015; Paul, 2015).

1.3.4.1.5   CropCam Unmanned Aerial Vehicle

CropCam is a sleek, small and lightweight agricultural drone that can be packed into a small brief case. It can be transported to any location and launched swiftly in a matter of few minutes (CropCam, 2012). It is a mini-drone that helps in management of crops. It offers imagery of various natural resources, installations in the farm and so forth. It is most commonly used to map the crop and monitor its growth. The primary purpose depends on the cameras and sensors fixed on it. The flight path of this drone can be predetermined using GPS connection. Its flight path could also be controlled using remote control with wireless connection.

Technical Aspects: The average speed of CropCam in flight is 60 kmph. It withstands wind speed of 30 kmph in the atmosphere. Its electronic circuitry stops working below −20°C. As the drone flies at low altitudes, it offers excellent images of the crop. CropCam is 6 lb in weight and can be launched from anywhere using a catapult. It is 4 ft in length and the wingspan is 6 ft. Its endurance, that is, flight duration without brake is 53 min (CropCam, 2012, 2015; see Krishna, 2016).

1.3.4.1.6   Trimble's UX5 Agricultural Drone

This is a flat-winged drone that swiftly flies close to the crop canopy. It offers aerial images of the field. It is also used to obtain 3D map of terrain and soil. Digital data obtained while in flight can be relayed instantaneously to the ground station for use in variable-rate techniques. Most commonly, it is used to get normalized difference vegetative index (NDVI), plant chlorophyll and thermal imagery. It is useful in monitoring ranches and cattle (see Krishna, 2016; SPAR Point Group Staff, 2015). This drone has also been used to prepare work schedules and keep watch on general activity in farms, particularly, movement of ground vehicles such as tractors, fertilizer inoculators and so forth.

Technical Aspects: This flat-winged drone flies at 80 kmph above the crop fields, on the basis of predetermined flight path. It has 50-min flight endurance. This drone adapts well to high-speed aerial imaging. UX5 picks images from 180 ac (73 ha) per flight. This drone offers 3D images plus digital surface model of crop fields. It withstands wind speeds of 37 kmph while in flight. Landing is smooth and precise due to reverse thrust (Trimble, 2015a, 2015b). Trimble's UX5 is fitted with cameras modified to capture image at visual and NIR bandwidths. Such images help in detecting crop's health, pests, weeds, mineral deficiencies and potential soil-related problems like erosion.

1.3.4.1.7 *Agribotix Hornet Drone*

This is a flat-winged drone that could be used during precision farming. It is a low-cost drone with ruggedized and light body. It is commonly used to obtain routine images of crop fields and monitor work schedules. It is equipped with visual and IR sensing. High-resolution images of 3-cm ground resolution are possible. This drone covers about 160 ac per flight (Barton, 2015). The drone system comes with image processing software. Hence, we can relay crop growth and nutrient status data instantaneously depending on the software used. Software such as SST, Agleader, SMS and *SoilMapper* offer excellent help during precision farming (Agribotix, 2015a).

Technical Aspects: Hornet Drone is fully autonomous during takeoff, flight and landing. Its flight path can be predetermined. Ground control stations with computer connectivity and telemetry is also a possibility. It is energized using four rechargeable batteries for short flight and eight if used for longer endurance. The imaging system has Go-Pro-Hero camera (four numbers) that have high-quality visual lenses and NIR filters. This drone has two bays to fit the cameras. Additional R, G and B cameras could also be fitted. An Agribotix Field extractor software is used to select, prepare and upload images to Agribotix cloud processing service. The data processing solution that costs extra includes preparation of full colour maps of crops and their health. Agribotix's 'management zone map' helps to adopt precision farming methods. The software that helps in data storage, retrieval and analysis (Agribotix, 2015b) are also included.

1.3.4.2   ROTARY-WINGED DRONES

1.3.4.2.1   RMAX

Among the agricultural drones in use, RMAX models are slightly heavier at 80–90 kg·unit^{-1}. They are produced by Yamaha Motor Company of Japan. RMAX is used mainly to scout crops and natural vegetation. They are also used to spray pesticides, fungicides, liquid fertilizers and to spread tree seeds during replanting programs. RMAX's flight path can be controlled using remote controllers or even predetermined using computer connectivity. At present, RMAX is a popular agricultural drone in the rice, wheat and soybean production zones of Japan and other Far Eastern countries (RMAX, 2015).

Technical Aspects: RMAX is a roto-copter with a payload of 24–28 kg. The overall height is 108 cm and width is 72 cm. The fuselage has two containers (8 L×2) that carry liquid formulation, granules or seeds. This drone can be fitted with nozzles that respond to computer decision-support systems, to release pesticides/granular fertilizers at variable rates. The drone flies at 150–400 m above the crop canopy. Its flight duration, that is, endurance without refuelling is 60 min. RMAX has two-stroke IC engine (245 cm^3) that is energised using petrol. Most importantly, this drone carries visual, IR, NIR and thermal cameras to pick imagery of crops. Aerial images could be processed immediately using computer programs (RMAX, 2015).

1.3.4.2.2   Yintong

It is a Chinese design copter drone. Along with other models of drones, Yintong's agricultural drones are touted to help farmers in the agrarian regions of China. Yintong's agricultural drone is said to offer great advantages during precision agriculture. It is a hovering type drone which is light in weight and needs no special landing site. It draws power from electric batteries and is relatively small. This drone surveys, relays crop imagery and it is equipped with variable-rate nozzles for dusting and spraying. Application of pesticide using Yintong's drones is said to reduce the requirement of pesticide by 50% compared with spraying done by human scouts on the ground. This copter has been widely tested in China on vast stretches of wheat, rice, cotton and fruit plantations (Yintong Aviation Supplies Company Ltd, 2012; Krishna, 2016).

Technical Aspects: Flight weight including pesticide is 15 kg; flight altitude ranges from 0 to 1000 m above crop; flight speed is 5–10 mph, spraying speed is 0.5–2.3 L·min^{-1}; spraying perimeter is 2.8 m. This drone covers an area of 2.25 km^2·min^{-1}. Yintong drone holds about 5-kg pesticide in containers.

1.3.4.2.3   Venture Outrider and Venture Surveyor

Venture copters are highly versatile drones that are operated above crop fields. They move relatively slowly and can even hover for a longer time at a spot. This trait helps in monitoring crops in greater detail and in obtaining excellent close-up shots of crops. Brochures suggest that Venture Outrider and Surveyor both are rugged. So, they can be used in any harsh environment. They are autonomous, but remote-controlled navigation is also possible. These copter drones have slightly extended flight endurance of 45–60 min. They can cover up to 100 acres per flight. The payload usually consists of 20-megapixel visual and NIR cameras and aerial mapping facility. Instantaneous relay of digital data to sprayers is also a possibility (Volt Aerial Robotics, 2010; Paul, 2015). Sensors in the drone allow each and every image to be tagged with GPS coordinates. High-definition videos are also produced using Venture drones with sensors. These copters are suited to monitor disease/pest attack on crops.

Technical Aspects: Venture Surveyor has an overall dimension of $83 \times 83 \times 24$ cm. It weighs 3 kg per unit (6.62 lb per unit). The cruise speed is about 30 kmph. It can withstand ambient wind speed of 35 kmph. Flight endurance ranges from 15 to 30 min.

1.3.4.2.4   EnsoMOSAIC Quadcopter

This is a quadcopter developed especially for use in agricultural fields. However, it can also be used for deriving NDVI of natural vegetation and to monitor progress of work in mining zones or even to track city traffic.

Technical Aspects: EnsoMOSAIC copter is about 67 cm with a take-off weight of 3.7 kg. It supports a payload of 8.0 kg. The flight endurance is 25 min with full payload. The copter withstands atmospheric wind speed of 8 m·s^{-1}. The copter follows a predetermined path or it can also be guided using remote controller. The EnsoMOSAIC copter usually has a Sony Alpha 6000 (24 megapixel) and NIR cameras to obtain imagery of crops and terrain. EnsoMOSAIC copter also comes with 3D camera and software attached with it to map topography of crop fields (Mosaic-Mill, 2015).

1.3.4.3   PARACHUTE AND BALLOON DRONES

1.3.4.3.1   SUSI-62 UAV

It is a parachute-type drone. SUSI floats and drifts above the crops at relatively slower speeds than flat-winged/copter drones. SUSI is currently in use in Germany and Polish crop production zones. It moves autonomously over crops. It collects images and information about crop's water status, pests and fungal diseases (Thamm, 2011). SUSI-62 can be guided using iPads or ground computer station. Flight routes could also be predetermined (Thamm and Judex, 2006).

Technical Aspects: The frame of the SUSI is made of steel. It is 62 m^3 in size with a payload of 5 kg for locating sensors. The frame has four robust wheels. SUSI is powered by a two-stroke IC engine that sits in the payload chamber. SUSI parachute needs a small runway over a cliff or a location at an altitude to gain height. It moves in the atmosphere above crops at speeds ranging from 20 to 50 kmph. It withstands wind speeds of 10–20 km without distraction to its flight path. Most importantly, its payload contains digital single-lens reflex visual, IR and NIR cameras. Flight parameters and GPS video data can be conveyed within 6 km. This parachute drone has autopilot option (Thamm, 2011).

The above classification and list of agricultural drones mentioned utilizes engineering traits such as wing type, endurance, flight speed, sensors and a few other aspects to classify the drones. Flat-winged and copter types are the most common way to identify and classify further. However, within the realm of agricultural crop production, we can classify drones on the basis of actual agronomic procedure for which they are meant. The sub-classification includes

    a.   Seeding (aerial) and seedling-monitoring drones;
    b.   Canopy and growth-stage-scouting drones;
    c.   Crop protection drones that detect disease/pests and spray plant protection chemicals;

d. Crop-monitoring drones that collect data regarding NDVI, leaf chlorophyll and plant-N status. Such drones could also be utilized to apply liquid fertilizer-N, if the fuselage has such a facility with nozzles;

e. Harvesting drones detect the crop maturity. They map the entire field of over 10,000 ha and provide accurate digital data/imagery for combine harvesters to operate. Sometimes, digital data could be applied to the robotic combine harvester. These machines later offer yield maps to farmers; and

f. Data drones are those vehicles that provide periodic data about the soil-type variations, seeding trends, seedling emergence, crop growth parameters, grain maturity and so forth. Such drones add vast amount of data to the 'big data bank'. Many of the computer-based decision-support systems rely on data banks.

1.4   UTILITY OF DRONES

1.4.1   DRONES ARE VERSATILE IN USAGE

Drones have been shrewdly used by us in a variety of situations and with great advantage. In many cases, drones are better than manual operations in terms of efficiency. Drones have been expertly utilized in aspects such as surveillance of natural resources, observing weather conditions, recording natural disasters, study of general topography, forests, wildlife migrations, monitoring civilized inhabitations, waterways, railroad, tramways, buildings, military equipment, electric transmission systems, oil pipelines, traffic movement and guidance. Drones are used efficiently for large-scale seeding aimed at developing natural vegetation. They are also needed for monitoring of crops periodically and spraying chemicals in the fields.

Let us consider few examples of drone usage. Several of them may become routine and essential in future. Many of these tasks are performed with great ease and accuracy by the small flying machines called micro-drones. There are actually several compilations about different uses of drones, which relate to human endeavour. Gogarty and Robinson (2012) have listed a few conspicuous uses of drones. They are mostly related to drone's usage in the military warfare, in monitoring naval vehicles and offshore inspection of installations, in border security, policing of military zones, cities and towns, patrolling and inspection of large industrial units, in management of mines, in handling hazardous chemicals (e.g. radioactive wastes), in the surveillance of transport vehicles and their movement and so forth. Recently, Frey (2015) has listed over 192 sets of *possible* uses for drones in different professions. Drone usage has been classified by him into different aspects as listed below:

Early Warning Systems: Drones are efficiently used in early warning systems. At present, drones help in earthquake warning networks; hurricane warning systems; tornado warning and in following its trail; in initiating sirens to indicate hailstorms; avalanche warning and reporting, flood alerts; tsunami alerting systems, forest fire alerting systems and so forth. Monitoring of power lines, oil pipelines and water supply systems, and early warning of brake down, if any, is also possible using drones. Drones are highly efficient in terms of economy, because human scouts and security staff involve high costs. For example, Chakravorty (2015) has compared use of drones for monitoring oil pipelines with human scouts and manned aircrafts. It seems drones cost much less and offer uninterrupted surveillance of pipelines. Drones are now used for industrial inspections. Their flight path above the installations could be preprogrammed and simulated (SkySquirrel Technologies Inc., 2015).

Emergency Services: Emergencies caused by several factors could be handled using drones, particularly, to keep watch and provide early warning. A few examples to quote are drones fitted with thermal sensors to judge volcano eruptions, avalanche and so forth. Fire in big buildings could be detected using periodic thermal imagery of specific buildings. Drones with IR cameras help in early warning of forest fires. Drones are used to detect lost pets, endangered species and provide information on their situation (Frey, 2015; Hinkley and Zajowski, 2011; Huang et al., 2011).

News Reporting: It seems that news agencies have adopted drones with great advantage. They have been using drones to get pictures and even audio tapes of events occurring in parts of a town, city or a region. Drones provide accurate imagery of events rapidly and often instantaneously. Sometimes, drones have also been used to relay a live event using perfect computer controls and broadcasting devices. Reports about happenings in remote areas, accidents or other events could be relayed rapidly using drone imagery. Drones are extensively used to provide pictures of atmospheric conditions and terrain. Time-lapse pictures of weather patterns are highly useful. There are several events such as public protests, street happenings and games in stadiums that are picturized and relayed using drones.

Delivery Drones: Drones could be used effectively in the delivery of postal parcels and letters. Medical prescriptions could be rapidly picked and delivered to exact location with great ease using drones. They avoid any kind of traffic jams on the road. Grocery delivery also seems a clear possibility using drones with appropriate payloads and navigation facilities. Farmers use drones for variety of purposes that involve delivery of packages containing seeds, fertilizer, pesticide and other essential items. Drones can be effectively used to deliver hazardous chemicals. Farmers even deliver a certain load of their harvest of fruits such as fresh peaches, tomatoes and watermelon to the points of sale or use.

Business Activity Monitoring: Drones are effective in monitoring different kinds of construction projects and help the businessmen to take appropriate decision on investment and cash flow. Drones could keep regular watch on the speed with which a business deal is completed, for example, tracking goods movement and delivery.

Gaming and Entertainment Drones: Drones literally help us in playing chess games at 3D level, compared with 2D games. Drone racing, obstacle courses and drone matches are a few other types of games. Regarding entertainment, Frey (2015) states that there are comedian drones, magician drones and, interestingly enough, drone circus is also a clear possibility. Drones are also used for movie making (Warwick, 2014).

Farming and Agriculture: Most recent trend in farming is to utilize drones for a range of agronomic procedures. They include, land survey, soil mapping, seeding, crop monitoring and detection of crop nutrient status using chlorophyll imagery. Drones are also used in yield forecasting; they play a crucial role in studying crop diseases, pest attacks and in spraying chemicals at variable rates. Drones have been deployed in precision farming. Drones, interphased with ground robots, can be of immense value to farmers adopting precision techniques to cultivate crops. Drones with multispectral imagery could also be interphased with satellite imagery during precision farming (Gevaert, 2014).

Ranching Drones: Drones have a great role to play in maintaining ranches and their vegetation, cattle, horse and other animals. Drones could be utilized to monitor the movement of cattle herds and guide them to appropriate locations in the ranch. Drones are extensively used to seed pastures (aerial broadcasting of seeds), to monitor their germination and growth and periodically provide data on growth and nutritional status of established mixed pastures.

Police Drones: Drones are the recent and effective tools used by the police departments in several countries. Drones are used in curtailing drug movement, that is, drug sniffing. High-speed chases of vehicles on highways are being regularly conducted in North America and Europe using drones. Drones are good to keep a watch on the happenings in the neighbourhood and trace any commotion rapidly. Drones are now regularly in use in police departments of many counties of the United States of America. Drones have a big role in watching congregations and large gatherings of people, say, for political purposes, sports, cultural festivals and so forth. Drones that faithfully follow and keep watch on a specific vehicle are also available. Drones film and provide warning on traffic or obstacles present in the way. These drones are nicknamed as 'flying puppys'. There are drones that watch movements of people and vehicles in the surroundings and at the entrance of installations. They are being used by police departments in North America and Europe.

Real Estate Drones: Drones are apt in collecting pictures and information about progress and daily maintenance of big buildings and skyscrapers. It seems drones with thermal imagery are excellent in revealing the energy consumption and dissipation from buildings. They are helpful to municipalities that intend to watch energy usage trends and to make an energy audit of buildings (Frey, 2015). Drones with capability of thermal images in addition to visual and IR ranges could be used to detect fire flare-ups and heating problems in the homes, irrespective of season and time.

Library Drones: In future, drones could be utilized to borrow and return books from a library situated at a distance. Predetermined pathways of flight and programming helps drones to pick the books and deliver them at right spots.

Military and Spy Drones: As stated earlier, drones were first invented and deployed to accomplish tasks for military. Military functions such as dropping bombs of appropriate size, launching missile from air to destroy ground targets and to disrupt electronic communications of enemy military installations are conducted by drones. There are heat-seeking drones that are used to destroy military vehicles. Drones are also used for a peaceful function of delivering medical supply within in war zones. They are excellent bets to watch the borders between nations. They can be programmed to fly over most intruded regions and inform the headquarters that possess computer controls. Drones offer excellent details about barracks, military installations, movement of caravans and military regiments in the borders and other regions. They are highly useful as hyperspectral images and close-up images reveal great details about each and every military vehicle and personnel on the ground. Reports suggest that drones have been used effectively to monitor the regions suspected to be infested with landmines. They can keep track of each landmine that has detonated and map its location. Military operations could then be planned appropriately. Drones have been effectively used to detect and track movements of insurgents. In some nations such as the United States of America, military and civilian surveillance using drones has been encouraged because of the budget cuts to military. However, such surveillance may affect privacy of some regions close to borders, particularly, military installations and private residential areas (Epic.org, 2014).

Healthcare Drones: Drones can identify locations from where patients need to be picked. They pick and deliver medical supplies to remote areas. Drones could also be used to pick patients, depending on payload.

Educational Drones Including Science and Discovery: Drones have been commissioned to study archaeological sites located in North America and Mexico. Thermal images from drones have been useful in some cases. Drones have the potential to be used extensively in scientific studies, for example, to investigate the migratory trends of whales in high oceans, bird migration, forest health and logging effects, water currents in oceans, keeping a watch on *Aurora Borealis*, solar flare-ups and so forth (Frey, 2015). In Peru, small drones have been used effectively to picturize and study archaeological sites of *Incas*. The entire archaeological landscape of Peruvian coast north of Lima harbours 1300-year-old Moche civilization. It has been accurately mapped to few centimeters' resolution using drones. Some of the drone models are really very lightweight. They are made of balsa wood and carbon fibre. It seems several ancient human dwellings in Peruvian mountains have also been mapped accurately using such lightweight drones (Cisneros, 2013). Drones are used extensively in studying geographic locations such as hot geysers, volcanoes, natural resources (mineral prospecting), dams, small reservoirs and rivers.

Travel Drones: Some of the travel drones mentioned for future development and use are commuter drones, trucking drones and so forth.

Transport and Delivery Drones: Drones can land and release any item (goods) loaded onto them at a given spot in a town, village and so forth. Obviously, remote controls and/or preprogrammed flight pathways, decided by computer and GPS guidance systems, are adopted. In Southern China, mail service drones are being regularly used. They deliver about 500 small packets (letters and parcels) each day. Postal services in Germany and France are also employing drones, but only in some locations. It is based on feasibility and net advantage. They say it may not be long before U.S. Postal Services adopts drone-aided letter/parcel delivery in remote locations. Moreover, reports suggest that in spas and beach resorts, drone programs aim at delivering drinks and food at the spots predetermined or as guided. This program has been nicknamed 'Drone Booze Delivery' in some spas. In the transport industry, drones are now used to track the trucks, truck caravans and regulate their movement. For example, in South Carolina, companies that haul timber and large goods use drones to detect the progress of truck movements.

Drones in Sanctuaries: Drones could be used to surveillance the natural reserves and wild animal sanctuaries (Leonardo et al., 2013; Digirilamo, 2015). It seems rangers in South Africa are already using drones to watch and enumerate rhinos in their natural preserves.

Lighting Drones: Lighting is a necessity and crucial factor in variety of public work settings. Drone lighting is an interesting and useful aspect to general public and governmental agencies looking after the public structures and areas. Autonomous flying machines in the air actually position themselves accurately at required angles and emit photographic light at the patches that require lighting. We may note that illumination is crucial to photography, particularly, if the lux is less than sufficient. It is often difficult to place lighting at vantage points and operate at required brightness. Therefore, engineers at Massachusetts Institute of Technology and Cornell University in the United States of America have devised ground and aerial robots (drones) that move swiftly to a particular location and impinge light accurately on the required spots. They have used miniaturized copters (nano-copters) to reach the spots that are generally difficult to reach for humans. Drones spread light of exact colour and hue (Hardesty, 2014; Barribeau, 2015). The cameras on the drones (e.g. *Litrobot*) provide images and decision-support computers aid accurate movement of drones. Yet, they say, providing accurate signals, directing and automatic maneuvering of drones to vantage points is an aspect that needs to be mastered.

Lighting drones, particularly, micro- and nano-sized drones could be used at locations that do not receive natural light and are also not endowed with electric lamps such as inside the caves and tunnels.

1.4.2 DRONES TO STUDY NATURAL RESOURCES AND VEGETATION

Aeroplanes and satellites were among the earliest to offer aerial imagery of earth's natural resources. Satellite imagery covers vast stretches of earth's surface. However, resolution and clarity of images depend on sensors and natural weather conditions. Satellite imagery has actually improved over time in terms of resolution, accuracy and sharpness; yet, it falls short of great details, close-ups and rapid relay of pictures. This is possible with drone-aided imagery of earth's surface. Drones can fly at low altitudes above natural resources that need to be imaged and studied. Drones have been used to study natural vegetation, forests, forest plantations, farm land, rivers, rivulets, irrigation canals and so forth. High-resolution cameras fitted to drones usually relay excellent details of ground surface. For example, each shrub in a wasteland or seedling in a crop field could be imaged. Then, GPS coordinates could be used to identify them.

Regarding natural mineral sources, drones have been effectively used to monitor mining activity in several regions of the world. Drones fly into locations within open pit or underground mines. These locations are otherwise impossible for human scouts and skilled engineers/geologists to reach. Drones offer close-up shots of mineral distribution in the profile. Drones map the entire profile of the quarries. They show the distribution of specific ores using visual, IR and thermal imagery. To the engineers, drones offer accurate readings of mineral distribution. 3D images derived from drones help mine vehicles to negotiate the terrain safely. Drones have also been used to keep track of work flow in the mines, particularly in an open-pit mine. In addition, it is said that drones provide pictures of distribution of natural vegetation, drains, water sources, extent of soil deterioration and pollution, if any, in the surrounding regions of the mine. This has direct relevance to environmental quality. Flora around the mining zone could be monitored effectively by using drone imagery (CivicDrone, 2015). Perhaps, we can keep a clear watch and study the effect of pollutants, if any, on the flora and then fix up the thresholds for mine waste dumping zones.

Micro-drones have been used with great advantage to study the shrinkage and expansion of arid regions and deserts. The scarce vegetation could be observed, marked and quantified with better accuracy compared to satellite-based techniques. Arid zones support a certain degree of agricultural activity as well. We have to understand the variability related to natural topographical settings, cropping pattern, water resources and agronomic procedures adopted. Many of these investigations could be performed using drones with greater efficiency. Recently, Gallacher (2015) has reviewed the utility of agricultural drones in monitoring deserts of Middle-East Asia and adjoining regions. Drones are capable of regularly monitoring deserts, particularly dune shifts, natural vegetation and crops. Drones have also been used to study the high mountain ranges, steep terrains, glaciers and their movements, ice caps, ice melting processes and flow rates.

Linehan (2013) has reported that, in near future, drones will find use in natural resource management on a daily basis. Based on trials by U.S. Geological Survey's report, he states that almost each and every forester will fly drones over the pine stands to ascertain the growth and productivity. They further state that world population cuts and burns about 26 billion trees in a year but, replants only about 15 billion per year. Hence, tree-planting programs that adopt drones to drop seeds in replanting zones and those to be reclaimed to vegetation are being popularized. Seeds of perennial trees, agroforestry species and perennial grass species could be sprayed using drones. Drones with facility to carry seeds under the fuselage are needed. Seed planting could be made accurate using variable-rate nozzles. Aerial imagery and computer decision-support systems that use digital data to plant seeds as per directions are essential. Seedling establishment could be monitored effectively as well.

Lucieer et al. (2012) have discussed the utility of micro-drones in observing ice caps and icy terrain that support green patches of moss. These bryophytes are photosynthetic and add to carbon content of the ecosystem. Incidentally, red algae and lichens could also be surveyed. Their density could be assessed accurately using drones. These drones have a reach that allows them to picturize even steep and difficult to navigate terrains in the Antarctic regions. For example, Lucieer et al. (2012) have reported that a small multi-copter drone with R–G–B multispectral and thermal IR photography, 3D sensors and facility for over-laying imagery could be effectively used. Such drone copters obtain information on density, thickness and spread of moss in the Antarctic area. Further, they have stated that drones could obtain data about the effect of climatic changes, such as temperature and water, on growth and spread of moss in the Windmill Islands of Eastern Antarctic region.

Drones are also used in swarms, particularly when area to be covered is large, but survey or spraying job needs to be completed swiftly. For example, drones are used to survey the vast natural geographic regions, particularly during prospecting mineral-rich regions for mining. 'Drone swarms' are also used while seeding of forest species, that is, replanting forests. Drones are flown in swarms to hasten the process of surveying and obtaining imagery of large expanses. Drone swarms have to be coordinated very accurately using ground control stations (Burke and Leuchter, 2009). Autonomy of individual drone is also to be kept intact. It is done using a microcontroller equipped with video camera. Drones could be flown in complex formations and different strategies could be used to cover the vast fields efficiently.

1.4.3   THE USAGE OF DRONES DURING AGRICULTURAL CROP PRODUCTION

At present, 'agricultural automation' is among the popular topics of discussion. Agricultural automation primarily reduces human drudgery or sometimes removes it altogether. Improvement in agricultural automation has been recently reviewed in greater detail by Zhang and Pierce (2013). Agricultural machinery and their automation add to accuracy. It reduces errors associated with human fatigue. It also adds to energy and economic efficiency of agricultural operations. We believe that, in future, agricultural automation may involve drones to a greater extent (Krishna, 2016). Such drone-aided automation may creep into general surveillance of the entire farm, specifically, to study topography, soil characteristics and crop growth. Drones could conduct agronomical operations, assess irrigation potential, conduct disease/pest survey, weed survey and herbicide spray. Drones are becoming quite common for general surveillance and scouting of crops. Most common advantage stated is that drone usage increases crop production efficiency. Drones save time of scouting and spraying. Economic returns on investment are supposedly excellent with drones. Drones are easy to use in agricultural fields and add data that are useful while making computer-based decisions. Drones help farmers in periodic supervision of crops, particularly for nutrient deficiencies and disease/pests. We should also note that drones used in crop fields are fail-safe. They return to original place where they started the flight in case of an emergency (Precision Drone LLC, 2013).

The relatively tiny unmanned airplanes that fly closely above the crop canopy provide details, particularly about growth, leaf colour, chlorophyll content and leaf N status. These aspects are expected to revolutionize farming procedures and labour needs of farms. *Scouting the crop swiftly to help farmer with much needed digitized data to conduct agronomic procedures such as fertilizer supply, irrigation, herbicide and pesticide spray may make drones more popular than any other agricultural gadget* (Ghose, 2013). Incidentally, crop scouting anywhere includes few simple steps. First, survey the crop field and analyse the local vegetation, crop species and terrain in general. Then, we identify pest, diseases and weeds that occur in the fields. Next, we assess soil fertility constraints, if any, and the severity of crop's retardation. Finally, we compile general observations and devise a plan to fly the drone repeatedly to obtain digital data and to adopt remedial measures. We execute the plan by operating drone with most appropriate sensors (SenseFly, 2013).

Drones are also employed by farmers to spray entire fields. It could be done on the basis of a generalized recommendation (not variable rates) of pesticides/herbicides. In such cases, knowledge about factors such as wind speed, accuracy of spray in terms of quantity and drone's speed are essential. Drones may often leave certain areas without spray, for example, in the corners of field not marked well, at higher altitudes, areas masked by trees or other structures and so forth. These aspects could reduce profitability. Hence, there are now computer programs with appropriate algorithms that allow such areas to be covered properly. They carefully consider wind speed, its direction and the drifts it creates, location, topography, spray volume and so forth (Falcal et al., 2015). Of course, a thoroughly mapped field could be sprayed using precision techniques. In other words, adopting variable-rate spray nozzles and computer decision-support systems will avoid such losses.

Hill (2015) reports that the current trend among farmers with large areas of field crops is to switch to drone-aided monitoring. Agronomic procedures are decided using drone-derived multispectral data. Farmers prefer to go for drones, instead of leaving the vast patches of field crops to natural vagaries, diseases and pests. They say drones allow them to keep a close watch of the entire farm. For example, in the lettuce production zone, a few farmers have expressed satisfaction with drone-aided monitoring. They state that even a single lettuce plant could be marked using GPS coordinates. They could be monitored using drones that fly closely over the crop, say, at 70 ft above the crop canopy (Hill, 2015). Farmers could monitor the field crops sprayed with different concentrations of nutrients (N and P). Multispectral images could also be used to apply variable rates of phosphorus to crops. Actually, crop monitoring tells us about the soil's fertility and water status indirectly. Therefore, several of the agronomic procedures could be processed accordingly.

Let us now briefly mention a few reports about drone usage from different agrarian regions. Recent updates from farmers in Saskatchewan in Canada suggest that drone imagery shows up variations in green colour, on large acreages of canola. Drone imagery has helped farmers in applying fertilizer-N precisely into areas that need the nutrient. Top dressing of fertilizer-N using NDVI maps has produced a better crop of canola. In some areas, drone imagery, particularly high-resolution close-up shots, have helped farmer in locating fungal diseases. So that appropriate spray schedule could be organized (Ilnsky, 2015; Precision Drone LLC, 2013). Interestingly, farmers in Saskatchewan have used drones to monitor the progress and accuracy of combine harvesters in the vast stretches of canola. Skilled computer technicians guide the combine harvesters to chop the straw and separate grains by using drone-derived images. This is done instantaneously using *Wi-Fi* electronic connectivity.

Crop scouting as frequently and accurately as possible is almost a necessity during crop production. In Canada, drones are used to scout soybean stretches for nutritional status, growth pattern and disease/pest occurrence. Several types of drones have been utilized to accomplish the scouting of soybean crop (Ruen, 2012a, 2012b).

Drones are in vogue in Michigan's farmland. They are used to derive high-definition pictures of crops by using cameras that operate at visual, NIR and IR bandwidths. Drones are also equipped with thermal imagery that offers farmers with knowledge about water status of crops and surrounding vegetation. Spectral data from drones, actually, help farmers in applying nitrogen and water at spots that need them the most. Such applications are based on the spectral maps and digital signals to variable-rate applicators. Cameron and Basso (2013) state that in Michigan, drone companies offer high-resolution images of crop fields to farmers. Digital imagery and data are given prior to application of nutrients, water or pesticides. A few farmers have also used imagery of topography and soil type prior to planting crops in their farms. It helps farmers in creating 'management blocks' that are essential for precision farming procedures.

In Indiana, USA, drones are being deployed within the large stretches of maize or soybean crops. It is mainly to make an aerial survey of the terrain. Drones identify submergence (flooding) that occurs in scattered spots, usually, after torrential rains. Drones also trace gully erosion, pest attack and disease incidence (Agweb, 2015). It is said that drones offer excellent images of crops that are tall, say a late-stage crop of maize, which hitherto was not possible easily. At the bottom line, drones cost less than human scouts, yet offer useful data to farmers in Indiana.

Glen (2015) points out that agricultural drones are destined to surpass other farm vehicles and gadgets in terms of number and usage mainly because they are versatile and lead to higher crop biomass/grain productivity. Actually, field drones of various sizes, capacities and with ability to perform different agronomic tasks are on the rise. Agricultural drones are particularly useful in conducting precision farming methods, such as crop imagery, obtaining digital data about soil fertility variations, also soil/crop water status. The crop growth and nutrient status is deciphered through visual and NIR imagery. Such data is then relayed to variable-rate applicators, either instantaneously or using chips, discs or pen-drives. Most farmers are now using auto-steer and variable-rate seed planters, fertilizer applicators and crop dusters. Drones that fly past above the crop, actually, offer the hard data for variable-rate equipment (Glen, 2015). In a way, agricultural drones are hastening and driving famers towards automation, and more of electronic controls in farms. They could lead us finally to a situation that we call 'push-button agriculture' (see Krishna, 2016).

We ought to use drone technology shrewdly so that they offer relief, greater accuracy and ease to farm workers, while conducting crop production practices. They should also offer economic advantages. If not, capital investment, researcher's time and farmers' efforts too may go futile. Sometimes, it may just offer a kind of variation from previously known time-tested procedures. For example, if sunflower crops with 10,000 seedlings ha^{-1} have to be pollinated using highly sophisticated small-sized drones that hop on to each flower, take a sufficient time to release pollen and pollinate the heads, then it is cumbersome.

Perhaps, there is no need to mimic nature's complicated processes in exactness. It takes longer duration compared with simple dusting of pollen collected from panicles/heads, which is the present efficient method to obtain enhanced seed set. The situation is similar with many cereal crops that release lots of pollen. In fact, just bagging the panicle with paper bags after dusting pollen from a different genetic population has worked wonders for plant breeders. We really don't know if pollens remain viable for such long periods when small drones fly past each flower. Above all, what is the energy and fiscal costs of using small bee-sized drones? It is much greater than collection of pollen from heads and allowing rapidly flying copter drone (e.g. RMAX R50; Autocraft) to do the needful pollen dusting. For example, drones cover 80–100 km^2 of sunflower crop in a matter of 30–50 min within their endurance period. Therefore, dusting pollen, a method known to us since long, seems better than devising minutely small drones. We may mimic nature accurately only where necessary.

Most commonly listed uses of drones in agriculture are crop scouting, crop growth monitoring, spraying and dusting. However, we ought to realize that during each flight that a drone takes above the crop canopy, it collects and allows farmers to archive relevant crop data. Such data can be retrieved and used in decision-support systems. These aspects are difficult and cumbersome, if conducted using skilled human scouts. Human errors due to fatigue and inaccuracies creep into data sets as well.

Field trials on vineyards in Sonoma county of California has shown that a copter drone such as RMAX can spray liquid formulations or water on foliage. It can conduct at a very rapid rate compared with ground vehicles steered by farm workers. Researchers at Oakville, California report that copter drones have relatively short flight endurance. They can spray water/pesticides for only 15 min at a stretch. However, the drone covers about 18–24 miles of vineyard per hour. It amounts to spraying 12 ac of grapes per hour. Given that a large number of human scouts and farm workers are needed to cover similar area in the same period, drones are extremely useful and economically impressive (University of California, Davis, 2013). Again, field trials indicate that grape farmers tend to suffer 5–10% loss in fruit yield due to improper scouting and lack of timely application of nutrients, water and pesticides. However, if drones are used to rapidly assess crop-growth status all across the large farm of say 10,000 ha, remedies could be applied timely. As a result, losses get reduced remarkably.

Drones have been repeatedly employed to assess NDVI, leaf chlorophyll and N status (SkySquirrel technologies Inc., 2015).

Brazilian Agricultural Research Agency has initiated programs to adopt drone technology to study forest, forest clearance trends, forest soils, soil-related problems such as large-scale gully erosion, river sedimentation and pest status (EMBRAPA, 2015). At present, in Brazil, cost of drones depends on size and sophisticated sensors attached to it. The cost ranges from 3240 to 6490 €. It is expected to reduce, in due course, as drone industry produces drone vehicles in larger numbers. Forecasts suggest that, soon, we may expect farmers producing soybean, maize and wheat in the Brazilian Cerrados, to benefit from usage of drone technology.

Cisneros (2013) has reported that lightweight drones made of indigenous balsa wood and carbon fibre, and attached with low-cost visual and NIR sensors have served the Peruvian farmers excellently. For example, reports emanating from International Potato Centre, Lima, Peru state that drones could collect data about potato crop swiftly and cover large areas. The high-quality crop images pertaining to NDVI, photosynthetic area, leaf chlorophyll content and plant-N status could be useful (Cisneros, 2013).

Drone technology has reached even the remotely located forests, plantations and crop land of Australia. For example, a project known as 'Terraluma' operates to introduce drones into different aspects of forestry and agriculture. It aims to study the growth pattern of forest plantations, natural vegetation and climate change effects in Tasmania. Drones allow them to estimate forest biomass, canopy cover, leaf area index (LAI) and so forth. In fact, drones are used right from the stage one, that is, to detect forest/shrub growth that needs to be cleared. It helps to initiate field crops such as wheat (Lucieer, 2015). Monitoring wheat crop's growth, LAI, crop-N status and maturity is done using drone imagery. They expect drones to become more common in the Tasmanian forest and farm belts.

In the present context, the ultimate aim is to adopt drones that are remarkably versatile, particularly in their ability to study the agricultural terrain and crops. We are still somewhere in the very early stages of gaining advantages from agricultural drones. However, we may realize that there are now companies and governmental agencies that are getting ready to offer a series of drone-related services to farmers. Along with satellite-guided systems, drones could provide digital data to support precision farming activities. Drone services that suit large-scale production of grain/

fruits and experimental observation of the crop have been offered in North America. Drone companies that evaluate crop cultivars are an example.

Precision Hawk Inc., a drone company located in North Carolina, USA offers to collect data about crop phenology at various stages, during a crop season (Precision Hawk LLC, 2014). Farmers need this service absolutely. Crop hybrids could be compared using phenological data. Drones are also useful in canopy profiling, monitoring crop canopy temperature, leaf area and chlorophyll, monitoring spore dust and pollen collection in a crop field. Perhaps, drones could also be used to measure the CH_4 and CO_2 emissions from the canopy. Use of drones for crop scouting and recording data is immensely easier, accurate and economically advantageous. In summary, using drones during experimental evaluation of crops, and while conducting agronomic procedures is a good proposition. It avoids tedious data collection and analysis by farm technicians. Field trials that quantify economic advantages accrued due to adoption of drones are needed.

In a large farm, hiring skilled farm scouts and collecting accurate data about crop growth, its nutritional status, locating water logging or drought affected patches and grain maturity status could be difficult. Drones are now capable of providing farmers with regular data on various aspects, such as crop species sown, genotype, plant count in a field or management block, canopy cover, LAI, soil moisture, drought affected patches, if any, crop growth stage, plant height, leaf chlorophyll and nitrogen status, and disease/pest attack if any. A generalized crop yield (panicle) monitoring done by drones could fetch useful data prior to harvesting (Precision Hawk LLC, 2014). In specific cases, drones could effectively trace out bird and insect pest damage and offer timely information to farmers. Drones are quick in their action and provide data rapidly. In addition, they offer a rough idea about crop status, replanting requirements if any, drainage data and irrigation needs. Water shed planning prior to planting could be done using detailed 3D aerial imagery from drones. Drone imagery using oblique shots helps farmers to know crop height and maturity (Precision Hawk LLC, 2014; Drone Life, 2014).

Farmer enquiries in Idaho's potato belt suggest that drones are used in deciding replant locations, in detecting floods/drought affected regions and in spotting cattle in ranches. Drone costs are based on sophistication. But paying 30,000 US$ was still economically worthwhile as farmers reached break-even in just a few seasons (The Des Moines Register, 2014).

Let us consider farmers' response about micro-drones and their utility in the corn belt of Iowa. They say, flat-winged small, low-cost drones are excellent propositions for any farmer with small or large farm. Flight plans for the drone (e.g. *eBee*) using *eMotion 2* software allows farmers to simulate the drone flight over the farm, prior to actual usage. It allows a certain degree of flexibility, if remote controllers are used later. Computer stations and mobiles could be used to regulate flight path and collection of images. Most importantly, some of these micro-drones come with postflight image processing software. For example, *Postflight Terra-2* software helps farmers to process the ortho-images and swiftly decide on remedies. No doubt, drones offer imagery rapidly compared with hiring skilled human scouts. The aerial image gives a total view of the farm almost instantaneously which is simply impossible with human scouts (Labre Crop Consulting, 2014). Such postflight image processing software is being improvised incessantly. For example, SenseFly Company of Lausanne, Switzerland has revised its *Postflight Terra-2* software to offer 3D images (Aasand, 2015). Yet another use of drones that has great consequences to farming operations is fertilizer and irrigation input. Drones also help in farmer's decisions about yield goals, each season. Next, drones make it possible for farmers to over-lay multi-year data about soil, crop growth pattern and yield variations in a field. This is possible because they collect aerial images of crops at regular intervals. Farmers can arrive at more appropriate decisions regarding supply of inputs, costs and economic advantages (Labre Crop Consulting, 2014). Similarly, there are other models/programs such as 'AgEagle'. They allow rapid automation and processing of postflight data. Processed data could be handled on an iPad rather immediately (Precision Farming Dealer, 2014).

1.4.3.1 DRONE USAGE ON DIFFERENT CROP SPECIES

At present, farmers cultivating several different crop species have explored, tested, tried and even evaluated the advantage of drones during crop production. Farm companies have also evinced interest and many now routinely use drones. Drones are actually amenable for use during production of several crop species. Drones are now fairly common among farmers specializing in major cereals such as maize, wheat, rice, even sorghum, a few legumes and oil seeds. Drones have been excellently adapted for use on plantation crops such as grapes and citrus grown in North America and Europe (Bailey, 2013; Garcia-Ruiz et al., 2013; Table 1.1). There are indeed innumerable crop species whose production zones are still to be invaded by drones. Major cereals that are produced in larger expanses by commercial companies have been exposed to drone usage. Drones seem to offer immediate and higher profits by reducing inputs and labour requirements. They aid rapid accomplishment of complex and tedious tasks related to crop scouting, mapping and variable-rate application. Farm companies tend to use least number of farm workers if they adopt drones. Drones are actually used for a range of agronomic procedures, therefore, perhaps in near future, skilled farm worker requirements will get to rock bottom and negligible. Few crop species that are already exposed to drone technology are listed in Table 1.1.

1.4.3.2   DRONES IN CATTLE RANCHES

Drones have an excellent role to play in the surveillance and upkeep of cattle ranching yards and monitoring cattle herds, both during day and night. Sensors with night vision and IR cameras allow the ranchers to track the cattle herd all through day and night. Drones provide ranchers with accurate status report of green pastures and insect/disease attack, if any. Cattle ranch monitoring using human labour is costlier by many times compared with watching the herd using drones and computer monitors. The computer software adopted also stores the entire happenings in the ranch. Farmers can refer to images at any later date. This aspect is not easily possible with human scouts. Human scouts do not offer aerial view of large areas of ranch in one go (Grassi, 2014). Reports from Central Great Plains of the United States of America suggest that drones could be effectively used to assess pasture growth, its nutritional status, also drought and disease affliction, if any. Moffet (2015) believes that forage and pasture management will be accurate, easier and profitable if drone technology could be adopted and standardized.

Agricultural drones are getting ever popular in the Australian continent. One of the recent reports suggests that drones are used for pest control in ranches. A private agency, 'The Bluebird Agroecosystems', has developed drones of 3-m wingspan with dual camera system. The drone keeps regular vigil of invasive pest animals such as dogs, boars, rabbits, pigs, deer and so forth. Experimental trials indicate that depending on the computer decision-support system used, drones could pin-point occurrence of even two

TABLE 1.1   Drone Usage on Different Crops—A Brief Report.

Crop species	Location, agronomic procedures and reference
Wheat (*Triticum aestivum*)	Cordoba, Spain: Drones are used to collect aerial data/imagery to decide wheat production procedures. Aerial data of soil is used to decide ploughing schedule and 'management block formation'. Leaf area, leaf chlorophyll content and plant-N status data from drones is used to supply fertilizer-N accurately. Liquid fertilizer-N could be sprayed using drones fitted with variable-rate applicators. Soil and crop water status is detected using thermal near-infrared (NIR) imagery. Such data helps in deciding irrigation schedules (Jensen et al., 2007; Torres-Sanchez et al., 2014). Drones are also used to quantify vegetation portion in wheat fields. Low flying drones are able to transmit sharp imagery suitable for use in precision farming. Such imagery is also used to distinguish between weeds and wheat seedlings in the field (Torres-Sanchez et al., 2014, 2015)
Rice (*Oryza sativa*)	Rice belt, Japan: Drones are used in phenotyping rice fields. It includes measurement of plant height, leaf area, crop's growth rate and biomass accumulation trend. Drones are employed to spray pesticides, either based on blanket recommendations or using nozzles with variable-rate applicators (e.g. RMAX 50) (Thenkabail et al., 2000, 2002; Nicas, 2015). Aerial spraying of pesticides and disease control chemicals using drones fitted with pesticide tanks is gaining acceptance (Bennett, 2013; Tadasi et al., 2014; Giles and Billing, 2014)
	Xinjiang, Northeast China: Drones are used to collect aerial images. Crop surface images are collected to prepare 'crop surface models' and compare it with current field data. Agronomic procedures such as fertilizer supply (split dosages), using liquid sprays could be decided, by comparing present crop status with known 'crop surface models'. Drone imagery is also used to decide water supply rates. It is usually based on infrared (IR) and thermal imagery from drones (Bendig et al., 2013; Gnyp, 2014)
Maize (*Zea mays*)	Davis, California, USA: Drones are used to collect aerial images useful to prepare field layout and 'management blocks'. Aerial imagery of soils is used to decide on ploughing schedules and then to collect data about crop growth rates. Drones are used in experimental farms to periodically observe several hundred genotypes of maize, in one go, using aerial imagery. Drones are exceedingly useful in phenotyping and collecting information regarding specific traits of maize genotypes (Raymond-Hunt et al., 2010)
	Los Cruzes, Mexico: Drone technology introduced at the International Centre for Maize and Wheat is called 'SkyWalker Project'. Here, drones are used at regular intervals to collect data about several hundreds of maize genotypes exposed to different inputs, soil amendments, plant protection chemicals and many other agronomic procedures (Mortimer, 2013)

TABLE 1.1 *(Continued)*

Crop species	Location, agronomic procedures and reference
	Illinois, USA: Drones are used to survey for crop damage due to root worm. Crop patches showing root worm attack are then treated with plant protection chemicals, using variable-rate nozzles fitted to drones (AUVSIAdmin, 2012).
	Cordoba, Spain: Drones detect weed patches in maize fields so that appropriate herbicides at correct concentration/quantity could be sprayed only at locations showing weeds (Pefia et al., 2013; Torres-Sanchez et al., 2015). Here, drones replace farm scouts
	Hohenheim, Germany: Drones are used to obtain detailed 3D imagery of maize crop. Such images are then used to develop 'crop surface models'. Such models are essential for deciding on several agronomic practices. Further, datasets derived adopting red, green and blue sensors during early and mid-season are also used to predict maize grain yield (Geipel et al., 2014)
	Pretoria, South Africa: Drones are utilized to monitor crops for health, NDVI and adoption of precision techniques, particularly to apply fertilizers (Jager, 2014)
	Southern Highlands, Tanzania: Drones have been utilized to monitor crop growth and health. Arial imagery is procured periodically to detect maize rust incidence. The digital data is used to spray plant protection chemicals (Agape Palilo, 2015)
Soybean (*Glycine max*)	Missouri, USA: Drones have been used to collect data on soybean crop grown in Missouri State. Major thrust is to obtain digital data, to develop maps, to scout soybean crops periodically, to count seedlings, to assess crop establishment and to dust crops with pesticides (United Soybean Board, 2014)
Sunflower (*Helianthus annuus*)	Missouri, USA: Drones fitted with sensors that operate at visual and NIR bandwidths have been used to collect data about leaf chlorophyll and leaf-N status. Such data helps in estimating fertilizer-N needs of the crop (Aguera et al., 2011)
Groundnut (*Arachis hypogaea*)	Rajasthan, Gujarat, India: Drones are deployed to identify and provide close-up imagery of groundnut crop and disease affliction, if any. Digital maps are used to spray pesticides and disease control chemicals (The Economic Times, 2015)
Strawberry (*Fragaria* × *ananassa*)	Passo Fundo, Brazil: Aerial imagery of strawberry fields derived from drones is used to assess an array of crop characteristics such as growth, fruit yield and its quality. Drones are used to alert farmers about various agronomic procedures required to be performed by them (Rieder et al., 2014)

(Continue...)

TABLE 1.1   *(Continued)*

Crop species	Location, agronomic procedures and reference
Citrus (*Citrus sinensis*)	Lake Alfred, Florida, USA: Drones are used to collect hyperspectral images of citrus groves to study canopy, tree growth, leaf chlorophyll, leaf-N status, soil water distribution and fruiting. Drones are also used to spray pesticides on to citrus trees. Drones can provide aerial data of each and every tree in the farm using Global Positioning System tags. Drones are effectively used to detect citrus greening disease, that is Huanglongbing disease (Lee et al., 2008; Garcia-Ruiz et al., 2013)
Grapes (*Vitis vinifera*)	Central Italy, Italy: Drones are used to collect digital data/imagery of grape vines. Such imagery is used to estimate leaf area, leaf chlorophyll content, leaf-N status. Thermal imagery is used to detect variation in plant water status. Fertilizer and water supply is decided based on aerial images from drones (Primecerio et al., 2012)
	Napa Valley, California, USA: Drones are used to collect 3D images. Such images help to gauge ripeness of fruit (berries) bunches. Such 3D images are also useful in detecting soil erosion, loss of top soil and fertility. It helps in adopting soil conservation procedures immediately (Bailey, 2013)
	Tasmania, Australia: Drones are used to collect data on leaf chlorophyll, leaf-N status, soil moisture distribution, and to prepare 'digital surface models (DSMs)'. Such 'crop surface models' are quite handy to compare with present crop and quickly decide on appropriate agronomic procedure. DSMs are usually compared prior to supply of fertilizer-N and irrigation water (Turner et al., 2012; Hall et al., 2011; Hall and Louis, 2008; Lamb et al., 2001; 2013)
Coffee (*Coffea arabica*)	Hawaii, USA: Coffee plantations could be studied using a drone fitted with visual, NIR and thermal NIR sensors. Further, coffee beans could be detected and their ripening status can be identified using drone imagery (Herwitz, 2002; Herwitz et al., 2004)
Peach (*Prunus persica*) and Olive (*Olea europaea*) Orchards	Cordoba, Spain: Drones are used to detect water stress affected peach and olive trees. Drone images help in assessing leaf chlorophyll (Zarco-Tejada et al., 2009). Drones have also been utilized to detect incidence of Verticillium disease and to map the affected regions within each plantation (Calderon et al., 2013)
Oil Palm (*Elaeis guineensis*)	East Malaysia, Malaysia: Drones are used to detect and mark the zones affected with Gonoderma disease. The digital imagery from drone is then used to spray plant protection chemicals (Shafri and Hamdan, 2009)
Eucalyptus (*Eucalyptus globulus*)	Tasmania, Australia: Drones are used to study tree growth, canopy closure pattern and biomass accumulation. Drones are also used to study soil erosion and deterioration of fertility (Terraluma, 2014)

TABLE 1.1 *(Continued)*

Crop species	Location, agronomic procedures and reference
Pasture and Rangeland	San Diego, California, USA: Turf grass management using drone-derived aerial imagery has been attempted. Thermal imagery is used to detect moisture distribution in soil and pasture grass/legume (Stowell and Gelentr, 2013)
	Southern USA: Aerial imagery is used to document pasture growth pattern, soil moisture, nutrient status and disease/pest attack, if any. Drones are also used to spray seeds during replanting. Pesticides and disease control chemicals are also sprayed using drones (Ahamed et al., 2011; KSU, 2013; Rango et al., 2009)

wild dogs in a very large flock of sheep in the ranch. This is indicative of accuracy of drone images and computer-based decision-support systems. Sharp shooters placed on the drone could then eliminate pests (Precision Farm Dealer, 2015).

1.4.4 DRONES IN PRECISION FARMING

Precision techniques are currently in vogue in several different crop-production zones. At present, adoption of precision methods is frequent in the agrarian regions of developed world. Most of the agronomic procedures are amenable to precision techniques. We should note that highly sophisticated farm vehicles and electronic sophistications are not mandatory, to practice all precision procedures. A few of the precision procedures are possible with lesser levels of sophistication. Now, let us focus our attention to agricultural drones and their role in precision framing. Agricultural drones are quite handy and exceedingly rapid compared with several other classical approaches. They are easy to adopt during precision farming. They offer excellent economic advantages by scouting the entire field rapidly. They avoid excessive expenditure on skilled labour. Grassi (2014) lists at least five different uses of drones during precision farming. They are as follows:

a. Midseason crop scouting for growth traits, height, leaf area and chlorophyll content. Drones with multispectral sensors allow farmers to obtain NDVI. Drones cover an area of 50 ac in a matter of 15–30 min. This same activity requires several skilled farm workers to methodically move in the crop fields and note the readings carefully. Then, the data has to be skillfully collated and maps prepared for the farm workers so that, later, farmers can apply amendments at variable rates. We may note that drone imagery removes human errors and fatigue-related aspects. The digitized maps and electronic signals required for autonomous/ semi-autonomous variable-rate applicators (fertilizer or pesticides or weedicides) are transferred to the farm vehicle in a matter of seconds. Also, there are drones that are endowed with computer-based decision-support systems. They are capable of variable-rate application of chemicals, simultaneously.

b. Precision irrigation involves movement of centre-pivot systems in the field at prescribed rates. It is dependent on soil moisture variation and crop's need. Precision irrigation equipment could be effectively monitored using agricultural drones. Grassi (2014) states that crops such as maize, sugarcane or tall sorghums require careful inspection of centre-pivot and nozzles from above the crop canopy. Inspection of nozzles and sprinklers using drones is gaining in popularity in the North American cereal belt.

c. Precision technique aimed at eradication of weeds first involves careful survey (land-based) or airborne imagery using drones. Drones with multispectral imaging facility are employed to judge the spread of weeds and their intensity. The computer decision supports are generally endowed with ability to identify weeds using their spectral signature. Often, spectral signatures of weeds common to an area or cropping system are available. They allow us to read the drone images properly. The NDVI data and post-flight image processing helps the variable-rate herbicide applicators with necessary electronic signals (Grassi, 2014).

d. Agricultural drones are gaining acceptance during variable-rate supply of pesticides and fungicides. Drone imagery helps in the identification of disease/pest-attacked area. It may also depict the intensity of infection. The digitized maps and data could then be channelled to variable-rate applicators. There are a few models of drones with facility to carry chemicals under the fuselage, as well as variable-rate nozzles. Such drones utilize spectral data instantaneously to apply pesticides/fungicides. They carry out the task at a rapid pace and are highly profitable economically (e.g. Yamaha's RMAX).

e. Fertilizer supply to crops is generally based on soil sampling, chemical analysis of major and micronutrients and preparation of soil fertility maps that depict variations. It is followed by application of fertilizers using farm tractors fitted with variable-rate applicators. Previous data about crop yield and multi-year data are also used to judge the fertilizer needs of a crop. Yield goals are invariably considered. However, at present, agricultural drones are finding use in obtaining NDVI maps during a crop season. The in-season data collected by drones could be electronically channelled to variable-rate applicators. In fact, drone-generated crop-growth maps are used efficiently during distribution of fertilizer (nutrients). Grassi (2014) states that application of all three major nutrients, namely N, P and K could be accurate and highly refined, if farmers adopt drones during precision farming.

Crop phenotyping using drones is a recent technique. It is adopted by sophisticated farms and agricultural experimental stations. It deals with selection/breeding of crop genotypes for grain productivity. Farmers may require rapid phenotyping of standing crops so that computers could compare data sets and arrive at remedial measures and fertilizer input schedules. Crop scouting and measuring a series of crop characteristics related to growth and grain formation could be first cumbersome. Without doubt it is a time consuming and costly procedure if done by human scouts. In such situations, drones that fly above the crop canopy rapidly are highly useful. Drones acquire sharp digital imagery of crop phenotype. Crop phenotyping is now routinely done using drones in some parts of Canada, the United States of America, Spain and Australia (Perry et al., 2014). Drones such as CropCam, eBee and Quad-Copters have been used for crop phenotyping.

Yield data and maps showing variations in productivity, in each grid cell or management zone, is an essential item of precision farming. In fact, farmers depend immensely on crop yield data and yield maps of previous years. Grain yield data maps from multiple years are overlayered, then schedules for fertilizers and water are decided. In North America, there are several precision farming companies that deal with crop yield maps and forecasts. They use drones to capture crop images at different stages and to arrive at accurate forecasts. For example, Pioneer and Dupont Inc. do provide farmers in Canadian provinces such as Alberta, Saskatchewan and Manitoba with drone imagery and video of crop yield, mainly soybean, corn and canola. Drone imagery helps in seeding crops using precision techniques. It is actually based on soil fertility and productivity trends (ProSeeds Inc., 2015).

As stated earlier, there are fantasies, suggestions and projects still on drawing board about several types of agricultural drones. In future, farmers may derive innumerable uses from drones. One of them is about 'mosquito-sized drones' with miniaturized visual and NIR cameras. The cameras are placed at the tip of the antennae. *They believe such minutely small drones will allow us to carry out the principles of precision farming, a stage further.* Using mosquito-sized drones, farmers will be able to read the growth pattern, leaf nutrient and water status of each and every plant in the field. A few hundreds of such mosquito-sized drones are expected to swarm. They throng the individual seedlings of maize, wheat or sunflower and bring home data for variable-rate applicators to act on. At present, it seems a bit cumbersome to see those many drones working just to achieve a little more accuracy. Such high accuracy may not be essential for a large farm. Factors such as cost of production of such small drones, their intro-duction in the fields at correct locations, managing their movement in the crop field among plants and retrieval for reuse may all be too difficult. We may have to debate and evaluate the logistics, economics and envi-ronmental aspects of several ideas and the particular models of drones proposed (see Krishna, 2016).

Drones are among most useful gadgets while collecting and storing data about effects of various agronomic procedures, nutrient inputs, water supply and pesticide spray schedules on the crop productivity. Drones collect data that can be easily applied on the 'management blocks'. Smaller grids within the 'management blocks' too could be treated using digital data obtained by the drones. Now, consider the minute drones that collect data about each plant or specific leaf 3, 4 or 7 and large data sets about chlorophyll, water status, disease/pest attack and leaf biomass. During treatment of crops using precision farming, such minutely accurate data sets are overlayered on each other to derive recommendations using computer decision systems. Perhaps, it will be really a highly accurate farming method. It will lead us to a kind of 'micro-precision farming'. Such micro-precision techniques may be apt while dealing with individual trees within fruit plantations and during experimental evaluation of individual plants in green houses.

Huang et al. (2008) stated that development of electronically controlled nozzles that suit the sprayers placed on fully autonomous UAVs (drones) is an essential step, particularly, if drones are to be adopted during precision farming. The drones with nozzles that are controlled using digital data, computer decision support and electronic signals, and that are aligned well with GPS coordinates are available. They could be used during crop dusting. Such drone crop dusters, indeed, cover large areas of crop land in short time. Drones cover a large area during a single or couple of flights. They dispense large quantities of pesticides and fertilizer liquid formulations based on variable maps and digital data supplied (e.g. RMAX; see Krishna, 2016).

Precision farming enterprises need a series of services involving satellite and drone imagery of a crop zone (Trimble, 2015a, b; Precision Hawk LLC, 2014). Detailed knowledge of terrain, soil fertility variations and crop productivity data from different locations within the farm is almost essential. Many of these precision farming aspects are conducted by specialized drone technology agencies. For example, in Hawaii, drone agency begins with offering farmers or farming companies with details (3D) about terrain, soil-type distribution, possible 'management blocks', general surveys of cropping systems, crop growth and nutrient status and so forth (Drone Services Hawaii, 2015). Similarly, drone companies in other regions of the United States of America, for example in California, offer a range of services and prescriptions. They usually suit the precision farming approaches adopted by farmers producing grapes, vegetables and cereals (Trimble, 2015a, 2015b). These drone agencies aim at lowering cost of production and improving productivity. Drones have also been utilized to monitor vineyards for various aspects such as growth, its vigour, disease/pest attack if any, droughts/flooding conditions that may occur in the farm and so forth. In Italy, for example, *VIPtero* is a drone model assembled in 2012. It is used to judge vineyard vigour and several aspects related to site-specific management (precision farming) (Primecerio et al., 2012; Costa-Ferreira et al., 2007). *VIPtero* collects a sizeable amount of data and imagery during each short flight. Heterogeneity of vineyard growth and fruit yield can be studied very accurately. They say, the data from drone is very useful during formation of 'management blocks'.

Farmers need accurate information about crop's nutrient status. Fertilizer supply depends much on the variations in crop nutrient status. Drones with sensors that operate at visual, NIR and IR bandwidth are used to obtain data about crop LAI, NDVI, chlorophyll content and water status (Aguera et al., 2011; Reyniers and Vrindts, 2006; Krienka and Ward, 2013; Walker, 2014, Innova, 2009). Spectral reflectance from the canopies of crops such as sunflower could be obtained, using micro-drones (e.g. md 4–200). Such data could be compared with those obtained using handheld chlorophyll meters and ground stations (Aguera et al., 2011). We should note that most of the leaf-N is localized in the chlorophyll. Therefore, appropriate calibrations are done based on leaf chlorophyll content. It helps farmers to apply fertilizer-N at variable rates after considering the yield goals. Over all, there are several reports showing that, digital data and crop imagery obtained using drones are highly useful during application of fertilizer-N at variable rates.

We may note that satellite imagery also provides a certain degree of details about cropping systems, variations in NDVI and crop growth stage. The resolution of the image is typically in the range of $20–100$ cm·pixel^{-1}. Such accuracy may not suffice. Satellite images are affected by cloudiness and could be hazy. On the contrary, one of the major advantages of drones is that they can fly closely just above the canopy of a crop or 10–50 ft above the crop canopy (Aerial Drones, 2015). The cameras on drones focus the crop from very close range compared with those situated on a satellite. The resolution of the imagery collected using drone offers great details about plant health, leaf area, colour and chlorophyll status. Variations in NDVI, plant chlorophyll and N status, disease/pest attack can be accurately assessed. Digital maps could be supplied to variable-rate applicators. Hence, drones are highly useful during precision farming. In fact, accurate digital data is obtained from above the crop. Drones cover the entire field in one stretch. It allows farmers to accurately plan and even simulate movement of variable-rate applicators in the field.

While describing the uses and definite advantages of agricultural drones, Lyseng (2006) has stated that combines with yield monitors offer useful information to farmers. They offer details about the grain productivity variations within a plot/field. The data depicts the end result of agronomic procedures adopted. However, during crop evaluation in the agricultural experimental stations, we actually need periodic data that depicts the progress of crop, during a season. Such data is not easy to obtain using human scouts. Drones are the right choices so that, farmers could collect, archive and make a comparative study of the entire field. They can compare crop species/genotypes during the entire season (Lyseng, 2006).

Caldwell (2015) has a few take-home statements about drones and their role in agrarian regions of the world. He suggests that drones are most recent gadgets that may be used for adoption of precision farming techniques worldwide. Their ability to offer accurate imagery of crops is immediately useful. Accurate spatial data provided by drones allows equally accurate application of nutrients, water and herbicides. Most importantly, drone technology is relatively very quick and useable by farm workers. Farmers need only a short training in flight control and image processing methods. It is the small flight endurance, small storage tanks, higher cost of computer software and drone vehicle itself that may be constraints. Even then, general opinion among Drone Industry Businessman, experts and farmers are alike. They say drones and related software could be produced in large numbers. It is economically efficient to operate drones and relinquish farm workers.

Drones are generally touted as small flying machines. They could be extremely quicker, accurate and provide useful information to farmers, in a timely fashion. At the same time, drones are highly economical as they reduce cost on human skilled farm workers and scouts. Drones are known to scout and apply pesticides on a few 1000 ha in a matter 1 h. Therefore, farm workers may get replaced in a big way due to adoption of drones. However, we ought to realize that simultaneously, drones create large number of jobs that require computer skills. Technicians with skills to regulate drone's flight path, read the data and prescribe appropriate agronomic procedures will be needed. For example, at present, there are several start-up companies that produce agricultural drones and offer computer-based decision-support services. They are expected to create several thousand jobs and exchequer to drone companies. Drones, therefore, replace a few farm workers. However, simultaneously, they create jobs that need a different set of skills. Drones definitely remove human drudgery out in the crop fields.

Overall, drones have appeared on the scene within the agrarian regions of different continents. No doubt, drones took several decades to traverse the distance from military barracks to farms. At present, drones have been deployed in smaller scale. Their usage in farms is sporadically distributed, depending on geographic region. Whatever is the current rate of acceptance in the farm world, we should realize that drones are capable of accomplishing exceedingly useful tasks which farm companies require. Drones are destined to reduce dependence and expenditure on farm labour, improve accuracy and accomplish farm operations swiftly and in time. Net results from drones are higher profit and reduced energy consumption. Therefore, if not right now, very soon drones are going to dominate farms. They are expected to flourish.

Thus far, this chapter has offered a glimpse about agricultural drones and their various facets of use in agricultural farms. In Chapters 2–9, a range of topics dealing with agricultural drones has been discussed in greater detail. They relate to soil and its fertility management; agronomic procedures wherein drones are effectively utilized; precision irrigation; weed detection, estimation of its intensity and distribution as well as spraying herbicides; detection of pests/diseases that afflict crops, marking their spread and spraying with plant protection chemicals; study of natural vegetation and climate change effects; and economic aspects of adoption of drones during large-scale farming. A summary is provided in the last chapter.

KEYWORDS

- **agricultural drones**
- **drone technology**
- **unmanned aerial vehicles**
- **sensors**
- **precision farming**

REFERENCES

Aasand, E. SenseFly Improves eBee Ag UAV. 2015, pp 1–2. http://www.uasmagazine. com/articles/1057/sensefly-improves-ebee-ag-uav/ (accessed May 16, 2015).

Aerial Drones. Benefits of Farming with UAV. Precision Agriculture. 2015, pp 1–5. http:// www.aerial-drones.com/uav-applications/precision-agriculture/ (accessed April 3, 2015).

Agape Palilo. Monitoring and Management of Maize Rust (*Puccinia Sorghi*) by a Drone Prototype in Southern Highlands, Tanzania. 2015, pp 1–15. http://www.academia.

edu/8063999/MONITORING_AND_MANAGEMENT_OF_MAIZE_RUST_ DISEASE_Puccinia_sorghi_BY_A_DRONE_PROTOTYPE_IN_SOUTHERN_HIGH- LANDS_TANZANIA (accessed May 30, 2015).

Agribotix. Agribotix's Drones and Data Services for Precision Agriculture. 2015a, pp 1–5. http://agribotix.co (accessed Sept 8, 2015).

Agribotix. Hornet Hardware Specs. Agribotix Company, Boulder, Colorado, USA. 2015b, pp 1–2. Info@agribotix.com (accessed Sept 8, 2015).

Aguera, F.; Carvajal, F; Perez, M. Measuring Sunflower Nitrogen Status from an Unmanned Aerial Vehicle-Based System and an on the Ground Device. Conference on Unmanned Aerial Vehicle in Geomatics, Zurich, Switzerland. *International Archives of the Photogrammetry, Remote Sensing and Spatial Information Sciences* 2011; 38, pp 1–22.

Agweb. Drones Eyed for Farm Use. 2015, pp 1–4. http://www.agweb.com/article/drones-eyed-for-farm-use/ (accessed May 22, 2015).

Ahamed, T.; Tian, L.; Zhang, Y.; Ting, K. C. A Review of Remote Sensing Methods for Biomass Feedstock Production. *Biomass Bioenergy* **2011,** *35,* 2455–2469.

AIRX$_3$ Visual Solutions. Sky is the Limit: Mining. 2015, p 14. http://www.arix3. com/#!mining/cm89 (accessed Nov 24, 2015).

Arjomani, M. Classification of Unmanned Aerial Vehicles. The University of Adelaide, Australia. 2013. http://www.personal.mecheng.adelaide.edu/mazair/arjoomandi/aeronautical%20engineering%20projects/2006/group9.pdf (accessed Sept 1, 2015).

AUVSIAdmin. Corn and Soybean Digest: Put Crop Scouting on the Auto-Pilot. Corn and Soybean Digest. 2012, pp 1–3. http://increasinghumanpotential.org/corn-and-soybean-digest-put-crop-scouting-onauto-pilot/ (accessed Sept 8, 2013).

Bailey, P. Wine Grape Drone Flying Over California Vineyards. Western Farm Press. 2013, pp 1–4. http://www.westernfarmpress.com/grapes/wine-grape-drone-flying-over-california-ineyard (accessed Sept 15, 2015).

Barribeau, T. The Litrobot is an Autonomous Lighting Drone: A Quad-Copter that'll Provide you with Perfect Lighting, no Matter how you or your Subject Move. 2015, pp 1–4. http://www.popphoto.com/gear/2014/07/litrobg.2013ot-autonomous-lighting-drone (accessed July 10, 2015).

Barton, J. Agribotix Software Picked for ACCO's Solo UAV. Agribotix Inc., Boulder, Colorado, USA. 2015, pp 1–3. http://www.precisionfarmingdealer.com/articles/1648-agribotix-software-picked-for-agcos-solo-uav (accessed Sept 22, 2015).

Bendig, J.; Wilkomm, M.; Tilly, N.; Guyp, M. L.; Bennetz, S.; Quing, C.; Miao, Y.; Lenz-Weidman, J. S.; Bareth, G. Very High Resolution Crop Surface Models (CSMs) from UAV-Based Images for Rice Growth Monitoring in Northeast China. *International Archives of the Photogrammetry, Remote Sensing and Spatial Information Sciences* 2013; 22, pp 45–50.

Bennett, C. Drones Begin Decent on US Agriculture. Western Farm Press. 2013, pp 1–3. http://westernfarmpress.com/blog/drones-begin-descent-us-agriculture (accessed Sept 6, 2013).

Bowman, L. UAV Drones for Farmers and Ranchers. Agriculture UAV Drones. 2015, pp 1–2. http://www. agricultureuavs.com/uav_drones_for_farmers_ranchers.htm (accessed May 23, 2015).

Burke, A.; Leuchter, S. *Development of Micro UAV Swarms*. Autonome Mobile Systeme; Springer Verlag: Heidelberg, Germany, 2009; p 217.

Calderon, R.; Navas-Cortes, J. A.; Lucena, C.; Zarco-Tajeda, P. J. High Resolution Airborne Hyperspectral and Thermal Imagery for Early Detection of Verticillium wilt of Olive Using Fluorescence, Temperature and Narrow-Band Spectral Indices. *Remote Sens. Environ.* **2013,** *139,* 231–245.

Caldwell, L. Unmanned Aerial Vehicles-How can They Help Agronomists? Lachlan Fertilizers Rural. 2015, pp 1–3. http://www.grdc.com.au/GRDC-Update-Papers/2015/02/ (accessed May 16, 2015).

Cameron, L.; Basso, B. *MSU Lands First Drone.* Michigan State University, East Lansing, Michigan, USA. 2013, p 1. http://msutoday.msu.edu/news/2013/msu-lands-first-drone/ (accessed June 17, 2015).

Chakravorty, S. *Oil Field Drones: Monitoring Oil Pipelines with Drones.* Angel Publishing; Baltimore, USA, 2015; pp 275.

Cisneros, L. J. In Peru, Drones Used for Agriculture, Archaeology. AFP News. 2013, pp 1–4. http://phys.org/news/2013-08-peru-drones-agriculture-archaeology.html (accessed Aug 28, 2016).

CivicDrone. Study of Soil: A Drone for Mapping and Quarries. CivicDrone Inc. 2015. http://www.civicdrone.com (accessed May 30, 2015).

Costa-Ferreira, A. M.; Germain, C.; Homayouni, S.; DaCosta, J. P.; Grenier, G.; Marguerit, E. Transformation of High Resolution Aerial Images in Vine Vigour Maps at Intrablock Scale by Semi-Automatic Imaging. *Proceedings of 15th International Symposium GESCO* 2007, pp 1372–1381.

CropCam. CropCam Unmanned Aerial Vehicle (UAV). 2012, pp 1–9. http://www.robot-shop.com/en/cropcam-unmanned-aerial-vehicle-uav.html (accessed July 22, 2015).

CropCam. Crop Analysis at Finger Tips with CropCam. 2015, pp 1–8. http://www.cropcam.com/ (accessed July 22, 2015).

Digirilamo. Drone Herders: Tanzanian Ranger and Researchers use UAVs to Protect Elephant and Crops. 2015, pp 1–5. http://wildtech.mongabay.com/2015/05/drone-herders-tanzanian-rangers-and-Researchers-use-UAVs-to-Protect-Elephant-and-Crops (accessed Feb 20, 2016).

Dobberstein, J. Technology: Drones Could Change Face of No-Tillage. Precision Farm Dealer. 2013, pp 1–7. http://www.no-tillfarmer.com/pages/Spre/SPRE-Drones-Could-Change-Faceof-No-Tillng-May-1-2-13 (accessed July 26, 2014).

Drone Life. Precision Hawk: Designing the Future of Agricultural UAS. 2014, pp 1–7. http://www.dronelife.com/2014/04/29/precisionhawk.agriculture-drone/ (accessed Aug 5, 2014).

Drone Services Hawaii. Drones in Paradise; Services. 2015, p 107. http://www.droneserviceshawaii.com/services/ (accessed June 23, 2015).

EMBRAPA. The Drones land in Brazilian Agriculture. EMBRAPA, Brasilia, Brazil, 2015, pp 1–2. http://www.freshplaza.com/article/120401/The-drones-land-in-Brazilian-agriculture (accessed May 1, 2014).

Epic.org. EPIC: Electronic Privacy Information Centre. 2014, pp 1–2. http://epic.org/privacy/drones/ (accessed March, 14, 2014).

Falcal, B.; Pessin, G.; Filho, G. P.; Furquin, G.; Carvalho, A.; Ueyama, J. *Exploiting Evolution on UAV Control Rules for Spraying Pesticides on Crop Fields,* Proceedings on Conference on Engineering Applications of Neural Networks. 2015, pp 1–3. DOI: 10.1007/978-3-319-11071-4_5 (accessed June 19, 2015).

Floriano, D. Swiss Researchers Design Super-Sharp 'Insect Eye' for Drones. 2015, pp 1–2. http://sputniknews.com/science/20150813/1025687505.html (accessed Oct 18, 2015).

Frey, T. Future Uses for Flying Drones. 2015, pp 1–37. http://www.futuristsspeaker.com?2014/09/192-future-uses-for-flying-drones/ (accessed May 5, 2015).

Gale Encyclopaedia of Espionage and Intelligence. Unmanned Aerial Vehicles. 2014, pp 1–32. http://www.answers.com/topic/unmanned-aerial-vehicles (accessed July 24, 2014).

Gallacher, D. Applications of Micro-UAVs (drones) for Desert Monitoring: Current Capabilities and Requirements. Zayed University, Dubai, 2015. http://www.academia.edu/11338070/Applications_of_micro-UAVs_drones_for_desert (accessed June 7 2016).

Garcia-Ruiz, F.; Shankaran, S.; Maje, J. M.; Lee, W. S.; Rasmussen, J.; Ehsani, R. Comparison of Two Imaging Platforms for Identification of Huanglongbing-Infected Citrus Orchard. *Comput. Electron. Agric.* **2013,** *91,* 106–115.

Geipel, J.; Link, J.; Claupein, W. Combined Spectral and Spatial modelling of Corn yield based on Aerial Images and Crop Surface Models acquired with an Unmanned Aircraft System. *Remote Sens.* **2014,** *11,* 103335–103557.

Gevaert, C. Combining Hyperspectral UAV and Multispectral FORMOSAT-2 Imagery for Precision Agriculture Applications. Thesis, Centre for Geographical Information Systems, Lund University, Solvegetan; 2014, No 32 pp 83.

Ghose, T. Drones and Agriculture: Unmanned Aircraft may Revolutionize Farming, Experts Say. HuffPost Science. 2013, pp 1–4. http://www.huffingtonpost.com/2013/05/20drones-agriculture-unmanned-aircraft-farming_n 3308164.html (accessed March 20, 2014).

Giles, D.; Billing, R. Deployment and Performance of an Unmanned Aerial Vehicle for Spraying of Speciality Crops. International Conference on Agricultural Engineering. 2014, pp 1–7. http://www.eurageng.eu (accessed June 10, 2015).

Glen, B. Rise of the Field Drones. Precision Agriculture of the Future Employs Robots to Work Fields and it's Here Now. *The Western Producer.* 2015, pp 1–12. http://www.producer.com/2015/02/rise-of-the-field-drones (accessed June 19, 2015).

Gnyp, M. C. *Evaluating and Developing Methods for Non-Destructive Monitoring of Biomass and Nitrogen in Wheat and Rice Using Hyperspectral Remote Sensing.* University of Kuln: Germany, 2014; pp 1–188.

Gogarty, B.; Robinson, I. Unmanned Vehicles: A (Rebooted) History, Background and Current state of the Art. *J. Law Inf. Sci.* **2012,** *21,* 1–18.

Gounden. The Other Side of Drones: Saving Wildlife in Africa and Managing Crime. South African Institute for International Affairs. Africa Portal Library. 2013, p 1. http://www.africaportal.org/search/site/the%other%20side%20of%20drones (accessed Sept 15, 2015).

Grassi, M. J. Rise of the Ag Drones. Precision Ag. 2013, pp 1–7. http://www.precisionag.com/data/imagery/rise-of-the-ag-drones/ (accessed July 19, 2015).

Grassi, M. Five Actual Uses for Drones in Precision Agriculture Today. Drone Life. 2014, pp 1–3. http://dronelife.com/2014/12/30/5-actual-uses-drones-precision-agriculture-today/ (accessed May 18, 2015).

Hall, A.; Lamb, D.; Holzapel, B.; Louis, J. Optical Remote Sensing Applications in Viticulture-Review. *Aust. J. Grape Wine Res.* **2011,** *8,* 36–47.

Hall, A.; Louis, J. P. Low Resolution Remotely Sensed Images of Wine Grape Vineyards Map Spatial Variability in Planimetric Canopy Area Instead of Leaf Area Index. *Aust. J. Grape Wine Res.* **2008**, *14*, 9–17.

Hardesty, L. Drone Lighting: Autonomous Vehicles Could Automatically Assume Right Positions for Photographic Lighting. Massachusetts Institute of Technology News. 2014, pp 1–12. http://www.dpreview.com/articles Cambridge Massachusetts, USA (accessed May 29, 2015).

Herwitz, S. R. Coffee Harvest Optimization Using Pathfinder-Plus (Solar Powered Aircraft). National Aeronautics and Space Agency, USA, 2002. http://www.nasa.gov/missions/research/FS-2002-9-01ARC.html (accessed Sept 2, 2014).

Herwitz, S. R.; Johnson, L. F.; Dunagan, S. E.; Higggins, R. G.; Sullivan, D. V.; Zheng, J.; Lobitz, B. M.; Leung, J. G.; Gallmeyer, B. A.; Aoyagi, M.; Bass, J. A. Imaging from an Unmanned Aerial Vehicle: Agricultural Surveillance and Decision Support. *Electron. Agric.* **2004**, *44*, 49–61.

Hill, C. *Drone Technology Brings New Perspective to Agronomy Decisions.* Eastern Daily Press. 2015, pp 1–4. http/www.edp24.co.uk/business/farming-news/drone_technology-_brings_new_perspective_to_ agronomy_decisions_1_4101360 (accessed July 6, 2015).

Hinkley, E.; Zajowski, T. USDA Forest Service NASA Unmanned Aerial Systems Demonstrations. Pushing the Leading Edge in Fire Mapping. *Geocarto Int.* **2011**, *26*, 103–110.

Homeland Surveillance and Electronics LLC. Agriculture UAV Helicopter Drones: UAV leasing solutions. 2015, pp 1–2. http//:www.agricultureuavs.com/faagrants_uav-permits_for-agriculture_real_estate-/ (accessed May 23, 2015).

Huang, Y.; Hoffmann, W. C.; Fritze Asabe, K.; Lan, Y. *Development of an Unmanned Aerial Vehicle-Based Spray System for High Accurate Site-Specific Application.* Drone Conference, Rhode Island, Providence, USA, 2008, p 1. DOI: 10.13031/2013.24628 (accessed June 19, 2015).

Huang, Y.; Yi, S.; Li, Z.; Shao, S.; Qin, X. *Design of Highway Landslide Warning System and Emergency Response Based on UAV.* Proceedings of 17th China Conference on Remote Sensing 820317. 2011, pp 1–7. http://dx.doi.org/10.1117/12.910424 (accessed March 1, 2016).

Ilnsky, B. Precision Drone: Easy to Use and Fly. Farm World. 2015, pp 1–5. http://www.farmworld.ca/drones/ (accessed June 17, 2015).

Innova, A. Estimating Crop Water Needs Using Unmanned Aerial Vehicles. Science Daily. 2009, pp. 1–5. http://wwww.sciencedaily.com/releases/2009/07/090707094702.htm (accessed March 15, 2015).

Jager, J. Flying in to Increase Crop Yields and Reduce Losses. Seed and Crop Services. 2014. http://www.SGS.com (accessed Aug 20, 2015).

Jensen, T.; Apan, A.; Young, F.; Zeller, L.; Cleminson, K. Detecting the Attributes of a Wheat Crop Using Digital Imagery Acquired from a Low Altitude Platform. *Comput. Electron. Agric.* **2007**, *59*, 66–77.

Juozapavicius, J. How weather Drones will Unravel How Tornados are formed. Huffington Post. 2013, pp 1–5. htttp://www.huffingtonpost.com/technology (accessed July 20, 2015).

Keane, J. F.; Carr, S. S. A Brief History of Early Unmanned Aircraft. *Johns Hopkins App. Tech. Dig.* **2013**, *32*, 558–593.

Kennedy, M. W. Moderate Course for USAF UAV Development. National Technical Information Service, US Department of commerce. 1998, pp 1–2. Info@ntis.goc (accessed August 10, 2015).

Krienka, B.; Ward, N. *Unmanned aerial Vehicles (UAVs) for Crop Sensing.* Proceedings of West Central Crops and Water Field day. University of Nebraska, Lincoln, 2013; pp 1–4.

Krishna, K. R. *Push Button Agriculture: Robotics Drones and Satellite Guided Soil and Crop Management*; Apple Academic Press Inc.: Waretown, New Jersey. 2016; p 405.

KSU. Small Unmanned Aircraft Systems for Crop and Grassland Monitoring. Farms Company. 2013, pp 1–4. http://www.Agronomy.k-state.edu/document/eupdates/eupdates04513.pdf (accessed Jan 24, 2015).

Labre Crop Consulting. Iowa Crop Consulting Firm Offers Ag Services from Drones. AG Professional. 2014, pp 1–4. http://www.agprofessional.com/news/dealer-update-articles/Iowa-crop-consulting-firm-offers-ag-services-from-drones-245201631.html (accessed Aug 5, 2015).

Lamb, D.; Hall, A.; Louis, J. Airborne Remote Sensing of Vines for Canopy Variability and Productivity. *Aust. Grape Grower Winemaker* **2001,** *5,* 89–94.

Lamb, D.; Hall, A.; Louis, J. Airborne/Space Borne Remote Sensing for the Grape and Wine Industry. Geospatial Information and Agriculture. 2013, pp 1–5. http://regional.org.au/au/gia/18/600lamb. htm (accessed Sept 6, 2015).

Lee, W. S.; Ehsani, R.; Albrigo, L. G. *Citrus greening Disease (Huanglongbing Disease) Detection Using Aerial Hyperspectral Imaging.* In the Proceedings of the 9th International conference on Precision Agriculture. Denver, Colorado, 2008, pp 32–39.

Leonardo, M.; Jensen, A.; Coopmans, C.; McKee, M.; Chen, Y. Q. *A Miniature Wildlife Tracking UAV Payload System Using Acoustic Biotelemetry.* Proceedings of the ASME International Design Engineering Technology Conference and computers and Information in Engineering conference. 2013, pp 1–4. http://aggieair.usu.edu/node/124 (accessed Oct 19, 2013).

Li, X.; Lee, W. S.; Li, M.; Ehsani, R.; Mishra, A.; Yang, C.; Mangan, R. Spectral Difference Analysis and Airborne Classification of Citrus Greening Infected Orchard. *Comput. Electron. Agric.* **2012,** *83,* 32–46.

Linehan, P. Drones and Natural Resources: White Pines. 2013, p 1. http://www.psu.edu/mt4/mt-tb.egi/367285 (accessed Oct 7, 2014).

Lucieer, A. About Terraluma Project. University of Tasmania, Tasmania, Australia. 2015, pp 1–8. http://www.terraluma.net/about.html (accessed Aug 8, 2015).

Lucieer, A.; Robinson, S.; Turner, D.; Harwin, S.; Kelcey, J. Using a Micro-UAV for Ultra-High Resolution Multi-Sensor Observations of Antarctic Moss Beds. *International Archives of the Photogrammetry, Remote Sensing and Spatial Information Sciences* **2012,** *39,* 429–442.

Lyseng, R. Ag Drones: Farm Tools or Expensive Toys? 2006, pp 1–7. http://www.producer.com/2006/02/ag-drones-farm-tools-or-expensive-toys/ (accessed June 20, 2014).

Ministry of Agriculture. Beijing Applies 'helicopter' in Whet Pest Control. Ministry of Agriculture of the Peoples Republic of China—A Report. 2013, pp 1–8. http://english,agri.gov.cn/news/dqnf/201306/t20130605 _19767.htm (accessed Aug 10, 2014).

Moffet, C. Unmanned Aerial Vehicles Advance Agriculture. Samuel Roberts Noble Foundation, Ardmore, Oklahoma, USA, 2015, pp 1. http://www.noble.org/ag/research/uavs-advance-ag (accessed May 11, 2015).

Mortimer, G. 'SkyWalker' Aeronautical Technology to Improve Maize Yield in Zimbabwe. DIY Drones. 2013, pp 1–5. http://diydrones.com/profiles/blogs/skywalker-aeronautical-technlogy-to-improve-maize-yields (accessed Sept 5, 2014).

MosaicMill. EnsoMosaic Quadcopter-Complete UAV system. 2015, pp 1–2. http://mosaicmill.com/products/complete_uav.html?gclid=CjwKEAjw96aqBRDNh (accessed July 20, 2015).

Newcome, L. R. *Unmanned Aviation: A brief history of Unmanned Aerial Vehicles.* American Institute of Aeronautics and Astronautics Inc.: Reston, Virginia, USA, 2004; pp 1–29.

Nicas, J. Yamaha Waits for FAA Approval on Agricultural UAS. The Wall Street Journal, New York, USA, 2015, pp 1–7. http://www.rpas-regulation.com/index.php/news-blog-archives/itemlist/tag/trimble (accessed Sept 15, 2015).

Nicole, P. Sustainable Technology- Drone Use in Agriculture. 2015, pp 1–7. https://wiki.usask.ca/display/~pdp177/Sustainable+Technology+Drone+Use+in+Agriculture (accessed July 6, 2015).

Ottos, J. UAVs are Next Wave of Agricultural Technology. Agrinews. 2014, pp 1–3. http://agrinews-pubs.com/Content/News/MoneyNews/Article/UAVs-are-next-wave/ (accessed May 16, 2015).

Paul, R. 2015 Wave Sight UAV—Volt Aerial Robotics. Volt Aerial Robotics, Chesterfield, Missouri, USA. https://www.youtube.com/watch?v=nwu5d8JamTg (accessed July 20, 2015).

Pefia, J. M.; Torres-Sanchez, J.; Isabel de Castro, A.; Kelly, M.; Lopez-Granados, F. Weed Management in Early-Season Maize Fields Using Object-Based Analysis of Unmanned Aerial Vehicle (UAV) Images. *PLoS One* **2013,** *8* (10), e77151. http://www.plosone.org (accessed Aug 30, 2015).

Perry, E. M.; Brand, J.; Kant, S.; Fitzgerald, G. J. Field Based Rapid Phenotyping with Unmanned Aerial Vehicles. 2014, pp 1–3. http://www.regional.org.au/au/asa/2012/precision-agriculture/79333_perrym.htm (accessed March 13, 2014).

Precision Drone LLC. Drones for Agricultural Crop Surveillance. 2013, pp 1–2. http://www.precisiondrone.com/drones-for-agriculture.html (accessed March 12, 2014).

Precision Farming Dealer. New AgEagle Rapid UAS Automation System Unveiled. 2014, pp 1–4. http://www.precisionfarmingleader.com/content/news-ageagle-rapid-rapid-uas-automation-system-unveiled (accessed Oct 21, 2014).

Precision Farm Dealer. Ninox Robotics to Launch Pest Control UAV in Australia. 2015. http://www.pcauthority.com.au (accessed July 21, 2015).

Precision Hawk LLC. Lancaster Platforms in Agriculture. 2014, pp 1–4. http://www.precisionhawk.com/i ndex.htm #industries (accessed Aug 5, 2014).

Primecerio, J.; Filippo, D.; Gennro, S.; Fiorillo, J.; Generio, L.; Matese, A.; Vaccani, F. P. A flexible Unmanned Aerial Vehicle for Precision farming. *Precis. Agric.* **2012,** *13*, 517–523.

ProSeeds Inc. See Regional Data for Corn, Canola and Soybean. Farmers Yield Data Centre. 2015, pp 1–3. http://yielddata.farms.com (accessed May 14, 2015).

Pudelko, J.; Kozyra, J.; Neirobca, P. Identification of the Intensity of Weeds in Maize Plantations Based on Aerial Photography. *Zembdirbyste* **2008,** *3*, 130–134.

Pudelko, R.; Stuzynski, T.; Borzeka-Walker, M. The Suitability of Unmanned Aerial Vehicle (UAV) for the Evaluation of Experimental Fields and Crops. *Zemdirbyste-Agriculture* **2012,** *99*, 431–436.

Rango, A.; Laliberte, A.; Herrick, J. E.; Winters, C.; Havstad, K.; Steele, C.; Browning, D. Unmanned Aerial Vehicle-Based Remote Sensing for Rangeland Assessment, Monitoring and Management. *J. Appl. Remote Sens.* **2009,** *3*, 1–15.

Raymond-Hunt, E.; Hively, W. D.; Fujikawa, S. J.; Linden, D. S.; Daughtry, S. S. T.; McCarty, G. W. Acquisition of NIR-Green-Blue Digital Photograph from Unmanned Aircraft for Crop Monitoring. *Remote Sens.* **2010,** *2*, 290–305.

Reich, L. Precision Hawk. Designing the Future of Agricultural UAS. 2014, pp 1–4. DroneLife.com (accessed April 14, 2015).

Reyniers, M.; Vrindts, E. Measuring Wheat Nitrogen Status from Space and Ground-based Platforms. *Int. J. Remote Sens.* **2006,** *27*, 549–567.

Rieder, R.; Pavan, W.; Carre Maciel, J. M.; Fernandes, J. M. C.; Pinho, M. S. A Virtual Reality System to Monitor and Control Disease in Strawberry with Drones: A Project. 7th International Congress on Environmental Modelling and Software. International Environmental Modelling and Software Society. San Diego, California, USA. 2014, pp 1–8.

RMAX. RMAX Specifications. Yamaha Motor Company, Japan. 2015, pp 1–4. http://www.max.yamaha-motor.Drone.au/specifications (accessed Sept 8, 2015).

Ruen, J. Put Crop Scouting on Auto-Pilot. Unmanned Aerial Vehicles Association, Corn and Soybean Digest: Exclusive Insight. Accession No 84562598. 2012a, pp 1–13. http://connection.ebscohoost.com/c/articles/84562/.html (accessed June 8, 2016).

Ruen, J. Tiny Planes Coming to Scout Crops. Drone Planes Take Aerial Imaging to a New Level. Corn and Soybean Digest. 2012b, pp 1–3. http://cornandsoybeandigest.com/tiny-planes-coming-scout-cros (accessed Aug 3, 2014).

Schwing, R. P. Unmanned Aerial Vehicles-Revolutionary Tools in War and Peace. USAWC Strategy Research Project. United States Air Force. 2007, pp 1–22. http://fas.org/irp/program/collect/docs/97-6230D.pdf/A/Acsc/0230D/97-03 (accessed July 21, 2014).

SenseFly. 'Swinglet cam' by SenseFly. A Parrot Company. 2013, pp 1–5. http://www.sensly.com (accessed July 1, 2015).

SenseFly Inc. Drones for Agriculture. 2015a, pp 1–4. http://www.sensefly.com/applications/agriculture.html (accessed Aug 16, 2015).

SenseFly Inc. eBee by Sensefly. 2015b, pp 1–4. http://www.sensefly.com (accessed Sept, 7, 2015).

Shafri, H. Z. M.; Hamdan, N. Hyperspectral Imagery for Mapping Disease Infection in Oil Palm Plantation Using Vegetation Indices and Red-Edge Techniques. *Am. J. Appl. Sci.* **2009,** *6*, 1031–1035.

SkySquirrel Technologies Inc. Unmanned Aircraft Solutions. 2015, pp 1–4. http://skysquirrel.ca/application.html/ (accessed May 20, 2015).

SPAR Point Group Staff. Trimble Targets Ag Sector with UAV System. 2015. http://www.sparpointgroup.com/news/precision-agriculture/trimble/trimble-targets-ag-sector. (accessed July 23, 2015).

Stowell, L. J.; Gelentr, W. D. *Unmanned Aerial Vehicles (Drones) for Remote Sensing in Precision turf grass Management.* Water, Food, Energy, and Innovation for a Sustainable World. Proceedings of International Annual Meetings of ASA, CSSA and SSSA, Tampa, Florida, USA, 2013, pp 1–2.

Tadasi, C.; Kiyoshi, M.; Shigeto, T.; Kengo, Y.; Shinichi, L.; Masami, F.; Kota, M. Monitoring Rice Growth Over a Prosecution Region Using Unmanned Aerial Vehicles: Preliminary Trial for establishing Regional Rice strain-Report **2014,** *3*, 178–183.

Terraluma. Applications selected case studies. 2014, pp 1–8. http://www.terraluma.net/ showcases.html (accessed Sept 22, 2014).

Tetrault, C. A Short History of Unmanned Aerial Vehicles (UAVS). Dragon Fly Innovations Inc. 2014, pp 1–2. http://texasagrilife extension.edu/ (accessed July 24, 2014).

Thamm, H. P. SUSI A Robust and Safe Parachute UAV with Long Flight Time and Good Pay Load. *International Archives of photogrammetry, Remote Sensing, Spatial Information Services* **2011,** *38,* 1–6.

Thamm, H. P.; Judex, M. The Low Cost Drone. An Interesting Tool for Process Monitoring in a High Spatial and Temporal Resolution. *International Archives of Photogrammetry, Remote sensing, Spatial information Science.* ISPs commission 7th Mid-term symposium. Remote sensing: From Pixels to Process. Enchede, The Netherlands 2006, 36, pp 140–144.

The Des Moines Register. Backers Say Drones will Prove Useful for Farmers. 2014, pp 1–2. Idahostatesman.com (accessed July 23, 2014).

The Economic Times. Drones to Help Rajasthan, Gujarat Farmers Detect Crop Diseases. 2015, pp 1–3. http://articles.economictimes.indiatimes.com/2015-02-19/news/59305367_1_groundnut (accessed May 19, 2015).

The UAV. The UAV-Future of the Sky. 2015, pp 1–4. http://www.theuav.com (accessed June 26, 2015).

Thenkabail, P. S.; Smith, R. B.; Pauw, E. D. Hyperspectral Vegetation Indices and Their Relationship with Agricultural Crop Characteristics. *Remote Sens. Environ.* **2000,** *71,* 152–182.

Thenkabail, P. S.; Smith, R. B.; Pauw, E. D. Evaluation of Narrowband and Broadband Vegetation Indices for Determining Optimal Hyperspectral Wave Bands for Agricultural Crop Production. *Photogramm. Eng. Remote Sens.* **2002,** *68,* 607–621.

Thomasson, A. Farmers of the Future will Utilize Drones, Robots and GPS. The Conversation. 2015, pp 1–10. http://phys.org/news/2015-03-farmers-future-drones-robots-gps.html (accessed June 28, 2016).

Trimble. Trimble UX5 Aerial Imaging Solution for Agriculture. 2015a, pp 1–3. http://www.trimble.com/Agriculture/UX5.aspx (accessed May 20, 2015).

Torres-Sanchez, J.; Pefia, J. M.; DeCastro, L.; Lopez-Granados, F. Multispectral Mapping of the Vegetation Fraction in Early-Season Wheat Fields Using Images from UAV. *Comput. Electron. Agric.* **2014,** *103,* 104–113.

Torres-Sanchez, J.; Lopez-Granados, F.; Pena, J. M. An Automatic Object-Based Method for Optimal Thresholding in UAV Images. *Application for vegetation detection in herbaceous crops. Computers and Electronics in Agriculture* 2015, 114, pp 43–52.

Trimble. Trimble Agriculture. Trimble Agricultural Division, Sunnyvale, CA, USA, 2015b, pp 1–4. http:// www. trimble. com/Agriculture/index.aspx (accessed May 20, 2015).

Turner, D.; Lucieer, A.; Watson, C.; Development of an Unmanned Aerial Vehicle (UAV) for Hyper Resolution Vineyard Mapping Based on Visible, Multi-Spectral and Thermal Imagery. 2012, pp 1–12. http://citeseerx.ist.psu.edu/viewdoc/summary?doi=10.1.1.368.2491 (accessed Sept 4, 2014).

United Soybean Board. Agriculture Gives UAVs a New Purpose. AG Professional. 2014, pp 1. http://www.agprofessional.com/news/agriculture-gives-unmanned-aerial-vehicles-a-new-purpose-255380931 (accessed April 23, 2014).

University of California, Davis. Drones in Agriculture. 2013, pp 1–5. http://ccivineyard. com/drones-in-agriculture/ (accessed May 23, 2013).

Volt Aerial Robotics. Venture: Multipurpose-Professional Grade. Volt Aerial Robotics, Chesterfield, Missouri, USA, 2010; pp 1–7.

Volt Aerial Robotics. Wave Sight. Volt Aerial Robotics. 2015, pp 1–3. http://www.voltaeri-alrobotics.com (accessed July 1, 2015).

Walker. M. Oregon Company Nabs Funding for Water Saving Farm Drones. 2014, pp 1–4. http://www.bizjournals.com/portland/blog/sbo/2014/03/oregon-company-nabs-funding-for.html?page=all (accessed Aug 14, 2015).

Warwick, G. FAA Approves UAV Use for Moviemaking. Aviation Week. 2014, pp 1–3. http://aviation week.com/business-aviation/faa-approves-uav-use-moviemaking (accessed June 21 2015).

Yamaha. RMAX-History. 2014, pp 1–4. http://www.rmax.yamaha-motor.com.all/history (accessed Sept 20, 2015).

Yan, L.; Gou, Z.; Duan, Y. A Remote Sensing System: Design and Tests. Geospatial Technology for Earth Observation. 2009, pp 27–38. DOI: 10.1007/978-1-4419-0050-0_2 (accessed July 11, 2009).

Yintong Aviation Supplies Company Ltd. A Precision Agriculture UAV. 2012, pp 1–3. http://www.china-yintong.com/en/productshow.asp?sortid=7&id=57 (accessed July31, 2014).

Zarco-Tejada, P. J.; Berni, J. A. J.; Suarez, L.; Sepulcre-Canto, G.; Morales, F.; Miller, J. R. Imaging Chlorophyll fluorescence with an Airborne Narrow-Band Multispectral Camera for Vegetation Stress Detection. *Remote Sens. Environ.* **2009,** *113,* 1262–1275.

Zemlicka, J. From the Virtual Terminal: Do your Homework Before Taking Flight with Drones. Precision Farming Dealer. 2014, pp 1–4. http://www.precisionfarmingdealer. com/blogs/1-from-the-virtual-terminal-post-/748-do-your-hoomework-before-taking-flight-with-drones (accessed Sept 15, 2015).

Zhang, Q.; Pierce, F. J. *Agricultural Automation: Fundamentals and Practices*; CRC Publishers: Boca Raton, Florida, 2013; pp 411

Zhao, Y.; Yang, J. X. The design of Mini Quad-copter Unmanned Aerial Vehicles Control Systems. *Adv. Mater. Res.* **2011,** 1–13. DOI: 10.4028/www.scientific.net/AMR.383-390.569 (accessed June 19, 2015).

Zufferey, J. SenseFly: Safely Capturing Data Anywhere, Any Time. 2012, pp 1–18. http:// www.sensefly.com (accessed Sept 12, 2015).

CHAPTER 2

DRONES TO STUDY NATURAL RESOURCES AND VEGETATION

CONTENTS

2.1   INTRODUCTION

Satellites help us in several ways to study natural resources, vegetation patterns and agricultural crop production trends. Satellite imagery was first utilized to study earth's features, such as terrain, topography, natural vegetation and agricultural crop production expanses, sometime in 1970s. Since then, tremendous improvements have occurred with regard to resolution, accuracy and processing of images. Beginning with LANDSAT satellite program in early 1970s, there are now innumerable satellites that help in collecting aerial data and directing global agricultural operations. Satellite imagery related to weather, terrain, water resources (rivers, lakes, canals), cropping systems, yield forecasting, insect and pest attacks, drought/ flood monitoring, soil erosion and desertification have been sought most frequently. A few of the satellite programs relevant to natural resource and agricultural crop monitoring are IKONOS, SPOT, QuickBird, IRS, NigeriaSAT etc. (see Krishna, 2016). Satellites have offered information on natural resources and crop production trends in many agrarian regions. Despite great strides in the sophistication of instrumentation, computer-aided processing and multispectral imagery, the satellite-aided mapping of crop production zones is still not an easy task. Spectral signatures of natural vegetation and crop species are utilized to identify and map the diversity, extent of spread and intensity of vegetation/crops (USDA-NASS, 2009). Satellite images cover very large patches of natural vegetation and agrarian regions, in one stretch. Hence, they are highly useful during macroscale decision-making. The advent of Global Positioning Systems (GPS), geographic information system (GIS) and specialized data banks has added to the accuracy, versatility and usefulness of satellite imagery, particularly in studying natural resources and conducting a range of operations related to agricultural crop production. Satellite-guided vehicles are revolutionizing the management of natural vegetation, forest plantations and field crops. Yet, it should be noted that satellite guidance has constraints with regard to revisit time, haziness of images if the atmosphere is cloudy, inability to focus sharply due to insufficient resolution if regions/ spots to be monitored are small and inability to offer close-up images. Several smaller geographic features may go undetected under satellite imagery. Satellite imagery has to be supplemented with close-range sharp and detailed images derived from airborne campaigns and/or drones.

Airborne campaigns have been utilized as an alternative to satellites as these offer high resolution photography. Such imagery is useful to agencies dealing with natural resource monitoring. Airborne remote sensing offers shorter turn around and helps in initiating aerial photography, at short notice and as many times as required. Yet, it is costly to operate aircrafts and man them with expert pilots. There are innumerable reports wherein, airborne remote sensing has provided high resolution spatial and multispectral imagery of natural resources and farming zones. The spectral resolutions of airborne campaigns have ranged from 0.5 to 2.0 m pixel sizes with 2–20 nm bandwidths in the 450–2500 nm spectral range (Berni et al., 2009 a, 2009b). Such aerial imagery can be utilized in studying vegetation, crop management, estimating leaf chlorophyll contents and detecting drought-affected regions. The Compact Airborne Spectrographic Imager offers images that depict chlorophyll content in the canopy/leaves (Berni, 2009 a, b; Moorthy et al., 2003; Zarco-Tejada et al., 2001; 2004 a, 2004b; Lucier et al., 2004). The Airborne Visible Infrared Imager provides information on leaf water potential. Airborne campaigns have also been used to obtain data related to dry matter accumulation and leaf area index (LAI) with minimized background effects (Berni 2009 a, b; Cheng et al., 2006; Jackson et al., 1977, 1981).

2.1.1 DRONES

Recent trend in monitoring natural resources, vegetation and agrarian belts is to adopt drones, that is, unmanned aerial vehicles (UAVs) that possess miniaturized sensors. They are rapid in turn around and offer very high resolution imagery, because of the proximity of sensors to the surface to be monitored (Berni et al., 2009 a, 2009b; Green, 2013; Vanac, 2014). Drones are slowly, but surely, becoming popular with the personnel involved in remote sensing and monitoring natural resources. They are gaining ground rather rapidly in the agrarian regions. There are reviews stating that drones have been evaluated for their efficiency and accuracy in offering aerial imagery. They are indeed capable of accomplishing a wide range of tasks related to natural resource monitoring and agrarian activity (Jin et al., 2009; Wang and Wu, 2010; Wang, 2010; Salami et al., 2014). Sensilize, a company in Israel, has combined drone-mounted sensors with analytical software. It is called 'Robin Eye' and it combines sensors that operate at eight different wavelengths, from visual (R, G and B) to infrared and thermal range. The multispectral sensor offers great details about natural vegetation and other resources (Leichman, 2015).

Let us consider a few different aspects of natural resources and agriculture where drones are deployed and are potentially useful. Drones are of great utility in mining regions. They allow us to trace and image areas that need attention regarding the intensity of mining activity. Drones are used throughout day and night to monitor activities, control vehicles in the mining region and guide storage of mined material in 'open-pit' mining locations. Let us now consider an example of 'open-pit' mining for phosphatic ores that is finally destined to become fertilizer for crops (phosphate rocks). Here, drones could guide many of the activities such as finding out the volume of phosphate deposits. High resolution imagery allows the detection of margins of minable area and to calculate the volume of ore available. Drones allow us to develop three-dimensional (3D) surface models of the open-pit mines, which in turn help in programming the mining and transport of ores accurately (AIRX$_3$ Visual solutions, 2015; MosaicMill, 2015d; CivicDrone, 2015; SenseFly, 2015; McCannon and Diss, 2015; Hagemann, 2014). Reports suggest that drones can also help in locating areas for dumping mine wastes. Actually, they can be flown over the 'mine waste dump' locations in order to detect the health of trees and vegetation. Excessive dumping of mine waste material may contaminate soil and deterioration may set in. Along with data from chemical analysis of soils, the drone images can help in relocating deposits, based on threshold levels of mined material in the waste dump (CivicDrone, 2015).

Drone-aided aerial survey and imagery of natural resources such as water, soil type, forests and crops are being conducted by several private agencies. They offer drone services to survey a particular region of interest. Mapping water resources and environment is done at a faster pace and at lower cost to farmers, using drones. Such facilities allow professional management of water resources. Major applications of drones with regard to natural resource monitoring, as quoted by many agencies are: (a) topographic mapping; (b) capture of images of natural features and other structures; (c) analysis of natural vegetation; (d) water level mapping in rivers and reservoirs; and (e) mapping environmental effects on soil resources, for example, monitoring soil erosion, top soil loss and soil contamination leading to suppression of vegetation etc. (Novodrone, 2014; UCMerced, 2014). Drones are preferred for data collection mainly because drone imagery offers high resolution in a relatively small execution time. This data is then analysed using appropriate computer software. Drones cost less and can reach areas that are dangerous and almost inaccessible to humans. Drones can fly frequently over natural stands of forest and crops to collect required data and deliver imagery quickly, so that, it could be processed. Drones cover on an average 50–250 ha in an hour. Drones are at least 50 times faster than ground surveys done by skilled technicians on similar sized locations (Novodrone, 2014; Trimble Navigation Systems, 2014; Trimble, 2015). The resolution and accuracy of images is higher than that offered by satellite-aided survey of natural resources.

Horcher and Visser (2009) believe that the size of drones and sensors used in the payload area are important aspects while monitoring forest vegetation. Experiments by USDA Forest Service at San Dimas, California, USA show that drones of small size can be effective in mapping and monitoring changes in forest, the invasive species and any other alterations to vegetation. They used sensors of 8 cm per pixel. 'Photo-stitch software' by Canon was used to arrive at sharp images of forest edges, streams, their course and floor vegetation, wherever possible. They further state that image processing facilities are almost mandatory in a drone for an efficient and accurate mapping effort. Incidentally, these small drones were also effective in determining timber thefts and log movement in the area, if any. However, we have to note that initial capital ranges at

40,000–45,000 US$ for drone platform and facilities at the ground station. Costs are also incurred on ground station to receive imagery and to process it using computers with appropriate software. Replacements may cost further 20,000 US$ based on the extent of usage. One of the observations made is that automated image processing and relay need to be streamlined.

Drones are excellent choice to study geographical features, topography and vegetation from a close range. This is attributable to their ability to fly at low altitudes above the ground. Also, because they and are fitted with high resolution cameras that possess capability for multispectral imagery. Further, processing using appropriate software offers 3D high resolution images (Cyberhawk, 2013; AnalistGroup, 2015). We should note that drones are among the best bets for surveillance and monitoring of disasters related to natural resources and agriculture. Drones are useful in imaging landslides, large-scale gully erosion of farm lands, avalanches, forest fires, dust storms, earth quakes etc. Such imagery from drones allows us to judge the extent of damage, locate the problem areas, note the topographic changes and the loss of vegetation/crop etc. Reports by National Aero-nautics and Space Agency (NASA) and US Geological Survey suggest that drones could be used in various aspects related to wild fire detec-tion, particularly, forest fires. Forest fires and aftermath have been mapped using drones (Cress et al., 2011). Often, disaster affected zones are simply not reachable by human scouts, but drones can fly very close to the spot and relay images. They can even carry out certain remedial activities, such as air-lifting etc. Both natural vegetation and crop production zones are equally vulnerable to environmental vagaries. For example, in a location, a thunder storm destroys crops and natural vegetation such as trees, shrubs and annuals, with same force of destruction. There are drones capable of imaging and mapping such disasters instantaneously. They relay processed images to ground stations or to ipads handled by farmers. For example, ENSO Mosiac disaster mapping and monitoring software offers processed images quickly (MosaicMill 2015c; DroneMapper, 2015; Precision Drone LLC, 2014; Precision Hawk, 2014). Another advantage with drones is that they could be flown without air strips. The copter drones are of course vertical-lift unmanned aircrafts. The resolution of the imagery often ranges at 5 cm. It is amenable for further detailed analysis on a computer (MosaicMill, 2015c; Precision Drone LLC, 2014).

In recent years, there has been a great interest in studying variety of geographic regions/locations, particularly the influence of climatic parameters and climate change effects. Let us consider a few examples. Natural resources, flora and fauna of Antarctica are an important attraction to study using drones. Vegetation in this region is sparse. It comprises mostly boreal and other cold tolerant species, such as mosses. Drones have been adopted in such remote terrain and ice sheets traced in Antarctica. Drone imagery has been utilized to monitor ice sheets, their shrinkage/growth, boreal vegetation, if any, and moss/algal growth (Turner et al., 2014). An Oktocopter was used to derive multispectral imagery of the icy terrain, boreal vegetation and moss growth. The extent of moss growth, its density and spread rate was monitored using the drone imagery. The images were sharp, well focussed and of high resolution, at 3 cm pixel^{-1}. Furthermore, drone-derived imagery could be applied to develop maps depicting spread of moss species. Particularly, to depict their fluctuations in response to temperature in the cold continent. In fact, impact of several other environmental parameters such as diurnal variations, irradiance, moisture and temperature on boreal vegetation and moss growth could also be studied using drone imagery. Drone imagery could help us in tracing and understanding climate change effects on several species of vegetation and fauna, in the remote continent. At the bottom line, drones are easy to fly over ice sheets, and avoid great risks and difficulty experienced by human scouts. Drone-derived images provide unmatched accuracy and clarity about ground conditions at relatively very low cost.

Drones are apt for adoption in tropics, semiarid tropics, arid regions, desert and temperate regions. They are apt for monitoring high intensity vegetation belts as well as low/sparsely vegetated areas. Shahbazi et al. (2014) state that UAVs (drones) have become popular with agricultural agencies and farmers, particularly those interested in remote sensing of natural resources and agricultural expanses. They are preferred when satellite-guided systems and airborne manned flights fail to offer decisive advantages. Their detailed review points out that drones offer robust digital data and orthomosaics that can be analysed and converted to high resolution images. They depict natural resources and crop fields accurately. Drones help in obtaining digital maps useful to national agencies that deal with variety of aspects related to natural resources and agriculture. They are sought for accurate 3D imagery of topography of crop fields and natural vegetation (MosiaicMill, 2015e; DroneMapper, 2015; Trimble Navigation Systems, 2014; Trimble, 2015).

Now let us consider a location with tropical savannah vegetation. Reports about natural resources, particularly, type of vegetation and cropping zones in the West African Savannahs have relied immensely on ground survey. Such surveys are done by skilled researchers and geographists. The careful and tedious mapping of vegetation types that dominate the region has helped us to understand a great deal about biomass and food grain generation in these savannahs (Coura Badiane, 2001; Junge, 2009; Krishna, 2008; 2015). Satellite-mediated imagery and aircraft campaigns have also provided details about vegetation and crops in Savannahs and Sahelian region of West Africa. In the context of this book, we may suggest that drones could be of immense help to several government agencies and farmers in this region in understanding the topography, terrain, soil types, vegetation changes and crop production potential.

Drones are efficient in terms of accuracy of images and cost incurred to obtain them. AnalistGroup (2015) state that topographic survey of land along with its vegetation using drones has almost become a separate discipline. The relief maps of sloppy terrain, land-slides, vegetation and elevation survey done with high resolution cameras are being sought frequently. The rate at which drones can survey and offer images is incredible. A small drone covers about 1 ha every 3–4 min in flight. Drones can be ready to fly in 10–25 min with all the accessories. The endurance of drone in flight is usually 25 min. Some of the computer programs, that is, 2D and 3D mappers, can provide photographs almost immediately thereby allowing the technician to analyse them on computer screens instantaneously (AnalistGroup, 2015).

Let us consider an example from temperate region. In the West Scotland, regions with peat bog and lush forest stand were imaged and mapped using drones. A small zone with predominantly moorland and bog was supposed to be converted to farm and small-scale industry. An area of 400 acres was surveyed using a light drone. About nine flights were required to image the entire stretch. Then, prepare 2D and 3D pictures of the terrain and its topography. The orthomosaics were processed using computer programs that offered images at 3 cm resolution. It seems such images taken periodically are invaluable to assess growth of forest plantation particularly, if the area is being developed as forest plantation. Drones' images were also used to assess environmental effects of the region and its vegetation (Cyberhawk, 2013).

2.2   DRONES TO STUDY NATURAL VEGETATION, FOREST PLANTATIONS AND AGRICULTURAL CROPPING BELTS

In this chapter, greater emphasis has been on studies related to natural vegetation (e.g. forest stands). Discussions on geological formations, mineral distribution and water resources are feeble. In fact, forecasts about regular use of drones to keep watch on natural stands of forests and plantations suggest that, in near future, drones will find a niche with foresters in North America. Linehan (2013) suggests that drones will be useful in surveying and monitoring growth of forest stands, particularly, pines and broad-leaved species grown in North America.

During past 3–4 decades, geographical aspects, such as the terrain, topographical variations, natural resources, forest vegetation and cropping zones have been studied using satellite imagery and airborne campaigns (Hansen, 2008; Getzin et al., 2012). Researchers have concentrated on accumulating data about forest plantations, their fluctuations and the extent of loss of forests in a given area, using satellite imagery. The aim is to understand the impact of deforestation and loss of natural vegetation on greenhouse gas emissions, loss of top soil, gully erosion and reduction in soil fertility. In some parts of the world, obtaining such data could be costly due to fiscal constraints. Of course, there are satellites that are easy to reach and obtain relevant data, for example, the LANDSAT system. However, there are others such as QuickBird or IKONOS that could be cost prohibitive to certain agencies (see Palace et al., 2008; Gardner, 2008). Reduced resolution of satellite imagery makes researchers to adopt ground-based survey and assess forest species, ground flora and fauna. However, ground-based survey is a time consuming and costly activity. It requires skilled scouts to note and map the biodiversity. Therefore, drones are a good alternative.

Drones have been used effectively to monitor the forested and natural vegetation, and to safe guard biodiversity (Jones et al., 2006; Watts, 2010; Lucier et al., 2014). Koh and Wich (2012) devised and developed a drone called 'Conservation Drone 2.0' in Indonesia. It is efficient in providing excellent images about natural resources, forest vegetation and biodiversity (Getzin et al., 2012). Let us consider the above study as an example and in greater detail. They used drones equipped with still-photograph cameras, like Canon IXUS 220 (4288-3216 pixels). The 'Conservation Drone 2.0' mentioned above could also be fitted with video cameras. The cameras could be customized to photograph at specific intervals and could also be calibrated regarding aspects, such as time-lapse for first exposure of cameras, to photograph natural forests. Focal length of cameras could be adjusted based on resolution aimed. Koh and Wich (2012) further state that, 'Conservation Drone' is a low-cost equipment, easily accessible to agencies situated in remote regions and those with limited fiscal resources. It covers about 25 km distance in flight of 15 min. They have reported that missions with this drone were successful in showing soil degradation, when it flew 100 m above the terrain. Drone imagery clearly showed up forest areas with human activity, loss of forest plantation and logging trends. Therefore, conservation practices could be prepared accordingly and the flow of rivers and canals could be monitored frequently. Again, drones flying just above 100 m provided imagery with higher resolution. They state that GPS-tagged photographs processed from orthomosaics were highly useful in understanding the effect of water resources, river flow, soil degradation and other changes. They could detect rampant logging and human activity on forest plantations. Drones also offered details regarding species diversity in natural vegetation. It has been reported that Koh and Wich (2012) flew 32 drone-based missions over the forest stands in the Indonesian forest zones. The drones provided higher resolution imagery and digital data. Also, the images were obtained at much lower costs than that required to buy satellite imagery.

Berni et al. (2009 a, 2009b) state that current sophistication of satellites and their high resolution sensors still falls short of the needs of high degree of details required during crop production. Details about crop canopy, its water status and diseases/pest build up, if any, are not clearly seen. In addition, spectral resolution offered is often less than that required for quantitative remote sensing of natural vegetation. Satellite-mediated monitoring is not an option if quick turn around and frequent revisits are required and we have to note that there is a need for short revisits and quick relay of imagery. Alternatives based on airborne methods such as manned aircrafts are still not feasible ideas; although, they may provide aerial imagery with high resolution and revisits at short notice. But, they are not preferred because of high complexity of operation, need for skilled pilots and prohibitive costs, in terms of capital and repeated costs.

Drones are employed to monitor natural resources and agrarian zones, mainly because the imagery and data derived using satellites and airborne campaigns lack high degree of resolution and accuracy. The spatial and spectral data from satellites is less useful to conduct day-to-day agricultural operations. In addition, the turn-around or revisit times of satellites, as stated above, are often unfavourable. Farmers have to wait for longer period, say, days/weeks (Berni et al., 2009a, 2009b). Berni et al. (2009b) point out that current satellite products have limited use during site-specific farming. Even though sensors in some satellites such as IKONOS or QuickBird offer high resolution, yet it is not sufficient to show up complete details. Further, thermal imaging is restricted to medium resolution sensors such as Terra-Aster.

In the study by Berni et al. (2009a), drones were flown over regions with natural vegetation and crops to obtain imagery using narrow-band multispectral sensors, including thermal sensors. Surface reflectance pattern and temperature were noted after making necessary corrections for atmospheric interferences. Biophysical parameters, such as normalized difference vegetation index (NDVI), transformed chlorophyll absorption in reflectance index, optimized soil-adjusted vegetation index and photo-chemical reflectance index were estimated. Based on above observations using drones, they could estimate the LAI, chlorophyll content and water stress. Water stress was actually detected using photochemical reflectance index measurements. They have concluded that estimates using sensors located on low-cost drones were comparatively more useful and timely both to farmers and other agencies dealing with natural vegetation.

A comparative analysis of drones, airborne campaigns and satellites and their utility in analysing natural resources is useful. Drones have particular advantages regarding the resolution of natural vegetation and crop land imagery (Candiago et al., 2015). Drones operate at 0.5–10 cm resolution, while aircraft photos offer 5–50 cm resolution and satellites are much hazier at 1–25 m. The field of observation and photography is smaller with drones. It is because they fly too close to the surface of the canopy of natural stands or crops. The field of view for drones is 50–500 m, for airborne photos it is 0.2–2.0 km and for satellites it is 0.5–5 km. In addition, drones do not require a pilot; however, a technician is required with expertise in flying different models of drones and directing their movement from a ground station. A pilot is a necessity for airborne campaigns. The cost for photography derived from aircrafts and satellites is relatively higher. Ultimately, agencies have to decide on the type of remote sensing vehicle. Economic advantages of drones too count while deciding which method to opt for survey of natural resources.

Often, in a natural setting, we find that the cropped zone, natural vegetation (herbs, shrub vegetation, trees) and waste land with scanty vegetation exist interspersed. There are drones that are specifically fitted with high resolution cameras allowing agencies to study such regions of mixed vegetation, map them accurately and monitor the changes that occur within each compartment of vegetation. There are actually private drone agencies that offer aerial imagery and advice to farmers mainly about happenings in the natural vegetation and crop fields simultaneously. If crop fields and natural vegetation are situated very close in the geographical settings, various climatic parameters and their influence are similar or at least inter-linked. In a day, drones may fly past as much as 10,000 ha of natural vegetation/crop land mixtures and provide orthomosaics for processing. Usually, 3D imagery is provided to farmers and governmental agencies that deal with natural resources (MosaicMill, 2015b; Trimble Navigation Systems, 2014; Precision Hawk, 2014). Incidentally, during a flight, drones could collect data about mineral resources, mines and mining activity, water resources and natural vegetation, simultaneously.

2.2.1   DRONES IN MAPPING AND MONITORING FOREST TREE VEGETATION

Reports suggest that there is a strong demand for information concerning forest structure and growth. Such information helps in planning forest development, wood production trends, logging and in optimizing wood flows from forested zones (Fritz et al., 2013). Knowledge about distribution of different tree species and quality of wood derived from them is almost essential to any one dealing with forest plantations. High resolution 3D information of forest stand offers greater detail (Frolking et al., 2009). A 3D image that offers planters with information regarding biomass accumulation is a necessity.

Knowledge about rate of increase and storage pattern of carbon in the forest stand is also useful. Several different methods have been adopted by forest planters to prepare maps and assess the status of their forest stands, the biomass accumulated and productivity (Asner, 2009). During past 5 years, interest in using drones to collect information about forest structure, tree species that cover the forested zones, their biomass accumulation trends and carbon storage has increased (Fritz et al., 2013; Wallace et al., 2012). Drones are being preferred against airborne Lidar-based methods (Fritz et al., 2013; Haala et al., 2011; Hyyppa, 2008). Recent studies suggest that drones could be used to reconstruct forest structure. According to Fritz et al. (2013), there is demand for aerial imagery (orthomosaics) of forest stands done using low flying drones mainly because their resolution is high. Most often, planters need aerial imagery of vegetation at various stages to compare and arrange logging schedules. Further investigation and standardization of drone-based procedures are needed particularly with regard to study of forest stands. Aspects that need attention are fixing flight path of drones above the forest stands, processing the raw imagery and utilization of spectral data.

Forest clearances, deforestation prior to initiation of food crops by farmers and gaps that occur due to rampant lumbering activity need to be monitored. The gaps in natural forest stands or plantations need to be regulated. These gaps are also created due to tree death, inter-tree competition, wind or storm, diseases, insect pests and senescence. Usually, human scouts are employed to mark the regions with loss of trees and gaps created for dwelling or crop production. Drones could serve the forestry agencies in efficiently tracing the gaps in a forest stand and detecting clearance made for crop growth by the forest dwellers (Getzin et al., 2014; see Plate 2.1). Let us consider an example pertaining to unmanaged temperate forest in the European loess region. Drones actually offer cost effective monitoring of forest gaps, tree health and productivity. The forest gaps are created naturally due to factors, such as destruction of trees due to storms, soil erosion or lack of tree growth. Forest gaps are also created by humans. Farmers adopting shift agriculture cut trees and create open spaces for crop production. It helps them derive food grains. Accurately detecting such gaps and the sizes of each gap is essential to forestry agencies. Drones could be used to image the small gaps, their frequency and loss of tree species. They could also be used to trace large gaps and their frequencies. Forest gaps larger than 100 m^3 were classified as large and smaller than 10 m^3 as small. Digital maps of forest stand showing the distribution of gaps that are accurately marked using GPS coordinates are highly useful to forest rangers and help during the course of forest management. Aerial imagery of forest gaps derived using drones could also help in detecting causes, such as diseases affecting trees, soil erosion, human activity etc.

PLATE 2.1 A drone scouting over natural vegetation in Kansas state, USA

Note: Drones could be used to monitor natural vegetation, droughts, fires, floods and fluctuations of natural vegetation and cropped zones.

Source: Tom Nicholson, AgEagle Inc., Neodesha, Kansas, USA

Now let us consider an example pertaining to temperate region. In Norway, peat bogs, mires and forest vegetation are being monitored using drones. The shifts in wetland, loss of wetlands and changes in botanical species within this region could be studied using low flying drones (Bazilchuk, 2016). The Norwegian project actually aims at restoring wetlands and their species diversity. They are using drone imagery as a method to collect data periodically. Drones are also used to assess greenhouse gas emission from wetlands. The digital elevation maps that drones provide are of great advantage to researchers and workers involved in restoration of forests (Bazilchuk, 2016).

Drones are gaining in popularity among government agencies dealing with forests. Forestry companies and plantation owners have also evinced great interest about drones and their utility in managing forest belts. They are using drones to obtain aerial imagery. There is wide range of drone companies that offer high resolution images at frequent intervals and almost immediately upon request (AIRX$_3$ Visual Solutions, 2015; Trimble, 2015; Precision Hawk, 2014; Plate 2.1). For example, EnsoMosaic imaging system for regular observation of forests helps in forestry planning, monitoring activity in forest plantations, wood procurement and wood movement (MosaicMill, 2015a). The hyperspectral imagery helps in mapping forest growth in great detail. Drone imagery could also be used to distinguish between broad-leaved trees and pine species. A digital library or data bank of various forest tree species traced in the region, along with their spectral signatures, is a useful idea. Drones are really rapid in terms of obtaining ortho-mosaics and offering processed images of forests and subregions of interest. Typically, drones collect images in an area ranging from 2000 to 50,000 ha. Such images can be picked at ground station and processed in a matter of hours or at best in a couple of days. This is something impossible if human scouts are employed. Simple cartography of forest itself is a tedious task if skilled technicians are asked to map the forests. Drone imagery has also been utilized to calculate data about volume of wood (MosaicMill, 2015a). Digital surface models (DSM) of forest vegetation prepared using drone imagery are useful. They are consulted while devising several of the procedures in the plantations. Drones offer 3D DSM of forests within a short span of time as the fly past the forests. Drone-aided forest management is more efficient in terms of labour requirement, cost of conducting various types of scouting and even applying remedial spray, if any. Over all, using drones to maintain forest plantations is a profitable proposition.

2.2.2 DRONES IN TREE PLANTING PROGRAMS AND AFFORESTATION

A recent report by BioCarbon Engineering, U.K. suggests that they have developed a drone-dependent system that helps the agencies involved in planting trees and developing forest vegetation (Markham, 2015). The system suits best when tree planting has to be accomplished rapidly. Traditional methods of tree planting, particularly, planting sets/explants or seed dibbling is a slow process. The above system takes seeds in the containers available in the payload region of drones. In addition, these drones try to obtain a detailed 3D view of the terrain. Such 3D images help planters to organize the drone's planting path carefully, after weighing the *pros* and *cons*. These drones actually offer precision planting of tree seeds. Such drones have been tested and proven as useful farm vehicles particularly during planting nutrient-encapsulated seeds, *Bradyrhizobium* treated leguminous seeds and fungicide/bactericide treated seeds. Even pre-germinated tree seeds are dispersed across natural areas by drones. The same set of drones can be used in monitoring seed germination and seedling establishment. Therefore, agencies related to developing forestry areas and natural vegetation should be inclined to adopt drones. Further, it has been opined that such 'rapid methods' involving drones are essential. No doubt, forest re-planting at a rapid pace could be done using drones. An important constraint to note is the viability of seeds and germinated sprouts.

2.2.3 DRONES TO STUDY NATURAL AND MANMADE PASTURES

Pastures are well distributed across different continents. They can occur naturally or can be man-made. A detailed knowledge about pastures is mandatory in regions supporting cattle and few other domestic species. Here again, drones could play an important role in the surveillance, upkeep and enhancement of productivity. Salami et al. (2014) have reviewed extensive possibilities that exist for drone usage. They emphasise on how drones are well suited to monitor and maintain range land, pasture and other similar vegetated regions. They say adoption of drones to monitor pastures has been possible mainly due to digital cameras/sensors. The high resolution of imagery offers excellent images of ground vegetation. Remote sensing of pastures could become much easier due to small drones as they could be launched at any instant to rapidly accrue data about pastures.

Pastures developed by farmers and those existing naturally are an important component of agrarian belts worldwide. They are essential for sustenance and production of domestic animals, such as cattle, sheep, piggery etc. Satellite imagery could be used if pastures are large and observations aimed at, fall within the limits of resolution. However, deciphering botanical diversity and dominant species within pastures may be difficult even with high resolution multispectral cameras. During recent years, drones have been tested and utilized to monitor and relay imagery of variety of vegetation, including pastures, and to study their botanical composition. The basic idea is to analyse the spectral reflectance properties of mono-species pastures, mixed pastures, mixed legume pastures etc. The NDVI data derived from alfalfa mixtures, for example, vary based on extent of different species encountered in addition to alfalfa. Spectral reflectance of mixed pastures could be utilized to analyse the effect of a wide range of factors such as season, precipitation, temperature, fertilizer supply, insect/disease control measures and grazing on pasture species and productivity (Kutnjak et al., 2014). Drones, both semi-autonomous and autonomous versions, cost much less to obtain aerial imagery of mixed pastures.

Now, let us consider an example that depicts usage of drones for the study of pastures and their botanical composition. In Croatia drones have been evaluated for their performance in judging pasture fields at the Experimental Station of Zagreb. The mixed pasture supported alfalfa sown at a seeding rate of 1.g m³. Alfalfa was the major component along with orchard grass (*Dactylis glomerata*) and Italian grass (*Lolium perenne*). A small flat-winged drone named *eBee* produced by Parrot Inc. of Switzerland was used repeatedly to image the mixed pasture (SenseFly, 2015). The drone was fitted with Canon IXUS 125 HS camera for visible range exposures and Canon IXUS 125 HS NIR for infrared imagery. The flight route was fixed prior to the launch of drone. The orthomosaics were processed using *Postflight Terra* 3D computer software. Mainly NDVI was noted on mixed pastures and individual species wherever possible. Sampling of individual pasture species and mixtures were done several times, as the drone flew over the pasture. According to Kutnjak et al. (2014), there was positive correlation between NDVI values noted and relative share of component legumes and grasses. In other words, NDVI indicated the extent of each botanical species or at least the dominant species in the pasture land. The accuracy of detection of each botanical species depends on its spectral reflectance properties and resolution of cameras. Incidentally, knowledge about pastures species and extent of legume and grass is essential. It allows farmers to get an idea about nutritional characteristic of the feed material. We have to note that drone imagery could also be used by focusing the cameras (sensors) at a particular wave length so as to collect data on NDVI specific to each grass/legume species that forms the component of the pasture. The above study shows the potential of drones, particularly, for use in large pastures of North America and Europe. It can help in monitoring the pasture lands from several different aspects of maintenance and production.

2.2.4   DRONES IN MONITORING WATER RESOURCES, FLOODS AND DROUGHTS

Drones could be effectively used to monitor reservoirs, fluctuations in water level in response to its usage, floods and drought conditions. Drones have been used to map the spread of noxious weeds that float and clog the large lakes and other water bodies. Algal blooms in reservoirs and ponds could be easily imaged using multispectral cameras fitted on drones (Wharton, 2013; Farms.com, 2013).

Reports from agencies dealing with water and other natural resources in the European Commission Area suggest that so far drones have been successfully utilized to monitor water resources in the reservoirs, influence of weather patterns on water flow in rivers, irrigation canals, storage in dams and ground water sources (Doward, 2012). A few other reports suggest that drones with visual, infrared and thermal imagery facility could be excellent in monitoring water bodies. Drones could detect water movement in the streams. Fluctuations in crop production zones due to changes in water flow within streams, canals and irrigation channels too could be monitored effectively. Influence of season on water storage in the underground aqua-ducts (e.g. Ogallala) has also been studied using drones. Melting of ice sheets and its effect on water flow has also been monitored using drones (Linehan, 2013; Jensen et al., 2014).

Water bodies, such as lakes and rivers that irrigate agricultural crops also support a range of nuisance green algae. If they occur as dense growth, then they may clog water and affect water quality. Monitoring such green algae is possible using drones fitted with cameras that operate at visual and NIR spectral band widths. Flynn and Chapra (2014) have reported that they conducted more than 18 drone missions over the water bodies infested with green alga—*Cladopora glomerata*. Digital data obtained through cameras were analysed and mapped automatically. The imagery correctly identified the patches of filamentous green alga in 92% cases. Results from drone flights also showed that green algal growth ranged from 5 to 50% in a given region within the water body. Algal growth peaked during summer season and declined later. It has been forecasted that, in due course, drone-aided imagery could be used to study a wide variety of floating and partially submerged flora on water bodies. Drones could be used to keep a watch over the undue growth, if any, in tanks, lakes, dams and rivers.

In tropical regions, aquatic weeds such as *Eichornia* and *Salvinia* are common. They grow luxuriantly and often clog water bodies and affect water quality. Such weeds have specific spectral properties. Therefore, they could be easily identified by cameras on drones. We could perhaps classify aquatic weeds of importance to the upkeep of water bodies using not just their botanical features, but including spectral properties. It then assures that weed growth on water bodies are correctly diagnosed for species diversity, extent of growth and dominance. Remedial measures to reduce or eradicate aquatic weeds will be easier, if aerial imagery and accurate digital data are supplied. Building a data bank of spectral signatures of water borne weeds is a good idea.

Researchers from the Chinese Academy of Agricultural Engineering report that small UAVs (drones) could be used to monitor large expanses of maize. Soil deterioration that occurs due to environmental vagaries could be detected (Zongnan et al., 2014). High rainfall events could be disastrous on maize crop stand. Scouting may be time consuming. It gets delayed if unskilled farm workers are hired. On the contrary, drones could fly over the entire field in a matter of minutes. They actually offer detailed images of soil loss, lodging, seedling loss and loss of grains. Drones have also been used to study effects of torrential rains and weeds on wheat crop including the lodging that may ensue in the fields (Liu et al., 2005; Hu, 2011).

Drones have a role to play in monitoring drought and ascertaining crop loss due to it, using aerial imagery. Drones are of great utility in surveying large areas of crop land that might have suffered drought effects. Droughts usually cause severe damage to crops. It is based on the intensity and period for which crops have been exposed to moisture dearth. Crop species too determine the extent of damage suffered due to drought. Usually in a large agricultural zone that has experienced drought, several crop species might have been grown. Drones with ability to identify crops based on their spectral properties and detect the drought stress suffered by each species are highly useful. Drought effects are also seen in patches based on soil type, its moisture holding capacity and variations experienced. Farm workers skilled in detecting droughts of different types and severity are required to detect and map the affected regions accurately. Human fatigue and skill related factors do affect the accuracy of such reports. Instead, if drones are used to make a few trips of flight over the crop fields, they can offer digital data and maps in a matter of few minutes. Drones give an overview of a large area of crop field along with drought affected zones. Such an aerial imagery is useful to farmers. Let us consider a few examples. Reports suggest that Californian agrarian regions have experienced drought for the entire past 2 years. Drought has affected natural vegetation (including forests) and crop land severely. Drone technology has allowed them to map the drought hit areas. Further, they are now trying to seed the sparse clouds that come over the region so that rainfall is created in the area that needs water most (Patrick, 2015). The use of drones to seed the clouds close to farm land has induced great interest among state agricultural agencies. In addition, they believe that drones, in general, have a great potential in improving our understanding about environmental effects on crops and alleviating them, wherever feasible.

2.2.5 DRONES TO STUDY WEATHER PATTERNS AND CLIMATE CHANGE EFFECTS

A drone could be small equipment flying over the farms, but it is expected to make large impact with regard to monitoring natural resources, agrarian regions and weather patterns that affect crop production trends. Drones could also have a strong impact on assessing weather related changes on water resources and on crop productivity (Wharton, 2013; Richardson, 2014).

Drones may become more common during collection of data pertaining to weather in agricultural belts and other areas with natural vegetation. Galimberti (2014a, 2014b) states that drones can provide rapid information and warnings about thunder storms and tornadoes. Reports and early warning could be relayed almost 60 min ahead. Previous timing was only 20 min ahead of a storm event. Earlier, researchers used radars, mobile instruments and balloons to collect data. Now weather research drones can provide high quality images due to the closeness of the sensors to the events unfolding. Reports by NASA (2013) suggest that weather drones could be useful in understanding, how tropical storms get strengthened. A robust drone is preferable compared to a costly aircraft as we have to send the vehicle into the storm. They say drones are much safer compared to other methods while dealing with strong storms and tornadoes. Galimerti (2014a, 2014 b) further states that drones could be handy in collecting data pertaining to general weather and storms, such as atmospheric pressure, temperature, humidity and wind velocity. Such measurements could be transmitted in due course or relayed immediately to weather computers. Regarding routine measurements, it is clear that drones offer advantage in terms of time lapse from collection of data to its relay to computers and other processers. Drones help in providing data to forecasters quickly so that they can analyse a series of parameters and develop weather models. Drones could be of utility while collecting data about climate change. Reports suggest that drones have been utilized to observe aerosol and gaseous emissions from agricultural fields. Periodic observations about stratosphere chemistry, troposphere pollution and air quality are possible using drones. Aspects such as vegetation changes, nutrients in coastal atmosphere, emissions from forest fires and emissions from agricultural fields, particularly CO_2, CH_4, NO_2 and N_2O are studied using drones. Drones are excellent in collecting air samples above the crop canopy. Hence, various gaseous parameters and water vapour could be deciphered. In addition to obtaining data from large expanses of crops/natural vegetation, drones could be handy in obtaining data about atmospheric changes just above an individual farm (Carr, 2004). At present, drones are becoming popular with weather experts. They are useful when used with other instruments, such as remote-controlled aircrafts, manned aircrafts and doppler radars placed at vantage locations in the farm or agrarian belts (Scott, 2011). Galimberti (2014a, 2014b) states that, on many occasions, drones could improvise the focus and sharpness of data from a region if satellite-aided data is already available but with insufficient resolution. In other words, drones could increase our accuracy of predictions about weather. We may note that drones have also been put to use to detect loss of ozone layer. For example, NASA has experimented with drones, to collect data relevant to changes in ozone layer (Oskin, 2013).

Weather experts in several different regions are trying to adopt drones to investigate parameters related to storms, tornadoes, torrential rains, droughts, dust storms etc. The idea, it seems, began with an effort to fly drones as 'Tornado Chasers' in the Central Plains of North America. A study at Oklahoma State University aims at collecting data related to precipitation pattern, storms and tornadoes using drones (Keber, 2013; Jouzapavicious, 2013; Jarvis, 2014). They say, drones could be used as instruments 'first to react' and respond to storms. They can offer most useful data and information about the strength and ferocity of storms to meteorologists. They aim at identifying which weather pattern will become a severe storm or tornado. Some of the drone machines tested in these laboratories weigh around 50 lb. They are made rugged, as much possible, to withstand storm strength. These drones are estimated to cost 10,000 US$. They could be owned by community or county weather agencies. Further, it has been stated that tornado warnings initially were before just couple of minutes, then it improved to 13 min owing to classical weather forecasting machines. Satellites further helped in advancing the warning period. Currently, with drones and other electronic surveillance equipment, farmers are able to know of imminent tornadoes about 13–20 min ahead of the strike. Farmers will have a chance to safe guard equipment, cattle, crop related material from destructive path of tornadoes. The general population too could move swiftly away from tornado's path.

Regarding floods, it is a routine practise to map the flood prone area and estimate the extent of damage caused in given farming zone. Such maps are sent to state agricultural agencies and national flood insurance programs. It is said that many of such flood mapping efforts are old and need updating (Shan et al., 2009). Previously, such flood maps were obtained via high resolution satellite imagery, manned aircrafts and human scouts. However, at present, we have the option of using drones. Drones could be highly specific to a region, a farm or even small group of crop fields that are prone to floods. Such maps could be updated and supplied to farmers. Drone imagery could be of great utility to farmers adopting remedial practices and also to insurance companies that have to compensate farmers who suffer flood damage. Aerial imagery obtained using drones could be digitized and sent via internet. They are highly specific, since flood maps are GPS tagged and of high resolution. Indeed, drones have a big say in monitoring, mapping and relaying post-flood images to farmers and agricultural agencies (Towler et al., 2012).

2.2.6 DRONES IN CLOUD SEEDING

Agrarian regions depend immensely on timely precipitation events. The extent of rainy period, intensity and total precipitation are all important factors. Droughts of lengthy period that cause irreversible damage to crop stands and productivity have to be tackled by one way or other, if grain yield levels are to be maintained. Irrigation is the common method. There are indeed several variations of irrigation methods adopted by farmers worldwide. Natural vegetation is also dependent on periodic precipitation events. In many regions, experts adopt cloud seeding methods so that natural vegetation and crops survive, grow and offer acceptable yield. Cloud seeding has been done using manned aircrafts. However, now, weather experts believe that drones could be handy instruments to induce rains through cloud seeding (Galimberti, 2014b). Cloud seeding is done using silver iodide or solid CO_2. Cloud seeding has been practised in United States of America as well as the Weather Department in China seems to be using cloud seeding rather frequently (Galimberti, 2014b). At present, there are a few states in North America, where in, weather experts are standardizing methods to adopt drones for cloud seeding. Drones could modify precipitation pattern at least for a short duration so that crops and natural vegetation grow better and alleviate water stress by seeding clouds in drought-prone regions eliciting rains (Galimberti, 2014b). It has been suggested that small drones can be effectively used to fly just below the base of the cloud and seed them. These could be helpful in micro-management of cloud seeding and the study of the effects of cloud seeding at comparatively lower costs.

2.2.7   MONITORING SOIL EROSION USING DRONES

Soil erosion is a worldwide phenomenon. It is one of the consequences of climate change and is a major factor that affects natural vegetation and crops alike. Natural vegetation including trees, shrubs and under-story species may suffer a great deal due to erratic climatic conditions. Heavy precipitation may cause breaches and induce soil erosion, leading to formation of gullies. Sheet and rill erosion induces loss of top soil and nutrients with it. Shrub vegetation, waste land and pastures are also prone to such maladies. No doubt we have to monitor the changes in the terrain, vegetation and its growth pattern periodically, and take note of ill effects of climate. Remedial measures should follow the suit.

In Australia, there are projects that are examining the usefulness of drones to monitor the natural forest stands, shrub vegetation and the coastal species. They are observing these same regions for soil erosion, loss of top soil and fertility aspects (Terraluma, 2014). They have assessed drone-derived images mainly to know if they are useful in ascertaining changes in the terrain, particularly, erosion and loss of surface soil at sub-decimetre level. The accuracy of drone-produced dense point clouds, using multi-view stereopsis techniques (MVS), was compared with other methods, such as differential GPS and total station. They report that geo-referenced point cloud (< 1–3 cm point spacing) were accurate to 25–40 mm specifically, if imagery was obtained from drones placed at ~50 m above the terrain. Such techniques could be useful in assessing coastal damage, vegetation loss and soil erosion (Terraluma, 2014). In Tasmania, drone-based monitoring has helped in assessing changes in terrain that supported salt marsh, shrub vegetation and coastal grasses.

Drones fly closer to the terrain affected by rains or dust storms or over-grazing, resulting in soil erosion. They can offer close-up view with greater details and at higher resolution. Satellite-mediated observation of sandy terrain and cropping belts too are conducted regularly. However, D'Oliere-Oltmans et al. (2012) have stated that drones could close the information gap that occurs, if satellites, aircrafts and ground-based observations are insufficient. For example, fixed-winged drones such as Sirius-1, equipped with visual and NIR cameras could offer images of high resolution. Such images depicted features of terrain and crops grown in Morocco. Clear images of terrain, soil erosion, leaching, loss of vegetation and poor crop stand could be obtained, using drones. Several types of erosion such as top soil erosion, gully erosion, sheet erosion and loss of crop stand could be periodically monitored. In fact, drone imagery can offer both 2D and 3D images of eroded region. Such images depict the size and severity of gully or sheet erosion etc. Usually, gully erosion occurs at different scales and shows marked temporal variability. Therefore, it is believed that frequent assessment of situation on the terrain, using small format aerial images, is useful. Gully retreat too could be monitored using 3D imagery and high resolution mapping obtained by the use of cameras mounted on drones (Aber et al., 2010; Marzolff and Poesen, 2009; Marzolff et al., 2011). Drones were utilized for both short- and long-term monitoring of gullies and surface soil erosion in the Sahelian region (Marzolf et al., 2011). In the Sahelian region, Marzolff et al. (2002) have experimented with kite aerial photography, to study gully erosion. This method was cost efficient.

Soil erosion in regions with natural vegetation and crops is caused by both, water and wind. The extent of erosion due to each factor varies depending on weather pattern, geographic location, nature of soil, crop-ping pattern and agronomic procedures adopted. To quote an example, in Eastern Europe, there are locations wherein water induces 50% erosion, but wind induces only 10%. In the Sahelian region that is prone to dust storms, wind is a major cause of soil erosion (see Krishna, 2008, 2014, 2015). Similarly, dust storms cause rampant loss of top soil in the Central plains of North America (Krishna, 2015). Several different methods are adopted to study the erosion pattern, understand causes and arrive at suitable remedies (Sarapatka and Netopil, 2010). In the present context, we may suggest that drones could be used frequently to monitor natural processes related to climate change. Drones with multispectral cameras provide excellent details about erosion (UCMerced, 2014).

Rivers flow at variable rates based on topography, season, volume of water brought by tributaries and catchment. Several other factors related to weather, terrain, soil profile structure and hydrological regimes also affect river flow. Knowledge about water from riverine sources is almost essential in all the agrarian belts. It is particularly important to monitor river flow in regions prone to floods, drought, erosion, dammed regions and agricultural cropping belts. Most often, satellite imagery has been used to monitor major rivers and their tributaries. Agricultural agencies conduct aerial surveys to understand water resources available for crops in the given zone. They also survey rivers for soil surface erosion and loss of embankment. Rivers are prone to overflow, causing inundation and erosion. Rivers remove large quantities of silt and deposit it at distant locations. Often, silts are rich in minerals and improve fertility of soils in locations where deposits accumulate. River monitoring can be done using human scouts, but it is tedious and costly to hire a scout. Often governmental and rehabilitation agencies need aerial imagery and accurate data rather quickly after a flood, or drought spell or for regular crop production. Drones, with their ability to take to sky quickly and in succession, as many times, procure detailed digital data and images. Hence, they are perhaps the best bets. Satellite imagery could be low in resolution, and not possible at all times. We have to wait for next transit and cloud-free conditions. Worldwide, rivers are being monitored regularly using satellites, manned aircraft campaigns and more recently using small drones. There are indeed innumerable examples possible, where drones could replace previously tedious and difficult methods. For example, in the Delta region of Bangladesh and India rivers are the reason for frequent floods, overflow, inundation and submerging of the cropped land. It is based on tides and rainfall pattern. River bank erosion is a major problem, since it affects crops. Human dwellings and crops are shifted to higher locations based on reports or warnings of impending floods. Such events can be regularly monitored. Prior information could be supplied to farmers in the delta (Krishna, 2015). Similarly, in Sahelian West Africa the embankment of rivers that flow over a sandy terrain is highly vulnerable to breaching (Descroix et al., 2012). It causes soil erosion. River breaching and loss of water resource is attributable to land clearing, overgrazing, change in hydrologic regimes, intensive cropping and torrential rains resulting in overflow. We have depended entirely on ground observations and satellite imagery, to take note of effects of erratic climatic pattern. However, drones with their ability for detailed high resolution images could be worthwhile. They could be used, along with satellite imagery that is suited better for observing very large regions. Drones could be used to study the impact of different sources of water such as seasonal precipitation, lakes, wells and rivers on the productivity of natural vegetation and cropping systems. It helps agricultural agencies in West African savannahs and Sahel to accordingly modify cropping patterns, adopt remedial procedures and improve production efficiency (Krishna, 2008, 2014, 2015).

Over all, drones are a better option and are destined to become a common instrument to study geological features, weather patterns, natural resources, soils, forest stands, shrub vegetation and most importantly crops and cropping patterns.

KEYWORDS

- **drone scouting**
- **geographic information system**
- **Global Positioning System**
- **weather research drones**
- **cloud seeding**

REFERENCES

Aber, J.; Marzolff, I.; Ries, J. B. *Small Format Aerial Photography: Principles, Techniques and Geoscience Applications;* Elsevier: Amsterdam/Netherlands, 2010; pp 256.

AIRX₃ Visual Solutions. Sky is the Limit: Mining, 2015, pp 14. http://www.arix3.com/#!mining/cm89 (accessed Nov 24, 2015).

AnalistGroup. The topographic Land Survey with Drone has Become a New Discipline, 2015, pp 1–14. http://analistgroup (accessed Nov 27, 2015).

Asner, G. P. Tropical Forest Carbon Assessment: Integrating Satellite and Airborne Mapping Approaches. *Environ. Res. Lett.* **2009,** *4,* 034–039.

Badiane, C. Senegal's trade in Groundnut. Economic, Social and Environmental Impacts, **2001,** pp 1–32. TED case studies number 646. http://www.Senegalcasestudy.com (accessed May 10, 2016).

Bazilchuk, N. Drones will Monitor Norwegian wetlands. *Front. Ecol. Environ.* **2016,** *14,* 64.

Berni, J. A. J.; Zarco-Tejada, P. J.; Suarez, L.; Gonzalez-Dugo, V.; Fereres, E. Remote Sensing of Vegetation from Uav Platforms Using Lightweight Multispectral and Thermal Imaging Sensors. In *International Archives of Photogrammetry and Remote Sensing Spatial Information Science* 2009a, 38, 722–728.

Berni, J. A. J.; Zarco-Tejada, P. J.; Suarez, L.; Fereres, E. Thermal and Narrowband Multispectral Remote Sensing for Vegetation Monitoring from an Unmanned Aerial Vehicle. *IEEE Trans. Geosci. Remote Sens.* **2009b**, 1–10. DOI: 10.1109/TGRS.2008.2010457. http://ieeexplore.ieee.org (accessed Sept 10, 2016).

Candiago, S.; Remondino, F.; Giglio, M. Farming Applications from UAV Images. *Remote Sens.* **2015**, *7*, 4026–4047.

Carr, P. Unmanned Aerial Vehicles: Examining the Safety, Security and Regulatory Issues of Integration into US Air Space, 2004, pp 1–55. http://www.ncpa.org/pdfs/sp_drones_-long-papers.pdf (accessed Sept 20, 2014).

Cheng, Y. B.; Zarco-Tejada, P. J.; Riano, D.; Rueda, C. A.; Ustin, S. A. Estimating Vegetation Water Content with Hyperspectral Data for Different Canopy Scenarios: Relationships Between AVIRIS and MODIS Indexes. *Remote Sens. Environ.* **2006**, *105*, 354–366.

CivicDrone. Study of Soil: A Drone for Mapping and Quarries, 2015. http://www.civic-drone.com (accessed May 30, 2015).

Cress, J. J.; Sloan, J. L.; Hutt, M. E. Implementation of Unmanned Aircraft Systems by the U.S. Geological Survey. *Geocarto Int.* **2011**, *26*, 133–140.

Cyberhawk. UAV Topography Survey of Proposed Wind Farm. 2013, pp 1–4. http://www.thecyberhawk.com/2013/03/uav-topographic-survey-of-proposed-wind-farms (accessed Nov 25, 2015).

Descroix, L.; Genithon, P.; Amogu, O.; Rajot, J.; Sighmonou, D.; Vauclin, M. Change in Sahelian Rivers Hydrograph: The case of recent Red floods of the Niger River in the Niamey region. *Glob. Planet. Change* **2012**, *98/99*, 18–30.

D'Oliere-Oltmans, S.; Marzolff, I.; Peter, K .D.; Ries, J. D. Unmanned Arial Vehicle (UAV) for Monitoring Soil Erosion in Morocco. *Remote Sens.* **2012**, *4*, 3390–3416. DOI: 10.3390/rs4113390 (accessed Nov 18, 2015).

Doward, J. Rise of Drones in United Kingdom Air Space Prompts Civil Liberties Warning. The Guardian. 2012, pp 1–6. http;//www.the guardian.com/world/2012/oct/07/drones-uk-civil-liberty-fears (accessed Sept 9, 2014).

DroneMapper. Geo-Referenced Ortho-Mosaic, DEM, DSM, NDVI and Point Cloud Generation. 2015, pp 1–4. DroneMapper.com/node/1.html (accessed June 4, 2015).

Farms.com Small Unmanned Aircraft Systems for Crop and Grassland Monitoring. 2013, pp 1–8. http://www.farms.com/commentaries/small-unmanned-aircraft-systems-for-crop-and-grassland-monitoring-61328.aspx (accessed Aug 29, 2016).

Flynn, K. F.; Chapra, S. C. Remote Sensing of Submerged Aquatic Vegetation in a Shallow Non-turbid River Using an Unmanned Aerial Vehicle. *Remote Sens.* **2014**, *5*, 12815–12836.

Fritz, A.; Kattenborn, T.; Koch, F. UAV-Based Photogrammetric Point Clouds-Tree Stem Mapping in Open Stands in Comparison to Terrestrial Laser Point Clouds. *International Archives of the Photogrammetry, Remote Sensing and Spatial Information Sciences* 2013; XL–1/W2, pp 141–146.

Frolking, S.; Palace, M. W.; Clark, D. B.; Chambers, J. Q.; Shugart, H. H.; Hurtt, G. C. Forest Disturbance and Recovery: A General Review in the Context of Space-Borne

Remote Sensing of Impacts on Above Ground Biomass and Canopy Structure. *J. Geophys. Res.: Biogeosci.* **2009,** *114,* 25–32.

Galimberti, K. Can Drones Offer New Ways to Predict Storms, Save Lives? 2014a, pp 1–4. AccuWeather.com (accessed Sept 23, 2015).

Galimberti, K. Drones offer New Horizon, Solutions for Weather Modification. 2014b, pp 1–3. AccuWeathr.com (accessed Oct 20, 2015).

Gardner, T. A. The Cost Effectiveness of Biodiversity Studies in Tropical Forests. *Ecol. Lett.* **2008,** *11,* 139–150.

Getzin, S.; Nusuke, R. S.; Wiegand, K. Using Unmanned Aerial Vehicles (UAV) to Quantify Spatial Gap patterns in Forests. *Remote Sens.* **2014,** *6,* 6988–7004.

Getzin, S.; Wiegand, K; Schoning, I. Assessing Biodiversity in Forests Using Very High Resolution Images and Unmanned Aerial Vehicles. *Methods Ecol. Evol.* **2012,** *3,* 397–404.

Green, M. Unmanned Drones may have Their Greatest Impact on Agriculture. 2013, pp 1–4. http://www.thedailybeast.com/articles/2013/03/26unmanned-drones-may-have-their-greatest-impact-on-agriculture.html#stash.c36uDpsT.dpuf (accessed Aug 3, 2013).

Haala, N.; Cramer, M.; Weimer, F.; Trittler, M. Performance Test on UAV-Based Photogrammetric Data Collection. *Proceedings of the International Photogrammetry Remote Sensing and Spatial Information Sciences* 2011; XXXVIII–I/C22, pp 7–12.

Hagemann, B. Drone Survey Tech Takes Off. 2014, pp 1–4. http://www.australianmining. com.au/news/drone-tech-investment-takes-off (accessed Nov 25, 2015).

Hansen, M. C. A. Method for Integrating MODIS and Landsat Data for Systematic Monitoring of Forest Cover and Change in Congo Basin. *Remote Sens. Environ.* **2008,** *112,* 2495–2513.

Horcher, A.; Visser, R. J. M. Unmanned Arial Vehicles: Applications for Natural Resources Management and Monitoring. 2009, pp 1–5. http://uavm.com/images/forrest_Management_UAV_for-Resource_Management.pdf (accessed Oct 31, 2015).

Hu, Z. Spectral Variation Characteristics of Wheat Lodging in Filling Period. *J. Anhui Agric. Sci.* **2011,** 1–8. http;//en.cnki.com.cn/Article_en/CJFDTotal-NYGU201419025 (accessed May 23, 2015).

Hyypa, J.; Hyyppa, H.; Leckie, D. Gougeon, F.; Yu, X.; Maltamo, M. Review of Methods of Small-foot Print Airborne Laser Scanning for Extracting Forest Inventory Data in Boreal Forests. *Int. J. Remote Sens.* **2008,** *29,* 1339–1366.

Jackson, R. D.; Reginato, R. J.; Idso, S. B. Wheat Canopy Temperature: A Practical Tool for Evaluating Water Requirements. *Water Resour. Res.* **1977,** *13,* 651–656.

Jackson, R. J.; Idso, S. B.; Reginato, R. J.; Pinter, P. J. Canopy Temperature as a Crop Water Stress Indicator. *Water Resour. Res.* **1981,** *17,* 1133–1138.

Jarvis, B. Drones are Helping Meteorologists Decipher Tropical Cyclones. Novnext. 2014, pp 1–3. http://www.pbs.org./wgbh/nova/next/eart/drone-meteorology (accessed Sept 21, 2014).

Jensen, A. M.; McKee, M.; Chen, Y. Procedure and Processing Thermal Images Using Low Cost Micro-Bolometer Cameras for Small Unmanned Aerial Vehicles. IEEE/IGARSS International Geosciences a Remote Sensing Symposium Quebec City, Canada, 2014; pp 34–45.

Jin, W.; Du, H.; Xu, X. A review on Unmanned Aerial Vehicle Remote Sensing and the Applications. Remote sensing Information. 2009, pp 1–8. http;//en.cnki.com.cn/Article_en/CJFDTotal-NYGU201419025 (accessed May 23, 2015).

Jones, V.; Pearlstine, L. G.; Percival, H. F. An Assessment of Small Unmanned Aerial Vehicles for Wildlife Research. *Wildl. Soc. Bull.* **2006,** *34,* 750–758.

Jouzapavicius, J. How Weather Drones will Unravel How Tornadoes are Formed. 2013, pp 1–3. TheHuffingonPost.com (accessed Sept 20, 2015).

Junge, B. Soil conservation options in the Savanna of West Africa: New Approaches to Assess Their Potential. International Institute for Tropical Agriculture, Ibadan, Nigeria, 2009, pp 1–7. http://www.slideshare.net/IITA-CO-conservation-options-in-the-savann-of-west-africa-new-approaches-to-assess-their-potential (accessed March 17, 2014).

Keber, P. Weather Drones: New Technology in Forecasting. The Weather Network. 2013, pp 1–3. http://www.theweathernetwork.com/news/articles/weather-drones-new-technology-in-forecasting/7772 (accessed June 8, 2016).

Koh, L. P.; Wich, S. A. Dawn of Drone Ecology: Low-cost Autonomous Aerial Vehicles for Conservation. *Conserv. Lett.: Trop. Conserv. Sci.* **2012,** *5,* 121–132.

Krishna K. R. *Peanut Agroecosystem: Nutrient Dynamics and Productivity;* Alpha Science International: Oxford, United Kingdom, 2008; pp 75–127.

Krishna, K. R. *Agroecosystems: Soils, Climate, Crops, nutrient Dynamics and Productivity*; Apple Academic Press Inc.: Waretown, New Jersey, USA, 2014; pp 546.

Krishna, K. R. *Agricultural Prairies: Natural resources and Crop Productivity*; Apple Academic Press Inc.: Waretown, New Jersey, USA, 2015; pp 246–312.

Krishna, K. R *Push Button Agriculture: Robotics, Drones, Satellite-guided Soil and Crop Management.* Apple Academic Press Inc.: Waretown, New Jersey, USA, 2016; pp 400.

Kutnjak, H.; Leto, J.; Vranic, M.; Bosnajak, K.; Perculija, G. Potential of Aerial Robotics in Crop Production: High resolution NIR/VIS Imagery Obtained by Automated Unmanned Aerial Vehicle (UAV) in Estimation of Botanical Composition of Alfalfa-Grass Mixture. *Proceedings of 50th Croatian and 10th International Symposium on Agriculture*, Opatija, Croatia, Section 5, pp 349–353.

Leichman, A. K. Drones-the Future of Precision Farming. 2015, pp 1–4. http://www.israel21c.orgheadlines-the-future-of-precision-agriculure (accessed June 21, 2015).

Linehan, P. Drones and Natural Resources: White Pines. 2013, pp 1. http://www.psu.edu/mt4/mt-tb.egi/367285 (accessed Oct 7, 2014).

Liu, L.; Wang, J.; Song, X.; Li, C.; Huang, W.; Zhou, C. The Canopy Spectral Features and Remote Sensing of Wheat Lodging. *Chin. J. Soil Sci.* **2005,** *03,* 1–9. http://en.cnki.com.cn/Article_en/CJFDTotal-NYGU201419025.htm pp 1-13 (accessed May 23, 2015).

Lucier, A.; Zarco-Tajada, P. B.; Rascher, U.; Bareth, G. UAV-based Remote Sensing Methods for Modelling, Mapping and Monitoring Vegetation and Agricultural crops. *Remote Sens. (Special Issue)* **2014,** 1–16 http://www.mdpi.com/jurnal/remotesensing/special_issues/UAVvegetaton%26crop.htm (accessed Sept 20, 2015).

Markham, D. How do you plant 1 billion trees a year? With drones of course. 2015, pp 1–10. http://www.treehugger.com/clean-technology/how-do-you-plant-1-billion-trees-year-course.html (accessed June 28, 2015).

Marzolff, I.; Poesen, J. The Potential of 3D Gully Erosion Monitoring with GIS Using High Resolution Aerial Photography and Digital Photogrammetry System. *Geomorphology* **2009,** *111,* 48–60.

Marzolff, I.; Ries, J. B.; Albert, K. D. Aerial Photography for Gully Monitoring in Sahelain landscapes. *Proceedings of Second Workshop of the EARSeL special interest group on Remote sensing for Developing countries.* Bonn, Germany, 2002; pp 18–20.

Marzolff, I.; Ries, J. D.; Poesen, J. Short Term vs Medium Term Monitoring for Detecting Gully-Erosion Variability in a Mediterranean Environment. *Earth Surf. Processes Landforms* **2011**, *36*, 1604–1623.

McCannon, T.; Diss, I. C. Drones Take Flight at Mining Sites Across Western Australia to Improve Safety and Efficiency. 2015, pp 1–3. http://www.abc.net.au/news/2015-09-02/drones-take-flight-at-mining sites-across-wa/6743394 (accessed Nov 23, 2015).

Moorthy, I; Miller, J. R.; Noland, T. L.; Nielsen, U.; Zarco-Tejada, P. J. Chlorophyll Content Estimation of Boreal Conifers Using Hyperspectral Remote Sensing. In *Proc. IEEE IGARSS—Learning from Earth's Shapes and Sizes*; 2010, Vol. I–VII, pp 2568–2570.

MosaicMill. EnsoMOSAIC for Forestry. 2015a, pp 1–2. http://mosaicmill.com/applications/appli_forestry.html (accessed May 19, 2015).

MosaicMill. EnsoMosai for Agriculture. 2015b, pp 1–2. http:/mosaicmill.com/applications/appli_agriculture.html (accessed May 19, 2015).

MosaicMill. EnsoMOSAIC for Disaster Mapping. 2015c, pp 1–2. http://mosaicmill.com/applications/appli_disaster.html (accessed May 19, 2015).

MosaicMill. EnsoMosaic for Mining. 2015d, pp 1–3. http://mosaicmill.com/applications/appli_mining.html (accessed May 19, 2015).

MosaicMill. EnsoMosaic for Topographic Mapping. 2015e, pp 1–2 http://mosaicmill.com/applications/appli_topography.html (accessed May 19, 2015).

NASA. NASA Weather Drones used to Determine How Tropical Storms Strengthen. 2013, pp 1–3. http://www.theepochtimes.com/n3/author/associated-press/ (accessed Sept 23, 2015).

Novodrone. Water and Environment Management. 2014, pp 1–6 https://novodrone.com/en/applications (accessed Nov 23, 2015).

Oskin, B. NASA Drone to Probe Ozone Layer Loss. 2013, pp 1–5.http://www.livescience.com/26161-nasa-drones-,ozone-study.html (accessed Nov 25, 2015).

Palace, M.; Keller, M.; Asner, G. P.; Hagen, S.; Braswell, B. Amazon Forest Structure from IKONOS Satellite Data and the Automated Characterization of Forest Canopy Properties. *Biotropica* **2008**, *40*, 141–150.

Patrick, P. Sustainable Technology–Drone Use in Agriculture. 2015, pp 1–7 https://wiki.usask.ca/display/~pdp177/Sustainable+pdp177/Sustainable+Technology+-+Drone+in+Agriculture (accessed June 23, 2015).

Precision Drone LLC. Precision Scout Drone. 2014, pp 1–17. http://www.precisionhawk.com (accessed Aug 5, 2014).

Precision Hawk. Lancaster Platform. 2014, pp 1–4. http://www.precisionhawk.com/index.html#industries (accessed Aug 5, 2014).

Richardson, B. Drones Could Revolutionize Weather Forecasts, but Must Overcome Safety Concerns. Washington Post, 2014, pp 1–2. https://www.washingtonpost.com/news/capital-weather-gang/wp/2014/04/25/drones-could-revolutionize-weather-forecasts-but-must-overcome-safety-concerns/ (accessed June 26, 2016).

Salami, E.; Berrado, C.; Pastor, E. UAV Flight Experiments Applied to the Remote Sensing of Vegetated Areas. *Remote Sens.* **2014**, *6*, 11051–11081.

Sarapatka, B.; Netpoli, P. Erosion process on intensively farmed land in the Czech Republic: Comparison of Alternative Research methods. *19th World Congress of soil Science, Soil Solutions for a Changing World*. Brisbane, Australia, 2010; pp 47–50.

Scott, W. Technology Used to Collect Weather Data. Bright Hub. 2011, pp 1–6. http://www.brighthub.com/environment/science-use-environmental/articles/107783.aspx (accessed Sept 20, 2014).

SenseFly. Drones for Agriculture. 2015, pp 1–14. https://www.sensefly.com/applications/agriclture.html (accessed October 10, 2015).

SenseFly. Using an Autonomous *eBee* Drone for Mining Exploration. 2015, pp 1–6. https//www.youtube.com/watch?v=d6t4nEKtZsY (accessed Nov 23, 2015).

Shahbazi, M.; Theau, J.; Menard, P. Recent applications of Unmanned Aerial Imagery in Natural Resource Management. *GISci. Remote Sens.* **2014,** *51,* 339–365.

Shan, J.; Hussain, E.; Kim, K.; Biehl, L. Flood Mapping and Damage Assessment–A Case Study in the State of Indiana. In *Geospatial Technology for Earth Observation*; Li, D, Shan, J., Gong, J., Eds.; pp. 473–495.

Terraluma. Applications Selected Case Studies. 2014, pp 1–8. http://www.terraluma.net/showcases.html (accessed Aug 28, 2016).

Towler, J.; Krawiec, B.; Kochersberger, K. Terrain and Radiation Mapping in Post-disaster Environments Using an Autonomous Helicopter. *Remote Sens.* **2012,** *4,* 1995–2010.

Trimble. Trimble Agriculture. Trimble Agricultural Division, Sunnyvale, Ca, USA. 2015, pp 1–4. http://www. trimble. com/Agriculture/index.aspx (accessed May 20, 2015).

Trimble Navigation Systems. Trimble UX5Aerial imaging Rover. 2014, pp 1–3. http//www.trimble.com/survey/ux5.aspx (accessed July 27, 2014).

Turner, D.; Lucieer, A.; Malenovsky, Z.; King, D. H.; Robinson, S. A. Spatial Co-registration of Ultra-high Resolution Visible, Multispectral and Thermal Images Acquired with a Micro-UAV over Antarctic Moss Beds. *Remote Sens.* **2014,** *6,* 4003–4024.

UCMerced. Unmanned Arial Systems. Mesa lab- Mechatronics, Embedded Systems, and Automation, Internal Report. University of California, Merced, California, USA. 2014; pp 1–5.

USDA-NASS. History of Remote sensing. 2009. http://nass.usda.gov/Surveys/Remotely_sensedData_Crop_Acreage/index.asp (accessed Oct 6, 2014).

Vanac, M. Drones are the Latest Idea to Improve Farm Productivity. The Columbus Dispatch. 2014, pp 1–4. http://www.dispatch.com/content/stories/business/2013/09/19/eyes-in-the-skies.html (accessed Sept 16, 2014).

Wallace, L. O.; Lucier, A.; Watson, C. Assessing the Feasibility of UAV-based LIDAR for High Resolution Forest Change Detection. *International Archives of the Photogrammetry, Remote Sensing and Spatial Information Sciences* 2012; 39-B7, 499–504.

Wang, Q. YAV applications I Remote Sensing Technology. Geomatics and Spatial Information Technology. 2010, pp 1–8. http;//en.cnki.com.cn/Article_en/CJFDTotal-NYGU201419025pp 3 (accessed May 23, 2015).

Wang, F.; Wu, Y. Research and Applications of UAS Borne Remote Sensing. 2010, pp 1–8. http;//en.cnki.com.cn/Article_en/CJFDTotal-NYGU201419025 pp 3–4 (accessed May 23, 2015).

Watts, A. C. Small Unmanned Aircraft Systems for Low Altitude Aerial Surveys. *J. Wildl. Manage.* **2010,** *74,* 1614–1619.

Wharton, C. Nevada Looks at 'Drones' for Economic Development and Natural Resources. 2013, pp 1–4. http://www.unce.unr.edu/news/article.asp?ID=1871 (accessed Sept 9, 2014).

Zarco-Tejada, P.; Miller, Harron, B.; Noland, T.; Goel, N.; Mohammed, G.; Sampson, P. Needle Chlorophyll Content Estimation through Model Inversion Using Hyperspectral Data from Boreal Conifer Forest Canopies. *Remote Sens. Environ.* **2004a,** *89,* 189–199.

Zarco-Tejada, P.; Miller, J.; Morales, A.; Berjón, A.; Agüera, J. Hyperspectral Indices and Model Simulation for Chlorophyll Estimation in Open-canopy Tree Crops. *Remote Sens. Environ.* **2004b,** *90,* 463–476.

Zarco-Tejada, P. J.; Miller, J. R.; Noland, T. L.; Mohammed, G. R.; Sampson, P. H. Scaling-up and Model Inversion Methods with Narrowband Optical Indices for Chlorophyll Content Estimation in Closed Forest Canopies with Hyperspectral Data. *IEEE Trans. Geosci. Remote Sens.* **2001,** *39,* 1491–1507.

Zongnan, L.; Zhgnxin, C.; Limin, W.; Liu J.; Qingbo, Z. Area Extraction of Maize Lodging Based on Remote Sensing by Small Unmanned Aerial Vehicles. Transactions of the Chinese society of Agricultural engineering, 2014, pp 1–13. http://en.cnki.com.cn/Article_en/CJFDTotal-NYGU201419025.htm (accessed May 23, 2015).

CHAPTER 3

DRONES IN SOIL FERTILITY MANAGEMENT

CONTENTS

3.1   INTRODUCTION

Agricultural drones may become one of the most useful agricultural instruments in near future. They may after all hold the key for global upkeep, surveillance and improvement of soils and their fertility. Drones' ability for detailed imagery of soil and crops is highly useful. Processing of the images using inbuilt software is possible. It allows farmers to perform soil fertility and crop management procedures with added ease and accuracy. At present, drones are making headway into farms worldwide to help farmers evaluate their field's soil and decide most appropriate management procedures. Several functions are done by drones at much lower fiscal cost and less human drudgery.

Tigue (2014) also points out that drones may be getting popular and serve the farming community in accomplishing several tasks more rapidly, accurately, repeatedly and at lower cost. However, it should be noted that drones affect the way we collect data about soils and crops to a certain extent, but they may not affect the kind of data collected from crops/soils in the farms. The sensor technology we adopt on drones is similar to those adopted by farm workers who carry hand-held leaf colour meters or sensors mounted on ground vehicles or those carried on aircrafts and even the satellites. However, drones have the advantage of rapidity, instantaneous repeatability, higher resolution, accuracy and lower cost to farmers. Above all, drones allow some extraordinary and rapid observations from vantage locations above the crop/field surface. Photography was never possible from that low altitude and vantage points above the crop canopy.

Drone technology is said to revolutionize the way information about soil fertility and its influence on crop growth is collected. Remedial measures to correct soil fertility deficiencies may become highly accurate and location specific even within a field. Drones are said to conduct survey for soil fertility deficiencies in a matter of minutes using sophisticated sensors. Drone imagery and digital recordings add to 'big data' pool. Further, the variable-rate applicators take instructions based on digital data collected via drones' sensors. These aspects are said to revolutionize application of soil fertility amendments (Taylor, 2014).

3.1.1 DRONES AID FIELD AND SOIL MAPPING

3.1.1.1 DRONES TO STUDY FIELD TOPOGRAPHY

Agricultural drones have found a niche for themselves when it relates to aerial imagery to study topography of crop land and detailed high-resolution mapping. They are used as a first step to survey the land and develop three-dimensional (3D) images showing contours and slopes. Such images are then used to decide management block formation, crop species, planting density and irrigation strategy, particularly in laying irrigation channels/pipes (Tara de Landgrafit, 2014).

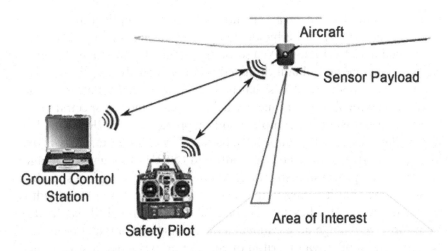

FIGURE 3.1 Diagrammatic representation of an agricultural drone system adoptable to study soil and crop parameters by using remote sensing methods.

Note: Area of interest could be a field soil type and characteristics of which such as texture (sand, clay, clods and gravel), moisture, crust formation, erosion effects and others need to be assessed.

Source: Prof Mac McKee, AggieAir, UWRL, Utah State University, Logan, UT, USA.

Drones with appropriate high-resolution multispectral sensors offer detailed soil maps. Such maps show the degree of soil weathering, texture, colour, surface features of the field, spots with soil erosion and organic mulches applied to soil (Microdrones GMBH, 2015). Drones have been useful in providing high-resolution multispectral images that show us the distribution of clods, gravel and coarse material. Drone-derived images are used to categorize and map the soil on the basis of textural classes. Drone images are used while marking 'management blocks' according to soil texture variation (Microdrones GMBH, 2015). Soil maps built using infrared (IR) sensors that show thermal variations are useful in judging moisture distribution in the surface soil. Figure 3.1 depicts a simple and easy arrangement of various aspects/instruments, including the drone. They are required to obtain aerial imagery of a particular location along with its soil features.

Farmers may actually prefer drone systems that are complete and compact to use. Farmers derive the necessary data depicting soils and their surface conditions and moisture status by using drones. The drone-aided soil mapping procedure has to be simple and easy to adopt. In general, drone-derived maps are utilized prior to application of basal fertilizer, pre-emergent herbicide sprays and during seeding. There are unmanned aerial vehicle (UAV) (drone) systems that are indeed simple, ready to operate and obtain digitized soil maps easily. Let us consider an example. There are UAVs that include a lightweight platform, a set of cameras with ability for imagery using a wide range of spectral bandwidths and software for complete soil mapping systems (Robota LLC, 2015; Plate 3.1). For example, 'Supernova' is a drone that has a five band [R, G, B, red edge and near infrared (NIR)], 5 megapixel imager and an image processing software. Another drone system is also called a 'Supernova' but it comes with sensors that operate at four bandwidths (R, G, B and NIR). They take pictures at 1.3 megapixel. It has facility for customized image processing at a local personal computer or ground station. The Supernova or Triton or Spectra drones make a streamlined and rapid system of mapping. As an end result, farmers have a drone system that just needs to be 'switched on' by power buttons. The predetermined path, cameras and computer software for image processing altogether offer the final product, that is, a soil map. The entire drone system is compatible with iPads or computer tablets having Windows 8. This allows farmers to import flight plans or newly/freshly prepare flight plans (Robota LLC, 2015).

3.1.1.2 DRONES IN SOIL FERTILITY MAPPING

Drones have been recently tested and adopted to measure a few different soil traits. The focus is to assess soil traits directly related to fertility. Soil fertility maps have been prepared by processing the ortho-images obtained during drones' flight over the fields. Drone imagery could be effectively overlayered with maps prepared by ground analysis. For example, soil survey maps, topographic maps showing the various undulations, hills, troughs and so forth and crop yield maps could be overlayered. Yield maps derived using Global Positioning System (GPS)-tagged combine harvesters too could be overlayered with those derived using the sensors on drones. Soil maps depicting maladies such as erosion, acidity, salinity/alkalinity, water stagnation, moisture deficit and so forth could also be overlayered with drone images depicting normalized difference vegetation index (NDVI), green normalized difference vegetative index (GNDVI) and leaf chlorophyll. In the recent years, soil electrical conductivity (EC),

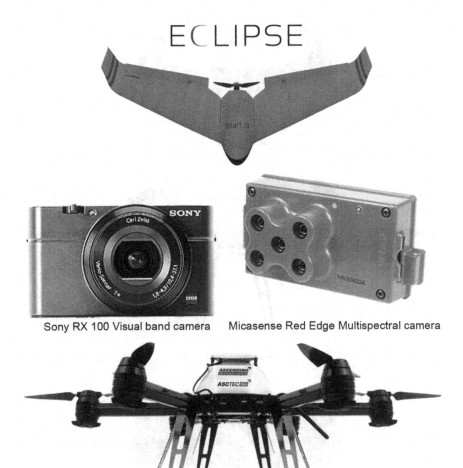

ECLIPSE

Sony RX 100 Visual band camera   Micasense Red Edge Multispectral camera

AscTec Copter

PLATE 3.1   Drones and sensors to study agricultural soils.

Note: Drones (Eclipse is a flat-winged model and AscTec is a copter) and sensors (Sony camera RX 100 visual, MicaSense RedEdge multispectral and Sony R7) are used to obtain images and maps depicting land, its topography, soil type, its colour, surface characteristics, thermal properties and vegetation (RedEdge). Such drone systems cover an area of 200–400 ac per flight of 50 min. They offer farmers with imagery at 1.1 inch (3 cm) pixel^{-1} to 3.14 inch (8 cm) pixel^{-1} depending on the sensor adopted.

Source: Antonio Liska and Domenica Liska, Robota LLC, Lancaster, Texas, USA.

Source for copter drone: Dr Guido Morgenthal, Technologien im Bauwesen, Ascending Technologien Inc., Krailling, Germany. Pictures of photographic cameras are from general websites that offer to market a range of products. http://zoom.cnews.ru/publication/item/47520/2.

(Continue…)

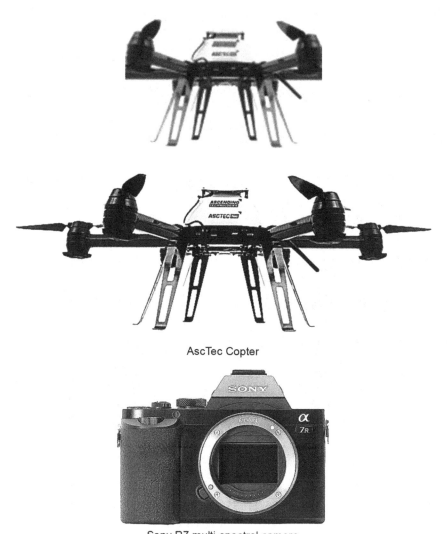

AscTec Copter

Sony R7 multi-spectral camera

PLATE 3.1   *(Continued)*

which is indicative of soil productivity trends, has also been used to overlay with maps depicting variety of other soil characteristics. The aim is to arrive at appropriate recommendations for formation of 'management blocks' and yield forecasts (Franzen and Kitchen, 2011). Overall, drones with appropriate sensors indeed offer maps depicting variations

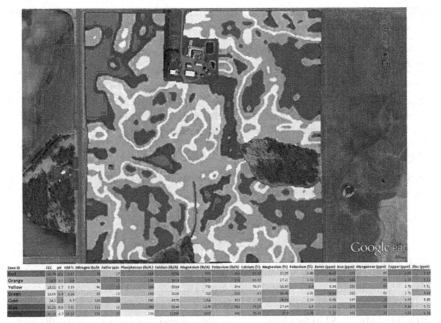

PLATE 3.2   A soil fertility map overlayered with topography and crop growth.
Source: Michael Dunn, Anez Consulting LLC, Little Falls, Minnesota, USA.

of a wide range of soil fertility factors (see Table 3.1). Such maps/digital data set could be overlayered, compared and evaluated. It helps to arrive at the most accurate soil fertility measures such as fertilizer formulation, its quantity, mulching, contour bunding and so forth (see Plates 3.2 and 3.4).

O'Leary (2014) has listed the potential uses of aerial images derived from drones that fly close to soil surface and crop canopy. We can actually procure maps depicting a very wide range of soil characteristics. Usually, soil colour which is indicative of soil weathering, inherent parent material, soil-C content and moisture status could be of good use to farmers. Farmers could overlay, compare and contrast soil maps and grain harvest maps of previous 5–10 years (if available) and then arrive at appropriate yield goals and nutrient/irrigation schedules. O'Leary (2014) further states that overlayering gives accuracy to farmers. Previously, farmers were growing crops by understanding fields' fertility and productivity by observing large areas of say 1.0 ha. However, with the advent of aerial digital imagery of soil types, the accuracy has increased. Description of variability of characteristics at high resolution, say a few centimetres, is

TABLE 3.1   Drones are Used to Study a Range of Field and Soil Characteristics Relevant to Crop Production: Few Examples.

Drone model/ Company	Cameras/Sensors	Field and soil characteristics	References
Topography, soil mapping, soil colour, texture, temperature, moisture			
Supernova/ Robota LLC, TX, USA or Eclipse/Robota LLC, TX, USA	RedEdge five bands—R, G, B, red edge, NIR, GEMs 4 band—R, G, B, NIR, 1.3 MP	Topography (2D and 3D). Soil mapping and locating ground sampling spots	Robota LLC (2015)
SUSI 62/ Geo-Technics, Berlin, Germany	Nikon 300 D; Canon D5 mark II	Field topography, gradients, soil types, soil temperature/ moisture. Management block formation	Thamm (2011)
AggieAir/ UWRL, Utah St. Univ. U, Logan, UT, USA	Canon visual (RGB), NIR and IR	Soil moisture (surface and root zone), surface temperature, VHI, VCI, NDVI and soil irrigation planning	Esfahani et al. (2015), AggieAir (2015)
eBee/SenseFly Inc., Chesseaux-sur-Lousanne, Switzerland	S110 for visual (RGB); SR110 for NIR; S 100 RE for RedEdge *Multispec 4c* for multi-spectral image; *ThermoMAP* for moisture and temperature	Aerial scouting, soil surface maps, soil temperature and moisture maps. Uses *Posterra* 2D and 3D software to develop colour images of field soil/crop	Sensefly Inc. (2016)
Trimble UX5/ Trimble Inc., CA, USA		Aerial scouting of fields, preparation of soil maps, thermal imagery/soil moisture. Surface soil moisture maps, NDVI, VCI, etc. Greenhouse gas–emission measurement	Trimble Inc., 2015a–d
Lancaster/ Precision Hawk, Inc., IN, USA		Aerial mapping of landscape and agricultural fields, seedling establishment, detecting gaps, estimating plant density, soil moisture, NDVI, VCI, leaf chlorophyll and plant-N status	Precision Hawk LLC (2016)

TABLE 3.1 *(Continued)*

Drone model/ Company	Cameras/Sensors	Field and soil characteristics	References
Monitoring soil erosion and fertility loss			
AscTec Humming Bird; Neo AscTec GMBH, Krailling, Germany	Sony Alpha 7R—36 MP; photo-package	Aerial mapping of agricultural fields for soil erosion	AscTec (2016)
AscTec Falcon 8 AscTec GMBH, Krailling, Germany	Panasonic Lumix DMC LX 3	Aerial imaging of fields for soil erosion, rills and gully erosion	Eltner et al. (2013)
Sirius 1/MA Vinci, Germany	Panasonic Lumix GF 1	To detect and monitor soil erosion in sandy Oxisol of Western Sahel	D'Oliere-Oltmans et al. (2012)
Crop scouting and fertilizer application			
md4-200/ Microdrone GMBH, Germany	ADC Lite Tetracam 520–920 nm bandwidth	Crop Scouting for NDVI, LAI, leaf chlorophyll content and plant-N status. Study effect of applied fertilizer-N on NDVI, leaf-N and canopy growth	Aguera et al. (2011)
Autocopter/ Autocopter Inc., NC, USA		Crop scouting, NDVI, LAI, plant moisture status, canopy chlorophyll content, application of liquid/granular fertilizers and foliar sprays of fertilizer-N	Effren (2014)
ESAFLY A2500-WH SAL Engineering, Modena, Italy	ADC Lite Tetracam; Aptina CMOS sensor	Crop scouting, measuring NDVI, GNDVI, SAVI, ortho-images of soil and crop in grape orchards	Candiago et al. (2015)
RMAX/Yamaha Inc., Japan	ADC Lite Tetracam Sony 7R; Canon S 100	Field soil and crop scouting, NDVI, LAI, plant moisture status, canopy chlorophyll content, application of liquid/granular fertilizers and foliar sprays of fertilizer-N	Yamaha Inc. (2015)

(Continue...)

TABLE 3.1   *(Continued)*

Drone model/ Company	Cameras/Sensors	Field and soil characteristics	References
Hercules-II Copter/check	EOS 30 D digital	Quantification of rice plant-N status., canopy reflectance and chlorophyll content	Zhu et al. (2009)
Aeyron, Scout/ Aeyron Inc., Ontario, Canada	ADC lite Tetracam; Photo 3S	Scouting crops. Comparing crop response with soils amended with organic and inorganic manures. Monitoring soil moisture	Zhang et al. (2014)
AgEagle/ AgEagle Aerial Systems, Kansas, USA	Sony QX 1; Canon S 100 NIR; MicaSense RedEdge	Soil/crop mapping; NDVI, soil fertility maps, variable-rate fertilizer supply maps; prescription maps for chemical applicators (pesticides)	AgEagle Aerial Systems Inc. (2016)
Vector-P UAV/ Intelli Tech Microsystems, Bowie, Maryland, USA	Kodak DCS cameras Fuji Fine Pix S3 PRO UVIR	Measurement of NDVI, GNDVI, leaf area index and biomass	Hunt et al. (2008)
MK-Okto copter/ Hisystems Gmbh, Germany	Panosonic Lumix DMC GF3	Development of crop surface models (CSMs) using stereo-images of rice CSMs are useful in deciding agronomic inputs to crops such as fertilizers	Bendig (2013)

Note: *NDVI* normalized difference vegetation index, *GNDVI* green normalized difference vegetative index and *SAVI* soil adjusted vegetative index. Soil adjusted values for NDVI are required if vegetation is sparse and reflectance from soil interferes with values noted by using sensors. In most cases, it is SAVI that is preferred for computation. *VCI* vegetative condition index, *VHI* vegetative harvest index (see Esfahani et al., 2015) and *LAI* leaf area index; Spectral Wavelength bands *R* Red, *G* Green, *B* Blue, *NIR* Near Infrared, *IR* Infrared, *GEM* global education management and *RGB* red, green, blue.

also possible. Thus, farmers can cultivate their fields by obtaining analytical data of each square foot. The size of grid cell/management zone could be made small and studied.

Soil type, textural variation and soil-N maps could be overlayered with grain harvest maps, and then the causes and effects/responses could be ascertained (Anderson, 2014). Farmers gain greater insights by analysing

PLATE 3.3   Aerial image from a parachute drone—SUSI-62 showing topography and soil type variations.

Note: Such topographic images showing soil variations could be of great help to farmers while deciding on cropping systems, planting and formation of management blocks. For example, slopes could be avoided by forming management blocks based on contour data. Soil erosion and loss of fertility could be reduced by adopting remedies such as bunds, mulches and tiled drainage canals appropriately and immediately.

Source: Dr. Hans Thamm, Geo-Technics, Berlin, Germany.

the effect of soil-N status on grain productivity maps, simultaneously (side-by-side) and accurately using computer programs. Fertilizer-N supply could be proportionately accurate if we adopt aerial imagery.

Repeated aerial surveys and mapping the soil fertility variation may be essential to decipher trends in crop productivity. A few reports suggest that observations using drones, particularly yield variation may be useful. For example, year after year, the same patches within a large field may tend to underperform. Also, sometimes same patches may yield high quantity of forage and grains. This is attributable to inherent soil fertility and its variations and fertilizer supply trends followed. In such a case, we may have to adopt precision farming techniques, particularly while applying fertilizer and water to obtain uniformity in soil fertility. In this way, we should be able to match the crop species and its nutrient needs with soil fertility levels. It is generally recommended to plant seeds at higher density in fertile regions within a field and to adopt relatively lower planting density in areas shown as less fertile. Further, it has been pointed out that to practise precision farming we must be in a position to measure soil fertility and crop growth variations using drones and their sensors. Quantification of soil fertility variation is essential. If possible, we may verify it with ground reality data (Farm World, 2014; Farms.Com, 2015). In its most simple form, the green colour recorded by multispectral cameras means healthy plants and high productivity crops, yellow foliage means stressed crops and red foliage means a severely hampered crop. Overall, fertilizer inputs should quantitatively match the soil fertility and crop growth maps.

3.1.2   DRONES IN LAND PREPARATION AND OPTIMUM TILTH: AERIAL IMAGERY

There is no doubt that land and field survey for determining topography of crop fields and deciding on actual timetable and intensity of ploughing are crucial. They are some of the earliest and most important soil management aspects. Ploughing needs a close look at soil conditions prior to running tractors with discs or tines. Excessively dry, sandy, gravely, cloddy or wet slushy conditions are not preferred. Ploughing is actually done to turn the soil and achieve a friable condition. This is to ensure that sowing could be performed with greater accuracy, at uniform depth and at optimum density of seeds. Flying drones above the open fields, prior to sowing, is becoming an essential procedure. It allows farmers to obtain information about soils and their condition at that instance when they are to send their tractors with discs and furrow markers into the field. Drones offer images of a large field (1000 ha) in great detail. It could be done within a matter of minutes prior to such agronomic procedures. This is something which is not feasible with human scouts and satellite imagery that lack in resolution. Manned aircrafts could turn out good imagery, but they are very costly and not repeatable often.

Farmers in Great Britain have started using drones to image their fields prior to sowing crops. Drones are easy to fly over large farms and obtain images of the terrain, water logging if any, dryness, clods and eroded areas. Farmers need not have to wait for satellite transit to obtain images of fields. At Bedfordshire, in the United Kingdom, farmers say that during sowing season, it is common to find clayey soils being inundated. Flooding makes it difficult or altogether impossible to run a tractor through the field. Ploughing and making furrows is just not feasible on a wet slushy field. Further, surveying and mapping zones affected by wetness and flooding is not easy. In addition, it takes time and funds to hire farm scouts to trace the spots accurately. Drones used currently are quick to deploy (fly) and obtain images so that farms could be ploughed accordingly (Crop Site, 2015). Farmers can obtain real-time firsthand views of fields covering entire field and soil conditions.

Ploughing needs optimum aeration and moisture status. A flooded area, wet patch or dry crusted zone may not be amenable. It may not yield optimum tilth if ploughed. Farms of just 200 ha could be difficult and costly to scout for soil conditions prior to sending tractors to plough the fields. Now, imagine these tasks to be done for a farm of 1000 ha. It is costly to hire several field scouts and the process is time consuming. Finally, we may not get very accurate estimates of soil moisture distribution and soil type descriptions even if we recruit skilled farm workers. Errors due to human fatigue will have impact on accuracy. Instead, we could adopt a drone with thermal cameras. Those get us maps that show up soil temperature and moisture distribution. Then, we can overlayer it with images captured by using sensors at visual range. It will then be easy for farmers to sit in front of a computer screen and decide the ploughing schedules. Farmers can pick areas to be ploughed immediately and those to be left for other time. Some plots may have to be discarded for planting all together. Thus, drones could help farmers in planning a ploughing schedule and also decide on the intensity of ploughing. Interestingly, in the near future, drone-derived imagery of the entire field and digital instructions could be fed to autonomous (driverless) tractors. Such autonomous tractors may operate in the fields, either singly or in swarms. Large fields could then be ploughed methodically covering each management block and as per farmers' needs (Gronau, 2016). The entire ploughing operation becomes accurate and easy.

3.1.2.1 ROLE OF DRONES IN NO-TILLAGE SYSTEMS

Drones surely have a role to play in the no-tillage systems that are adopted widely. No-tillage systems are popular among large farms of different agrarian regions. No tillage for couple of seasons or more, then a single disc ploughing is common. No tillage has its well-known and perceived advantages. They are as follows: first, it avoids expenditure on ploughing by using tractors. It reduces cost on turning soil, clod crushing and even ridging. No tillage avoids excessive disturbance to soil, lessens soil nutrient loss via erosion and soil oxidative processes. No tillage lessens greenhouse gas emissions. However, one major problem attached with no-tillage system is the eruption of volunteers and weeds. Weeds and its germination and establishment in between the rows have to be thwarted early. If not, weeds are often prone to outgrow and cover the crops (seedlings). A competing weed and its canopy can reduce incidence of photosynthetic irradiance on crop canopy. When fields are large and need quick scouting, then this exercise is done best by using drones. Drones could be flown periodically after planting (dibbling seeds), say, two or three times in a week, to ascertain areas infested with weeds. Later, postemergent sprays, using ground vehicles fitted with GPS connectivity, could be adopted. Unmanned weedicide applicators too could be used if digital directions are accurate and available from drones' sensors. Drone-derived digital maps could be inserted into ground-based weedicide sprayers. As an alternative, there are drones (e.g. RMAX) that pick aerial images of weed-infected zones in no-till fields and then they use the digital data instantaneously to guide variable-rate sprayers. Such drones are known to spray about 8–10 L of herbicide onto field surface in a matter of 15 min.

Drones could also be used to monitor application of mulches and their distribution on soil surface of a no-till field. Even prior to seeding, drones could be flown over entire fields of 1000–10,000 ha to judge the topography, landscape, soil type and vegetation. Based on drone-derived images, fields could be demarcated into 'management zones'. This step offers a certain degree of convenience and efficiency, particularly while operating drones to judge weeds, erosion, loss of fertility and so forth. Many of the points discussed above for no-tillage are also applicable to other types of soil tilth management.

3.1.2.2 *DRONE IMAGERY PRIOR TO SEEDING (PLANTING)*

Drones are of value to farmers right from the beginning, when they wish to plan, seeding the fields or even slightly prior to it. Drones provide an aerial view of entire field, so that plans for seeding specific crops, their seed rates, patterns and expected plant density could be developed. Drones may be of use prior to seeding, particularly, while forming 'management blocks' based on topography, soil type and its fertility and expected planting density. Previous data about soil productivity and maladies, if any, crop species that match the location and germination trends seem essential. They could be used along with the most recent drone imagery to form 'management blocks' (Shannon, 2014; Gallmeyer and Aoyagi, 2004).

Wheat production in Washington State in the United States of America has been already exposed to drone technology. Wheat seeds are sown deeper than the usual 10–15 cm soil depth to utilize soil moisture efficiently. The other aspect relates to formation of hard crust above seeds and the impedance it creates for seed germination. Aerial imagery using drones fitted with cameras that operate at visual range (R, G and B) are being adopted to scout wheat fields. They detect seed germination and hard crusts. Significantly, reports by Khot et al. (2014) suggest that ground reality data about germination rates of wheat seeds collected by using farm workers matched linearly with that collected by using drone images. The correlation values ranged from $r^2 = 0.78$ to 0.86 for different wheat types such as hard red and soft white wheat.

During rice farming, farmers may encounter a high degree of soil fertility variation and unevenness of soil surface. It leads to uneven germination and seedling establishment, particularly in direct seeded fields. Gaps in seedling establishment actually occur when tractors with seed broadcasters or dibblers move through fields with uneven soil surface. Drones could be useful in obtaining 3D maps of fields to classify them into level and uneven (medium or deep) fields. Soil surface could be leveled by referring to data (3D) from drone imagery. As an alternative, variable-rate seeding could be practised. Soil surface relief maps (3D) are obtained by using visual cameras on drones and digital data could be used in the variable rate technology (VRT) planters attached to tractors (UNSCAM, 2015).

3.2 DRONES AND PRECISION MANAGEMENT OF SOILS, THEIR FERTILITY AND PRODUCTIVITY

Precision farming aims at accurate management of nutrients in crop fields. It is primarily dependent on aerial imagery of entire field by using drones or manually by using human scouts. It then involves tedious soil sampling to arrive at appropriately accurate soil fertility maps. Such maps show up variation in mineral, nutrient and organic matter distribution. Aerial survey of field soil is the essential first step during drone-aided precision management of fields (USDA-NRCS, 2010; Krishna, 2013). Knowledge about productivity of previous crops, management zones, soil nutrient input trends, soil moisture variability maps from thermal imagery and residual nutrient status maps are most useful to farmers. Despite the great details depicted in the aerial images and the corroborating ground data, it is not always possible to guess grain yield or fix appropriate yield goals. Aerial images do not always depict the causes for soil fertility and crop productivity variations in grid cells, management blocks or entire fields. Yet, it is advisable to get soil/crop productivity maps with the highest resolution possible. Detailed study of a field by using drones' imagery is of course an essential first step.

According to technical reports by United States Department of Agriculture, basic requirements for precision management of soil fertility that mainly involves soil nutrient status assessment and removal of its variations are as follows (USDA-NRCS, 2010):

a. Identify within field variation and delineate management blocks by using drone imagery;
b. Prepare a nutrient budget for the field in question using data about inputs of nutrients via all sources such as inorganic fertilizers, organic manures, mineralization rates for soil-N, irrigation schedules, also consider cropping systems that add to soil-N such as legumes or residue recycling via cover crops;
c. Identify soil maladies such as salinity, acidity, top soil loss, erosion and so forth that may create variability in soil fertility;
d. Identify nutrient deficiency, sufficiency and excess levels that may affect crop growth;
e. Most importantly, determine soil variability and supply amendments such as fertilizers, gypsum, organic matter using drone imagery as a guide. Supply amendments at variable rates using appropriate computer programs and overlaying the soil maps.

During precision farming, aspects such as soil sampling, analysis of its nutrients and other relevant properties are essential. Soil sampling could be tedious if it is not organized well. Drones' imagery of field soils along with topography can be useful in deciding the soil sampling trends. Aspects such as number of soil samples to be picked per grid cell, number of grid cells, intensity of soil sampling prior to arriving at a composite sample and finally, analysis of soil nutrients could be based on imagery. Drone imagery can be helpful by offering field maps along with topographic and soil type variation, so that appropriate 'management blocks' could be created. No doubt, drone-derived images help in marking soil sampling locations accurately. There are at least two types of soil sampling done. They are grid sampling and management zone soil sampling. They help farmers to ascertain and authenticate drone imagery about soil fertility trends and variations.

Grid sampling, as the word means, involves formation of grids or cells covering entire field. These rectangular grids could be of different sizes. Larger grids of 1 ac are formed if the field to be covered is very large and extends into thousands of acres. Grid sizes could be smaller if the intention is to sample soils intensely. It leads to soil fertility maps of greater resolution and accuracy. Drone images could be highly helpful in providing an overview of the soil fertility situation. Aspects such as grid formation, number of grids to be sampled and number of soil samples per grid that goes to make a composite soil sample could be decided easily. An aerial photograph of the entire field is useful to farmers. Drones could also be used to assess the progress of the soil sampling process in the field, particularly when fields are larger extending into 1000 ac and more. Currently, there are automatic pilot-less soil samplers (see Krishna, 2016; AutoProbe Technologies, 2016). They need a digital image of entire field and preprogrammed instructions about the spots to be sampled. Digital images from drones could be used effectively, to feed the data or to remotely control the sampling (digging), by the autonomous soil-sampler (see Krishna, 2016). Digitized reports are almost essential to locate grid cells and then decide on soil sampling locations and prepare maps. Such maps could then be overlayered with data about several different soil factors such as nutrient distribution, soil pH, soil texture and crop growth trends of yesteryears (USDA-NRCS, 2010; Krishna, 2013, 2016; see Table 3-1).

Zone sampling involves picking samples of soil at spots within a previously marked 'management zone'. The size of the 'management zone' depends on variety of factors such as investment on labour, time at disposal, topography, soil type and soil fertility status. Soil maladies, if any, such as erosion, salinity affected area, soil moisture distribution and availability of drone imagery about the field in question also affect 'zone sampling'. Soil tests conducted on samples from a zone depends on previous knowledge about soil productivity and the exact problems a field or each location/management zone experiences. Again, intensity of soil sampling decides the accuracy and resolution of soil fertility maps prepared using ground reality data. Such maps could be overlayered with drone imagery. Drone imagery and ground reality data together help in deciding on soil sampling. It also helps to decide inputs to be channelled at variable rates using ground vehicles (USDA-NRCS, 2010; Krishna, 2016).

At this juncture, we have to note that aerial surveys adopted sampling procedure (grid cells or zones) and management blocks established have their impact on quantity of fertilizers applied. Nutrients are applied to a grid cell or zone based on the inherent soil fertility status and yield goal planned for the particular cell/zone. In case of grid sampling, each grid cell is handled separately using variable-rate methods. Similarly, if management zones are used, digital data about soil nutrient status is fed to variable-rate suppliers. Fertilizers are distributed appropriately within a management block. Drones are very handy and efficient in imaging. Digital maps showing the growth and grain formation trends in each grid cell/zone could be prepared. Drones flying above the precision field can relay images of each cell. Farmers can compare the crop growth with yield goals set for each grid cell. This aspect is not at all easy if human scouts are employed. It is cumbersome to survey each grid/zone and record growth patterns both during in-season and at final harvest. Usually, multiple nutrient supply pattern and yield goals are required if management zones are adopted (USDA-NRCS, 2010). Drone-aided precision soil nutrient management could become easy to accomplish. It may get practiced routinely in future. Drone-aided imagery offers instant visualization of soil and its conditions on computer screens. Therefore, it actually removes a lot of complications that otherwise occur during manual observation and recording of data for long stretches of a large field.

3.2.1 MANAGEMENT BLOCKS DEVELOPED USING CROP PRODUCTIVITY DATA

Soil fertility variation is directly related to and reflected as crop growth/ yield variation within fields. Final crop yields are often reflections of soil fertility status and fertilizer management strategies adopted by farmers. Therefore, yield maps obtained using GPS-connected combine harvesters are useful. In addition, aerial images of crops got using drones or satellites are of value during 'management block' formation. Yield maps obtained by using GPS-guided combine harvesters and satellite imagery showing crop canopy and NDVI values have been of use during management block formations. In fact, experiments that involve application of fertilizers have been tracked by using Landsat imagery. Drones too could be used to obtained high resolution images of crops and NDVI could be calculated for each grid or management block. Fertilizer (quantity) prescription based on variable-rate techniques was much lower than the normal blanket application procedures. For example, fertilizer-P requirement was 185 lb·ac^{-1} for plots treated normally. If grids were used, variable-rate application reduced the fertilizer-P needed to 164 lb·ac^{-1}. However, if management blocks were made using soil type and its characteristics then fertilizer-P needed could be further reduced to 134 lbs·ac^{-1}. Farmers applying fertilizer-P based on exact crop removal values and adopting variable-rate methods applied lowest at 121 lb·ac^{-1}. Similarly, fertilizer-K requirement reduced from 200 lbs ac^{-1} in fields under normal blanket application to 145 lb·ac^{-1} if grids are used. It reduced to 90 lb·ac^{-1} if management blocks based on soil type were used. Fertilizer-K reduced to 100 lb·ac^{-1} if crop removal values were used (GIS Ag maps, 2014). At this juncture, we may note that the above example used Landsat imagery that could be affected by noise, due to cloud or haze. Instead, a drone that flies close to canopy offers accurate data regarding NDVI and soil type details. Drone imagery has to be overlayered with yield maps obtained from GPS-connected combine harvesters. Overlayering has to be done prior to fertilizer prescription. Yield goals are also considered while applying fertilizers. It is interesting to note that variable-rate methods lead to savings on fertilizers ranging from 36 to 88 US$ ac^{-1} based on management block and type of fertilizer used.

3.2.2   DRONE AND VARIABLE-RATE TECHNOLOGY FOR FERTILIZER APPLICATION

Drones have been tested for their utility in detecting fertilizer requirements of rice crop. Field trials by using drones have shown that they could effectively help farmers in Italy to detect nutrient deficiency, fertilizer needs and apply (spray) fertilizers (Oryza, 2014). Agricultural drones have been useful in reducing costs incurred on fertilizers by 15% compared with fields not under drone surveillance. About 6 t less fertilizers were needed for a 160 ha field. In the normal course without drone technology, the same area required 41 t of fertilizers. Drones efficiently fly over crops and help farmers to check the fertilizer need rather accurately. Drones offer aerial imagery of both soil and crop characteristics. They transmit information packages from 'vigour sensors'. The vigour sensor data is decoded and utilized by computer decision–support systems to arrive at the exact quantity of fertilizer to be applied to soil. Farmers state that such a precision technique aided by drone technology helps them in reducing fertilizer input. It costs less compared with other techniques. Further, they could use this drone technology on crops grown in different types of soils. Of course, appropriate calibrations to decision-support systems were needed (Oryza, 2014).

Rice production in Southeast Asian nations such as Malaysia occurs under submerged conditions. Farmers often adopt blanket rates of fertilizer-N. They apply fertilizer-N at high rates to develop a good sized canopy. It finally offers commensurately high grain yield. However, this procedure may lead to accumulation of excessive N in the soil profile. Soil-N then becomes vulnerable to loss via emissions such as NO_2, N_2O and NH_3. Therefore, at present, farm agencies are asking farmers to adopt variable-rate applicators that avoid excessive supply of N to paddy fields. This procedure involves formation of 'management blocks' based on crop productivity maps of yesteryears. Soil maps depicting N availability are essential. We may adopt tedious soil chemical analysis (Kjeldahl's process) or hand-held leaf chlorophyll (or leaf-N) meters. In the present context, we may use drones that obtain images using multispectral sensors. Drone-aided maps show canopy/leaf greenness, leaf chlorophyll status and N deficiency, if any, and so forth. For example, a study by UNSCAM (2015) states that rice fields in Malaysia were first surveyed, simultaneously, by using both SPAD leaf meters and drone imagery. This is to develop plant/soil-N variation maps. The digital data was then utilized in the VRT applicators fitted on a tractor. Rice fields' boundaries were established by using drone imagery and a mapping technology known as 'Qmap'. Incidentally, tractors with VRT applicators have to be light for efficient use in flooded, well puddled and soft soil found in the Asian rice belt (UNCSAM, 2015). For comparison, blanket applications were done by using backpack motor blower to spread fertilizer-N uniformly. Blanket applications do not require detailed maps of soil/plant-N status. Spot applications of fertilizer-N granules done manually are costly. It needs farm labour, which in certain seasons could be difficult to obtain. Results suggest that variable-rate method based on aerial maps of rice crops' N status is efficient. This method utilizes relatively less quantity of fertilizer-N. The in-season N dosages get smaller. Adopting precision farming and drone imagery seems cost effective. It also reduces farm drudgery connected with spot placement of fertilizer. Right now, the above procedure may seem complicated, but with repetitions we should be able to accomplish it with ease.

Grassi (2014) states that drones have at least five major uses in precision agriculture. One of them relates to variable-rate soil fertility correction. Usually, farmers have consulted satellite maps, adopted ground-based measurement of soil, canopy and leaf-N status. Then, they have prepared maps manually or used computers. It helped them to build a variable-rate map of slightly low resolution/accuracy. However, we can refine the maps and fix variable rates of fertilizer (N, P and K) supply using multispectral images from drones (e.g., Agribotix). These drones are highly useful for in-season fertilizer-N supply to cereal crops. The variable-rate prescriptions, it seems, could lessen fertilizer-N supply to 40–50 $lb \cdot ac^{-1}$ if aerial imagery, NDVI and variable-rate fertilizer-N supply systems are adopted. About 60 $lb \cdot ac^{-1}$ was needed under blanket prescriptions. Overall, drone imagery could reduce fertilizer-N supply to soil, avoid its accumulation and reduce loss of soil-N. Consequently, it reduces cost on fertilizer, but at the same time improves crop yield.

3.3 DRONES TO ASSESS SOIL AND CROP NITROGEN STATUS AND TO APPLY FERTILIZER NITROGEN

Let us now consider a few facts, in background, about soil-N, plant/crop-N status and calculation of fertilizer-N requirement. Fertilizer needs are usually decided in relation to crop productivity and yield goals set by farmers. Perhaps, these are some of the most reviewed and rapidly updated aspects of soil fertility. Soil-N is among the most important fertility factors that affect foliage and grain formation. Nitrogen is needed by the crop in greater amounts compared with all other essential nutrients. Nitrogen is a highly variable soil fertility factor. It is also a relatively mobile element in soil. Soil-N variations occur spatially and temporally to a relatively greater degree. This is caused by both soil characters and differential removal of soil-N by the crop. Fertilizer-N is an important amendment added to soil, at relatively large quantities. Fertilizer-N supply has direct impact on crop productivity (see Krishna, 2002, 2015a, 2015b, 2015c; NDSU, 2012a and 2012b; FAO, 2007; Garcia, 2013; Alvarez, 2007; Rosas, 2011). Farmers, generally, first make grids, then sample the field soils and analyse soil-N content. This is done to arrive at appropriate fertilizer-N prescriptions. Soil-N status has been measured using chemical methods since few decades. They are tedious, laborious, sometimes complex and costly. They involve collection of a series of soil samples at various depths of profile. It is followed by equally complex soil analysis in laboratory. Soil chemical analysis in laboratories is time consuming. Often, it is not possible to obtain results showing soil-N maps quickly, as and when farmers need them.

Soil-N fertility could also be deciphered indirectly by measuring the plant-N status. Crop-N status could be measured indirectly by assessing leaf chlorophyll content. Plants store most of their leaf-N in the chlorophyll. Therefore, measuring leaf chlorophyll helps in understanding leaf-N status. Leaf colour and chlorophyll content could be quickly estimated using optical methods, that is, spectral reflectance measurements. Actually, there are several studies which prove that leaf-N and chlorophyll content are directly related. Measuring leaf/canopy chlorophyll content suffices, to provide us with, an accurate estimate of plant/crop-N status (Han et al., 2001; Reyniers et al., 2004; Reyniers and Vrinsts, 2006; Hunt et al., 2014). Crop canopy reflectance measured using ground-based optical instruments and satellites fitted with multispectral sensors has been useful. They have provided an estimate of crop-N status and vegetation index (i.e. crop health) (see Krishna, 2016). Basically, increase in reflectance is related to decrease in leaf chlorophyll. It could be caused by lowered leaf-N status. Enhanced reflectance at NIR band width means increase in leaf area index (LAI) and green biomass. At the same time, we may have to realize that, reflectance measurements done using satellites are affected by interferences. Such interferences are caused by natural vegetation (Suguira et al., 2005), soil brightness and reflectance, environmental effects such as haze, clouds and so forth. Usually, corrections for background reflectance are applied before arriving at values for such reflectance data.

Hand-held instruments (e.g. Konica SPAD Chlorophyll meter) to measure plant leaf colour, leaf chlorophyll and leaf-N are in vogue, in many farms (see Krishna, 2013, 2016). They are useful if the field is not large. Farmers will have to carefully scout every region of a large field/farm and arrive at plant-N status maps of appropriate resolution. Crop scouting for spots showing N deficiency or sufficiency or excess is costly. It is tedious to map them, if it is for a large field of several thousand hectares. It is immensely costly to hire skilled farm workers. Farmers find it difficult to obtain an overall view of the large farm for soil-N/plant-N status, whereas drone imagery and leaf chlorophyll maps give an excellent overview of the entire field, *at one glance*. Therefore, farmer may trace and point exactly to the locations that need fertilizer-N application (Peng et al., 1995, 1996).

At present, drones fitted with multispectral cameras are becoming popular. They could be playing a vital role in collecting background data for fertilizer prescription. Drones are used to obtain aerial imagery and collect data related to crop canopy, leaf area, chlorophyll content and leaf-N status. Drones fly close to crop canopy, say 100–400 m above the crops and offer high-resolution images. Such images could be effectively used to detect leaf/canopy reflectance, chlorophyll content and leaf-N. Drones could be flown repeatedly over the crop field to collect crop reflectance data. Such data are then relayed to ground computer stations that estimate crop-N. Then, fertilizer-N needed to be supplied could be decided, of course, based on yield goals. Drones, with their multispectral sensors, can be used as many times required throughout the crop season, to assess leaf-N. Aerial imagery and reflectance data derived from sensors on drones could also be compared and calibrated, using ground data from multispectral radiometers. If it permits, sensor data from drones could be compared with soil-N data from chemical analysis and so forth.

Let us consider an example that compares data from sensors on drones (mainly NDVI, leaf chlorophyll, leaf-N and plant-N status) with those obtained using ground-based multispectral radiometer (see Aguera et al., 2011; Reyniers and Vrinsts, 2006). Aguera et al. (2011) found that NDVI derived from both ground-based and drone-based sensors correlate with applied fertilizer-N and soil-N status. Crop response to applied fertilizer-N could be assessed using sensors' data. The correlation value for Tetracam data ranged from $r^2 = 0.530$ to $r^2 = 0.81$, and that for ground-based radiometer ranged from $r^2 = 0.48$ to $r^2 = 0.712$. Next, NDVI and leaf-N data from radiometer and ADC Lite Tetracam multispectral sensors on drones are highly correlated ($r^2 = 0.864$). We can, therefore, measure NDVI and leaf-N, using Tetracam images. Then, detect N deficiency/sufficiency in plants, instead of recourse to ground-based radiometer or tedious soil chemical analysis. Tetracam data about NDVI, leaf chlorophyll, leaf-N and plant-N could be utilized to calculate fertilizer-N requirements, of course, by using appropriate computer software. Drones are rapid and offer digital data that depicts variations in crop-N status swiftly, and as many times in a crop season. Sensor data could be utilized effectively to apply fertilizer-N at variable rates. Aguera et al. (2011) have concluded that crop-N measured using both methods, that is, ground and drone-based sensors, are highly indicative of fertilizer-N applied to sunflower crop in different plots. It is clear that drone imagery could be used to collect data, periodically, from a field experiment. We can assess fertilizer-N effects on crops. In this case, Aguera et al. (2011) have, in fact, assessed fertilizer-N effects on crop-N, growth and grain formation, using data from drone-based sensors.

Quemada et al. (2014) have examined the relationship between ground-based measurements of greenness, leaf chlorophyll, leaf-N and NDVI using optical sensors on instruments such as SPAD, Dualex and Multiplex and data from aerial surveys, using hyperspectral and thermal sensors. They found that aerial measurements were reliable and correlated significantly with ground-level data, regarding N deficiency and N sufficiency indices. Further, they say that comparing different methods of leaf chlorophyll, leaf-N and NDVI measurement helps in finding out, lacunae and usefulness of different techniques, particularly while developing fertilizer recommendations (Arregui et al., 2006; Blackmer and Schepers, 1994; Fox and Walthall, 2008; Tremblay et al., 2011). In most cases, chlorophyll concentrations have provided best correlation. They have been used as best indicators of plant-N status. However, leaf-chlorophyll estimations could be affected by interference from drought effects, phenol content and discolourations, due to factors such as insect/disease damage. We may also trace crops with non-uniform distribution of chlorophyll. These aspects affect accuracy of leaf-chlorophyll estimations (Fox and Walthall, 2008; Zarco-Tejada et al., 2001; Kyverga et al., 2012; Quemada et al., 2014). Hence, there are suggestions to estimate chlorophyll/phenol ratios, which could serve as a good indicator of leaf-N status. There are also suggestions to estimate canopy fluorescence changes at red edge wavelength bands to detect chlorophyll content. Then, apply it to develop fertilizer-N recommendations (Tremblay et al., 2011; Reed et al., 2002; Zarco-Tejada et al., 2012, 2013). High resolution fluorescence methods could be used to estimate plant water status. But, this method is yet to be thoroughly examined and adopted for leaf-N/canopy-plN estimations (Quemada et al., 2014). At present, agricultural consultancy agencies, who offer suggestions about basal and in-season fertilizer-N application, rely on previous spectral data or recent measurements of leaf chlorophyll, using visual (red, green, blue) and red edge sensors. The plant-N data are corrected to remove background effects of soil, water and undue discolourations of foliage-caused factors other than chlorophyll. In-season top dress of N is a key factor affecting N-use efficiency. It is a method that helps farmers to match fertilizer-N supply with crops' need. Hence, accurate estimations of N requirement, using sensors placed on drones is highly pertinent. Drones with sensors have been already used to collect such data useful for deciding split-N dosages (Scharf and Lory, 2002). This procedure may become routine in future.

We have to note that soil-N and fertilizer-N management procedures that aim at greater efficiency receive priority. Most recent fertilizer-N management method that promises to enhance efficiency is precision technique. Precision techniques involve series of sampling and analysis. This is done to prepare a field-soil map showing variations in soil-N status and N availability to crops. Soil-N dynamics is affected by phenomena such as organic matter recycling, incorporation of cover crops, chemical fertilizer supply; accumulation in soil profile, mobility in the profile, transformation due to soil enzymes, removal by crop roots; emissions as NO_2, N_2O, N_2 and NH_3; loss via top soil erosion, seepage, percolation and so forth. Therefore, crops' response to fertilizer-N supplied is affected by a series of factors listed above. Computer software that calculate in-season-N requirements do rely on crop growth, yield goals and several of the factors that affect soil-N dynamics. Drones with hyperspectral and multispectral cameras could be utilized to measure some of the factors that affect fertilizer-N efficiency in the field. Drones could be flown periodically to monitor crop growth rate and arrive at appropriate in-season fertilizer-N recommendations. As stated earlier, in-season fertilizer-N supply is among important procedures that affect fertilizer-N efficiency. Therefore, drones have a major role in measuring in-season NDVI trends and help farmers in deciding fertilizer-N schedules. In the farm and its surroundings, drones could be flown periodically to monitor and image drainage channels. This is done to estimate loss of N via drainage, if any. Eutrophication due to accumulation of N in drainage channels, ponds and lakes in the surroundings of crop fields could be easily detected through spectral images obtained using drones. Ultimately, if extrapolated to an entire agrarian region, drones in future could become important farm vehicles plying in the atmosphere. They could help farmers in regulating nutrient dynamics, particularly N in the ecosystem.

Agricultural drones are highly effective and useful to farmers who adopt precision techniques. Drones offer the all-important digital data required to be inserted into variable-rate applicators of fertilizer-N granules. Drones could also be utilized to spray liquid fertilizer-N or dust granules/powder at variable-rates (RMAX, 2015; Yamaha Inc. 2015). In some cases, drones could immediately process the crop-N data, using a computer decision-support systems placed in the payload area. Then, apply fertilizer-N at variable rates. Drones, indeed, could be highly valuable, rapid, cost effective and profitable to farmers.

Sugarcane is a cash crop. Its productivity is influenced to a certain extent by inherent soil-N status, N availability to roots and its efficient utility by crop. Fertilizer-N supply is practised, basically, to reach higher yield goals. Soil sampling by using grids or management blocks consequently developing a soil-N map is useful. Farmers also use leaf-N meters to monitor plant leaf-N status. Sensors placed at vantage points in the field also help farmers to prepare soil-N map (Amaraj et al., 2015). Usually, high soil-N and zero-N plots are used to calibrate sensors for plant-N status. During recent years, drone with sensors for biomass, leaf chlorophyll, leaf-N and plant moisture status is touted as better option. They help to decide fertilizer-N supply schedules to sugarcane crop. In addition, drones could also be used to supply liquid fertilizer-N formulation, as foliar sprays. Copter drones with fertilizer tanks could be efficient in supplying the nutrients as aerial sprays (Yamaha Inc. 2015).

Let us now consider an example from the Southern European agrarian region. Here, soil–nitrogen management is an important task to the farmers. Fertilizer-N recommendations have to be made carefully. Over or under supply of fertilizer-N could be detrimental to soil/crop. In this regard, Papadopoulos et al. (2014) have reported a method. First, a digitized soil map is prepared depicting a series of soil characters such as soil pH, nutrient distribution, (N, P, K), soil organic matter (SOM) and so forth. Such a digitized soil map could be used to prescribe site-specific fertilizer supply. Quantity of fertilizer-N to be supplied is decided by a computer software that considers a range of soil-related factors. It considers mainly the crop-N budget, soil-N balance, soil-N mineralization rates, soil texture, SOM content, $CaCO_3$ content, N loss due to emissions, seepage, percolation and so forth. Timing of fertilizer-N is dictated by the crops' demand at various stages of growth. Actually, drone-derived digitized field maps are used to decide fertilizer-N supply. Fertilizer-N application is done to each field parcel in a management block. Drone-aided imagery and estimation of NDVI was utilized to arrive at values of N sufficiency and deficiency indices (see Varvel et al., 2007). First, a fertile field with no soil-N limitation was established and sufficiency index (SI) was determined. The NDVI values were calculated with well-fertilized (soil-N non-limiting) plot as a standard. SI is calculated as a ratio of NDVI value measured in the field using drone sensors to the value known for well-fertilized field. Next, fertilizer-N supply was related to SI as follows: $SI = 0.655 + 0.002 \, (Nrate)^2 - 0.000003(Nrate)^2$; the correlation factor was ($r^2 = 0.60$). This equation shows that, to achieve maximum yield, field soils have to reach a value of 335 kg N ha^{-1} through variable-rate techniques. Incidentally, farmers in this area of Greece, normally use 320–350 kg N ha^{-1} to obtain best grain yield from wheat or triticale. Drones could be adopted to supply split dosages of fertilizer-N. Liquid formulations of fertilizer-N could be sprayed to canopy. It reduces fertilizer inputs. Fertigation using drip irrigation system is also a possibility to supply split-N dosages (Karyotis, 2006). Whatever is the procedure adopted, maps showing soil-N variations is a basic necessity. Such soil-N maps could be derived using sensors on drones.

Agricultural drones have been employed to aerially distribute seeds of several leguminous agroforestry tree species. This is an important tree replanting method. Seeds of leguminous trees are treated with appropriate *Bradyrhizobium* species. Then they are distributed at variable rates all across the landscape that is marked to support trees. This procedure actually adds to soil-N fertility, through biological N fixation by *Bradyrhizobium* in association with tree species. Strip-cropping of leguminous tree species and annuals is common. Drones may have a role to play in managing such cropping systems. Drone-aided seeding could be less costly compared with hiring farm workers. However, seed sprouting and seedling establishment may not be high. Hence, seed rate needs to be higher. No doubt, in future, drones may take over several aspects of establishment of agroforestry blocks, their surveillance and tree-management procedures. Such a cropping system finally aims at enhancing soil fertility (N) and its productivity.

3.3.1 DRONES AND SOIL ORGANIC MATTER

Soil carbon is a very important trait that affects physicochemical and microbial process in soil. Soil quality is dependent on total and organic C content. It is indicative of soil fertility and nutrient buffering capacity. Farmers and farm experts periodically estimate soil-C, as it immensely affects nutrient availability to crops. Soil carbon is usually replenished by using farm yard manure, composts and variety of organic wastes that get recycled during crop production. SOM improves soil aggregation and soil structure and acts positively on nutrient buffering capacity.

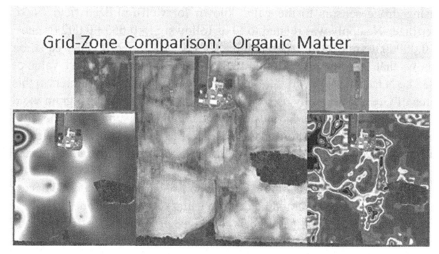

PLATE 3.4   Soil organic matter map showing its variable distribution.
Note: The SOM map has been overlayered with maps of field topography and crop growth.
Source: Michael Dunn, Anez Consulting LLC, Little Falls, Minnesota, USA.

Several types of agronomic procedures are adopted to protect SOM, sequester it in greater quantities in the soil profile and avoid its loss via erosion, seepage and top-soil loss. Crop residue applied to soil surface can be detected, using spectral reflectance measurement by sensors on drones. We can use either ground vehicles or drones. But knowledge about within-field variation of SOM is essential to adopt variable-rate organic manure supply. In this regard, Bartholomeus et al. (2014) have reported about possibility of estimation of within-field variation of SOM, using drones. They say, using drone-derived imagery (R, G and B band width) along with elevation data, it is possible to assess organic matter distribution on soil surface. Further, scouting soils to verify uniformity in soil-C using soil colour is a good idea. Soil colour could be measured using sensors on drones. Copters such as 'md 4–1000' fitted with appropriate multispectral cameras are used to assess soil humus content. Again, soil humus measurement is based on soil-colour saturation and reflectance. It allows us to compute and arrive at a value for soil-C content (Microdrones GMBH, 2015; see Plate 3.3).

Reflectance of crop residue (dry stems, twigs, leaf) occurring on soil surface, perhaps, could be utilized to ascertain the uniformity in the spread of organic matter on the soil surface. The drones' imagery has to be rectified for background soil reflectance. Prior knowledge about optimum spectral band width for a specific type of crop residue will be useful. Again, drone-aided survey for uniform spread of crop residue will save time and cost on scouting. It adds to accuracy of the agronomic procedure. Gaps in crop residue application could be identified using GPS coordinates rather accurately. Drone-based survey could be conducted periodically to check organic mulches. We can study their effectiveness in reducing soil erosion and rill formation in crop fields. Clearly, our efforts to detect and study crop residue distribution on soil surface depend on standardization of drone-aided spectral methods. The spectral data collected should distinguish soil and organic residues. Spectral reflectance characters of crop residues derived from different crop species, mixtures and other sources have to be known thoroughly. It is required to detect them with ease, using sensor placed on a drone.

3.3.2   DRONES AND MANURE SLURRY APPLICATION

Ground-based tractors fitted with variable-rate applicators (e.g. John Deere Tractor with GPS connectivity and variable rate applicators) are useful in distributing organic slurry/compost accurately, on to fields (John Deere Inc., 2015). There are reports suggesting that drones could also be used to apply compost slurry and other fluid forms of organic matter. It is done by filling slurry into the tanks attached to drone's fuselage and adopting variable-rate applicators. The computer software has to consider variation of SOM; note the nutrient content of organic matter (source) and arrive at appropriate dosage. Farmers could adopt a swarm of copter drones, if needed. This way, perhaps, spraying fluid organic matter formulation could be accomplished quickly and accurately.

3.4   DRONES TO MONITOR SOIL MOISTURE BY USING INFRARED IMAGERY

The expression of soil fertility is linked directly with soil–moisture distribution in the profile, particularly moisture that is available in the rooting zone. Most simple fact to be realized is that all essential nutrients required by a crop for optimum growth and grain formation are channelled to roots in dissolved state (in soil water). Then, nutrients translocate via roots to canopy and grains. Several of the agronomic procedures such as ploughing, seeding and, most importantly, fertilizer supplying and irrigation scheduling all depend on our knowledge about soil–moisture status. Therefore, soil specialists have bestowed greater interest in devising a series of methods to decipher soil–moisture status in the field. In the present context, we focus only on methods based on drones and their aerial imagery.

Soil moisture is among the most critical factors that affect crop productivity. Soil moisture affects a series of soil/plant interactions. Soil moisture availability varies immensely based on several factors related to soil type, crop species and environmental parameters. A thorough knowledge about soil moisture in the crop fields is almost essential to farmers, all the time, during a crop season. Soil moisture content at the surface and in the root zone is the main factor that affects crop productivity. The irrigation schedules are usually determined by using soil moisture data in the root zone. Soil moisture distribution pattern is to be known first, if variable-rate methods are to be adopted. Farmers generally adopt grid or management blocks to sample soils. They use the auger at different depths. Then, they measure soil moisture content in the sample. Such soil moisture maps could be fed to irrigation vehicles and variable-rate supply of water could be achieved (see Krishna, 2013, 2016). Historically, we have adopted tedious soil sampling, using grids of various sizes. We have then pooled the samples to arrive at a few representative composite samples. These composite soil samples are analysed for water content, using gravimetric method. It is usually done in a soil analysis laboratory. This procedure is tedious, time consuming and laborious, and costs are high. Further, results are not available immediately for the farmer to judge (see Brady, 1975; Jackson, 1973). Soil moisture content is a highly variable trait. Therefore, to obtain a clearer understanding, we need to sample soils intensely. It adds to work load of farmers. The data from gravimetric analysis have to be computed. Then, detailed soil maps showing variation in soil moisture distribution at various depths have to be drawn. We may have to use the classical cartographic methods. These maps will have to be then adopted in digital format to supply irrigation at variable rates, particularly if the farmer intends to adopt precision farming methods. Other methods, such as using gypsum blocks or imbedded sensors, also involve series of tedious measurements. Manual or computer-aided cartography is needed to adopt variable-rate irrigation techniques (Esfahani et al.,

2015). *In situ* measurement of soil moisture at different depths using such procedures can be costly, and it is generally exhaustive and slow. High-resolution maps will need grids of very small size. Consequently, soil samples from too many locations will have to be analysed. Precision soil fertility practise is therefore a labour-intensive and costly procedure.

Remote sensing, using IR (thermal) sensors along with data obtained about soils, using visual and NIR can be of utility. It can provide quantitative data about soil moisture. Surface soil moisture (SSM) content found in the top layer of 0–7 cm depth could be measured, successfully and accurately, using a combination of visual and IR cameras (Esfahani, 2015; Esfahani et al., 2014a; Kaleita et al., 2005; Yin et al., 2013). There are satellites such as meteorological satellite and European Remotes Sensing satellite that monitor SSM of large areas of 5–50 km² (Zaman et al., 2012). However, resolution could be less than that needed by farmers. Farmers' fields are generally relatively small. Drones come in handy, whenever high resolution thermal imagery developed, from a close range above the soil and crop canopy is needed. During recent years, thermal imagery of large cropping belts and most importantly individual farms/fields is on the rise. Such data are used by farmers to decide on irrigation schedules and in forecasting crop yield.

Farmers' fields may get affected by droughts, water loss and floods at various locations. Such detriments may affect fields at varying intensities for different lengths of time. They have to be identified at the earliest possible instance, and remedial measures should be adopted. If not, crop loss proportionate to severity of the malady will be surely felt in the form of reduced grain/forage yield. Scouting of farms accurately for soil–moisture dearth, droughts and flooded patches is essential. Drones have the ability for scouting and imaging the entire farm in one stretch. They offer an overview of the problem. Drone imagery could also help farmers in identifying locations that suffer droughts, marginal water scarcity and dryness of soil profile (AggieAir, 2015; Esfahani et al., 2014a and 2014b; Trimble Inc. 2015a and 2015b). Drone-derived imagery could also be used to identify flooded zones. It helps to take precautionary methods such as contouring, damming, draining the surface water and curtailing further irrigation.

Agricultural drones with ability to image the soil at high resolution, using multispectral sensors, are preferred. Such drones help to study soil temperature and moisture. Small drones, either copters or flat-winged models, attached with appropriate sensors are essential. Further, computer decision-support and decoding of ortho-mosaics are also required. They help to decipher variation in soil temperature and relate it to soil moisture. The IR cameras detect soil temperature and heat signatures of a particular soil within a grid cell, a management block or a field. The amount of heat reflected by soil is dependent on soil water content. Often the IR cameras are used in conjunction with multispectral cameras that offer soil maps at high resolution. The aerial images show the plant health, particularly leaf colour, LAI, NDVI and moisture content of canopy. The set of data obtained can be shrewdly analysed to arrive at soil moisture levels and plant growth status. A healthy plant accumulates more water in its leaf and canopy. Developing soil moisture distribution maps for the entire profile is not easy. Data about depth and spread of rooting for representative locations are needed. We can then correlate and find soil moisture distribution pattern in the soil profile. It is done using IR images and visual images of canopy and total plant water status (Goli, 2015; Tsouvaltsidis et al., 2015). Estimating moisture in the surface soil is important, because crop roots explore and extract a large portion of their water requirement from top layers (Esfahani et al., 2014b).

Soil moisture in the profile and the deep rooting zone of tree crops has direct impact on its foliage, leaf temperature, fruit bearing and final productivity. A period of water stress can immensely reduce fresh flush of leaves and fruit bearing. We may note that enhancing water-use efficiency of an orchard fruit crop is important. Water stress has its impact on many of the physiological processes including nutrient absorption. Drones fitted with IR cameras can map the foliage and canopy temperature of fruit crops. For example, Crawford et al. (2014) state that measuring leaf temperature of walnut and almond trees provides a good idea about water stress that the tree is suffering. Soil moisture has to be improved, if leaf temperature data of trees indicate so. Leaf temperature measured using IR sensors on drone correlates linearly with water stress experienced by certain woody fruit trees.

Soil moisture is among the most important factors that affect crop growth and productivity. Crops' response to soil nutrient inputs, soil biotic activity and series of physicochemical reactions in soil is dependent on optimum availability of soil moisture. Incidentally, both excessive and low moisture in the soil profile can be detrimental to crop growth. Large-scale droughts or floods and their impact on crop production are often monitored, using satellites. Satellite imagery has been useful in providing detailed information about soil moisture distribution in large agrarian belts. Major floods, droughts and dust bowls have been effectively forecasted, using satellite imagery. Farmers have been warned of impending weather-related maladies and their impact on soil/crop conditions. For example, most recently, a satellite named 'Soil Moisture Active Passive Satellite (SMAP)' placed by National Aeronautics and Space Agency of USA forecasts drought and soil moisture dearth mainly in large patches of landscape (NASA, 2014). It monitors soil surface regularly and offers soil moisture map. However, such a map has its limitations with regard to resolution, accuracy and ability to focus on very small regions. The resolution of above satellites is 50 km (Buis and Murphy, 2014). In such cases, it is preferable to first consult satellite imagery from SMAP. Then, for detailed analysis of soil moisture distribution, agricultural drones with NIR, IR and thermal band width cameras could be adopted. Satellite imagery only allows us to focus and identify regions broadly. We can then use drones and attain greater accuracy, while devising irrigation schedules. No doubt, scouting a large agrarian region using human scouts is time consuming and costly. The accuracy of maps produced could be of low resolution. Instead, a sequential use of satellite images from SMAP (NASA, 2014) or European Space Agency's 'Soil Moisture and Ocean Salinity Satellite Program' is useful. Then, venturing into detailed analysis of cropping zones or fields, using drones is worthwhile (see Krishna, 2016). Integrating satellite imagery with drone's activity and soil sensor networks is needed (Vellidis et al., 2008). It could be easier said than done, at present. But, eventually, we could be able to first search cropping zones with low moisture or those suffering drought, using satellite imagery. Then, quickly change to sharper focus and resolution, using lightweight drones. It is a kind of zoom-in and zoom-out technique, with regard to analysing soil moisture in crop fields. It seems irrigation schedules could be drawn by using data from satellite. Drone-aided thermal infrared (TIR) sensor also offers data on moisture in top layer of soil.

Acevo-Herrara et al. (2010) opine that UAVs have a great role to play in the aerial imagery of soils, particularly to study soil characteristics. Drones are attractive because they can be flown low over fields, as many times and in short intervals to study the same location in great detail. They are easy to deploy with very short runways or none (copters). Drones become essential when satellite imagery falls short of resolution and accuracy. In this case, Acevo-Herrara et al. (2010) predict that drones could be among common methods for obtaining high-resolution maps showing spatial and temporal changes in soil moisture. In fact, they have used a lightweight, airborne, L-band radiometer mounted on a small UAV. Drone technology included hardware and software to calibrate, geo-reference and to retrieve soil moisture data. Initial field trials show that results are promising. In due course, such methods may be needed to supplement data from satellites. We can also ascertain satellite/drone imagery, using ground reality data (see Lacava et al., 2012), if costs permit. Further, it has been suggested that in addition to retrieving soil–moisture data, during the same flight, drones can monitor other aspects of soil such as colour, texture, vegetation and weeds and so forth.

Parachute UAVs, that are slow in their transit above crop fields, do offer detailed aerial imagery. The imagery is obtained by sensors placed in the payload area. For example, 'Pixy' is an UAV that offers imagery about crops. It allows us to measure NDVI, leaf chlorophyll and plant-N status (Antic et al., 2010). Thamm (2011) has reported use of parachute-based UAV known as 'SUSI-62', to obtain aerial imagery of fields. The cameras (Nikon 300D, Canon D5 Mark II) in the payload cabin provide detailed picture of surface soil temperature. Such data could be utilized to assess soil moisture and schedule irrigation accordingly. Further, the thermal imagery also helps in tracing crop regions suffering from drought. Usually, the leaf temperature of drought-affected crop patches are higher by 5°C. Incidentally, imagery data along with soil/crop surface temperature maps have helped farmers in New South Wales (Australia). They could quickly identify drought-affected regions in a field and adopt remedial measures. Further, soil fertility deterrents such as erosion and top soil loss have also been identified, using aerial imagery from SUSI-62. The 3D elevation pictures about erosion effects are helpful, particularly, while adopting remedial measures to restore soil fertility.

3.5   DRONES TO DETECT SOIL MALADIES AND ADOPT REMEDIAL METHODS

Drones have been utilized to surveillance and offer images and data, pertaining to different maladies that afflict agricultural soils, for example, soil erosion, formation of gullies, loss of top soil, uncongenial soil acidity, high soil alkalinity and so forth. Some of these aspects are dealt in the following paragraphs.

3.5.1 DRONES TO DETECT SOIL EROSION IN AGRICULTURAL FIELDS

Agricultural fields are prone to suffer loss of soil fertility due to different types of erosion that occur at various intensities and continue to affect crop production, if unattended. Major factors that induce or create eroded soils are high-intensity rainfall events, rapid movement of water on soil surface, flooding of irrigation canals and small channels in the crop fields. High-intensity surface winds have the ability to create dust storms and dust bowls. Periodic dust storms, such as *Harmattan* in Sahel, for example, move fertile top soil from one location to other (Sterk et. al 1996, 1998; Michels, et al.1995; Anikwe et al., 2007). Soil erosion is among the major causes for loss of organic matter (Roose and Barthes, 2001; Bationo et al., 2006; Burkert et al., 1996) from the sandy oxisols. Soil erosion is a common malady in many other agrarian regions of the world (Bielders. et al., 2002a, 2002b; see Krishna, 2008; 2015b). It is believed that drones could play a vital role by forewarning the imminent soil deterioration.

Soil erosion is also induced by repeated ploughing and disturbance of soil structure (aggregation). Farmers may trace at least three different types of soil erosion in the crop fields. They are (a) sheet erosion which is removal of top soil due to excessive runoff that carries with it top soil and nutrients; (b) rill erosion is induced by ploughing and ridge formation plus movement of flood water in a ploughed field, after a rainfall event. If the ridges are unplanted or unprotected, it results in formation of rills that carry water and destroy land preparation. It leads to loss of top soil and its nutrients; (c) gully erosion is an important natural phenomenon resulting in land degrada-tion, soil erosion and loss of fertility. Gully erosion occurs when water that moves after rainfall events and strong winds consistently removes the soil. It causes small to large gullies over time. They are easily detected by using spectral sensors on drones (Marzolff and Poesen, 2009; Marzolff et al., 2002, 2011; D'Oleire-Oltmanns et al., 2012). Gully erosion varies spatially and temporally in a region. Drone-derived imagery of gullies and rills using 2D and 3D sensors could, actually, quantify the extent of expansion of gullies. Whatever be the type of soil erosion, all of them destroy soil struc-ture and reduce fertility. Farmers have to detect erosion at an early stage and take remedial measures. In addition to drones, blimps and kite-aided aerial imagery are also practiced to detect gullies and loss of soil fertility (D'Oliere-Oltmans et al., 2012). As stated earlier, it is tedious, laborious and at times difficult to cover the entire field using human scouts. However, timely detection of soil erosion and its correction is essential. Therefore, drones that offer aerial imagery of entire field are apt. Drones with high-resolution multispectral imagery help farmers in detecting erosion, right at very early stages. Sheet erosion, leading to loss of top soil after a rain fall event or rills that gain in size and length, could be detected within the crop fields. Drones are useful in detecting rills caused by ploughing and ridging. Bulldozing fields also creates rills. Small rills usually collect water in depressions and grow in size as crop season proceeds (D'Oliere-Oltmans et al., 2012). Farmers could easily point out and mark the areas affected by erosion, using processed imagery. Drones may be flown over fields immediately after a rainfall or a dust storm. Images covering 200 ha could be captured within a matter 15 min to 1 h. Such rapidity is not possible with human scouts. Moreover, resolution of satellites may be insufficient, if erosion is still rudimentary and eroded area is small. However, both small- and large-sized gully erosions can be easily detected by spectral imagery using drones. Let us consider a few case studies. In Italy, Bazzoffi (2015) has reported that UAV-geographic information system methodology is apt to detect soil erosion that afflicts crop fields. Aerial imagery can be used to restore eroded soils and adopt conservation practices such as mulching, formation of contours and bunds and so forth. Further, Bazzoffi (2015) reports that rill erosion that occurs in fields could be detected very accurately. They could be verified using ground data, with an accuracy of $r^2 = 0.87$. Drone imagery could be used to gain more detailed insight into rill formation, volume of rills, and eventually loss of soil and its fertility. Incisions caused to soil surface layers due to ploughing, formation of ridges and drains to collect runoff water may induce rill formation.

Next, drones with multispectral sensors can be useful to detect, quantify, analyse and model the soil erosion that occurs due to water flow through a gradient. River bank erosion could be identified with high accuracy using 3D images (UNEP-GEAS, 2013; USGS, 2013). In practical farming, field's gradient is an important factor that makes soils prone to erosion. Fields in the sloppy terrain, hills and mountains could be imaged. They could be assessed periodically, using UAV imagery and/or Light Detection and Ranging also known as Terrestrial laser Scanning (Neugerg et al., 2015; Eltner et al., 2013; Marzolff and Poesen, 2009; Ries and Marzolff, 2003). It is said, UAV could be used to generate Digital Terrain Models (DTM). The DTM could be verified with ground reality data (D'Oliere-Oltmans et al., 2012). Drone imagery derived could be compared instantaneously with DTM. Then, decisions on erosion control measures could be made. Further, detailed study of erosion, particularly, its spatial features, soil loss pattern and temporal changes in response to rainfall and wind is possible, using drones. Drones could be flown repeatedly and in short intervals to the same spots to collect the crucial data.

Bybordi and Reggiani (2015) have explained that drones have a big role in assessing soil characters during crop production. The same drone equipment may actually be used, to conduct different sets of analysis, based on season and importance of the soil trait. For example, imagery of topography and landscape to detect areas prone to soil erosion is relevant prior to sowing. After seeding, aspects like soil temperature, moisture and weed emergence could be important. Soil erosion detection using drones is done periodically, especially, immediately after a high-intensity rainfall event.

Dust storms in Sahel have dual effects on fields. High intensity wind removes soil nutrients via dust/sand grains. It leads to loss in fertility. At the same time, as dust storm settles at a different location, there, they add to soil nutrients (Bielders et al., 2002a, 2002b; Sterk et al., 1996, 1998; Michels et al., 1995; see Krishna, 2008; Krishna, 2015b). Therefore, whatever soil nutrient loss is reported in literature, it is actually the difference of nutrient loss and input due to sand drift and storms. Drones could be used to monitor local variations of the quantity of sand/dust removed and transmitted through the atmosphere. It seems two storms that occur in a crop season in Sahel results in loss of 76 kg Soil-C, 18 kg N, 6 kg P and 57 g K ha^{-1} (Sterk et al., 1996, 1998). In a matter of five storms, about 24–27 t ha^{-1} of soil could be displaced from a pearl millet/cowpea field in Sahel (Bielders et al., 2002a, 2002b). Such dust storms could aid spread of desert conditions. However, such events could be monitored accurately, using drone imagery and advance warning provided to farmers.

As stated above, soils in Sahelian West Africa are highly prone to erosion due to both wind- and water-related forces. The sandy soils with low aggregation and meager organic matter are exposed to storms. There are, in fact, innumerable studies conducted and reported about soil fertility loss due to dust storms. Soil erosion that results may reduce moisture and nutrients held in sandy soils. It accentuates crop loss. Bousset (2016) suggests that we should try as many shrewdly devised and planned innovations that reduce loss of soil fertility. We should restore crop biomass/grain yield. Drones are among the most recent devises that have yet to be tried, tested and used advantageously. The aim is to reduce soil degradation. Drones could be flown, periodically, in as many locations in the cropping zones of Sahel. They could bring in advance information on soil deterioration that sets in season after season. They say, in Sahel, soil erosion is a single factor that causes a loss of 30 kg soil nutrients ha^{-1} $year^{-1}$ (see Bousset, 2016; Gallacher, 2015; Gallmeyer and Aoyagi, 2004). Drones could be used effectively to surveillance desertification trends in the sub-Saharan Africa, as well as in other dry/arid zones. Drone-aided monitoring of terrain and crop fields could help farmers to plan the planting zones and soil-conservation procedures. Periodic monitoring could warn about imminent crop loss due to intermittent droughts common to Sahel. In addition, drones could be used to monitor seed germination, that is, establishment of seedlings and survival. Large gaps in seedling emergence could be filled (replanted) immediately after rains to take advantage of stored soil moisture.

Soil crusting is a problem that farmers experience immediately after rains. The soil dries and forms a crust above seeds. So, it suppresses seedling emergence and crop stand development. A careful scouting reveals gaps in seed germination and sprouting. In a large farm of 1000–10,000 ha, monitoring is tedious, and a time-consuming activity that costs to farmers. Sometimes, a swarm of farm workers may have to be employed to go out searching for soil crust formation and delayed emergence. Such scouting could be done effectively, using agricultural drones fitted with high-resolution multispectral cameras. Drones could capture images rapidly and offer details with GPS tags. Images showing areas affected with soil crusts and erratic germination could be acquired. Farmers could track each planting hill by using GPS coordinates, and monitor seedling emergence and establishment.

In drought-affected regions of tropics, for example, in sandy oxisols of Sahel and other semiarid/arid regions, soil temperature may reach 45°C or above. It is uncongenial for seedling growth. Soil particles, particularly sand grains of high temperature affect seedling survival and growth. Again, it is not easy to scout large areas for heat stress affected seedlings. Instead, we may adopt drones with visual, IR and thermal cameras to detect high-temperature zones in soil. Then, adopt remedial measures only at spots that show gaps in germination, caused due to high temperature and soil crusting (Bybordi and Reggiani, 2015). A single flight by a drone offers imagery depicting both soil temperature and soil surface moisture conditions. A drone covers 200 ha in a matter of 4–5 h flight over a cropping zone.

3.5.2 DRONES TO DETECT SOIL SALINITY PROBLEM

Soil fertility and its expression in the form of crop growth is affected by several physicochemical traits such as pH, aeration, alkalinity, salinity and acidity, soil aggregation, dispersion, soil crusting and so forth. Soils with alkalinity/salinity problems are wide spread across different agro-ecosystems. Soils classified as solonchaks are high in salts. It is due to aridity that induces precipitation and accumulation of salts. Such saline soils are widely distributed in arid and semiarid regions (FAO, 1998; Soil Survey Staff, 1998). High salt content limits plant growth. Sea coast and delta region too show up soils with salinity/alkalinity problems. Root activity, particularly, that related to uptake of major nutrient is hampered. Crop productivity is relatively low in salt-affected soils. Remedial measures include adopting irrigation excessively to drain out salts and planting crops that tolerate salinity (Beyer, 2003). Solonetz are soils that are high in sodium and are dispersed. Soil dispersion affects nutrient recovery and crop growth. Farmers in coastal plains, where salinity affected soils are conspicuous, have to scout using drones. Then, they have to pick and analyse soils. Later, follow it with remedial measures. They have to manually scout crop stand and measure soil characters related to salinity. Usually, agricultural agencies in different countries first adopt an aerial remote sensing through satellite imagery. This helps them to restrict their attention to salinity-affected regions. Then, soil sampling and chemical analysis is done to understand the severity of salinity-related problems. Agricultural drones are currently getting popular. They are used to assess a very wide range of soil/crop-related aspects. Drone with multispectral cameras could first digitally image the region and then focus on specific salt-affected farms. Usually, seed germination and seedling emergence are sparse, if salinity is beyond threshold. Crop growth traits such as NDVI, leaf chlorophyll and moisture status are depressed in saline soils. Crop productivity is lower. Such areas could be delineated, and separate 'management zones' could be formed. It helps while adopting remedial measures directed to reduce salinity effects. For example, in China, first they have obtained satellite (Quick Bird band 1–4)-aided imagery of saline soils in the coastal region. Then, detailed analysis of soil for EC, pH and salts have been conducted, only on well-focused regions and on just few soil samples. Farmers have later adopted remedial measures. Irrigation to flush out salts and planting salinity tolerant rice varieties are the common remedies (Guo et al., 2014). In the delta areas of South Asia, salinity is a common problem due to sea water encroachment. Such areas too could be delineated with greater accuracy, using drone imagery (see Krishna 2015c). Drone-aided aerial imagery of bare soil could be overlayered along with ground data pertaining to soil pH, salinity, EC, cation exchange capacity, Na, moisture, aeration and crop/vegetation status. Such a procedure is preferred prior to forming management blocks.

3.5.3   DRONES TO DETECT SOIL ACIDITY PROBLEMS

Soil pH (hydrogen ion activity) is an important character that affects series of physicochemical reactions in the profile. Foremost, availability of soil nutrients to crop roots is influenced by the soil pH. Soil pH preferred by crops differs on the basis of crop species. There is a wide spectrum of soil pH preference shown by crop roots. It ranges from acidic (4.5–5.0) to neutral (6.5–7.5) and alkaline (7.5–8.5). During practical farming, farmers often use amendments such as lime ($CaCO_3$) to adjust soil pH in the rooting zone. It allows crops to respond to soil environment, fertilizer and irrigation optimally. Most important fact is that soil pH is a highly variable trait. It is affected by soil parent material, cropping history and agronomic procedures, adopted on the field. Therefore, farmers have to first develop a map showing soil pH variations. Then, farmers could adopt precision techniques, to apply lime at variable-rates. They may use applicators guided and commanded by digital data and computer-based decision support systems. Farmers may also develop a soil pH map using the manual procedures of digging soil samples and measuring soil pH. They can also use one of the most recent farm instruments (e.g., Veris Technologies' soil pH detector) that are equipped, to measure soil pH, automatically on-the-go, then, arrive at a tangible digital map showing soil pH variations (see Krishna, 2013, 2016; Lund, 2011). The digital data/map showing soil pH variations has to be read, using computer decision-support system that considers several soil characteristics. Some of them are texture, moisture, crops' optimum pH and lime required to raise the soil pH. Such decision-support computers then direct the variable-rate applicators on drones. Drones release gypsum at exact rates on to the soil. This way, drones could hasten application of tons of lime on to soil. It could be accomplished in a matter of hours. A swarm of drones with predetermined flight path does the variable-rate application of gypsum for over 1,000 ha, in a matter of hours. It is not possible, even if tractors with variable-rate applicators are used. Drones are rapid and cost effective to apply lime/gypsum pellets on to soil (Yamaha, Inc., 2015; Yintong Aviation Supplies Company Ltd., 2012). Incidentally, another advantage with precision techniques is that quantity of lime added reduces significantly due to variable-rate inputs. For example, Shannon (2014) reports that at the Missouri Experimental Agricultural Station, Boone county, MO, USA, if blanket application of lime is 2 t ac^{-1}, total quantity required for the entire field is 294 t. But, it reduces to 95 t if variable rates are adopted.

Drones, in general, could be flown above large stretches, to detect uneven or depressed growth, particularly if soil pH has already been suspected as uncongenial. Remedial measures could be focused to only those regions. This way, again, it saves on lime input. Gypsum application to soils prior to sowing seeds is a common soil-management practice. In much of the cereal production zone of Great Plains, the Mollisols are amended with gypsum. In the Cerrados region, again, acid Oxisols are widespread but difficult to cultivate. Gypsum application to soils has to be done, prior to sowing (Lopes, 1996; Krishna, 2003; Krishna, 2008; Krishna 2015a). In these vast regions, drones could be effectively employed to apply lime/gypsum at variable-rates or at blanket rates. Drones could be apt to conduct targeted soil amendment with gypsum powder or granules (Donald, 2014). Drones could conduct the procedure rapidly and replace ground vehicles. A copter drone such as RMAX can spray or discharge 2.5 kg gypsum pellets or powder per minute. About 26 kg of gypsum pellets or any other fertilizer formulation could be stored in the two tanks that are attached to fuselage (Yamaha Inc. 2015; Donald, 2014).

A report by UNEP-GEAS (2013) states that lightweight drones called 'eco drones' are useful. They could be fitted with visual, NIR, IR sensors, meteorological sensors and greenhouse gas–detection sensors. These eco-drones can provide imagery of natural disasters (forest fire, loss of vegetation, land slide, volcanic eruption resulting in hazardous emissions) and environmental deterioration (e.g., soil erosion) (Hardin and Hardin, 2010; Hinkley and Zajkowski, 2011; Koh and Wich, 2012; NASA, 2013). Simultaneously, they can provide meteorological data such as wind direction, speed, temperature in the ambient atmosphere, humidity and pressure (NOAA, 2008, 2012). The eco-drones can provide data on greenhouse gas emissions, particularly CO_2, CO, CH_4 and NH_3 (Khan, 2012; Nowatzki et al., 2014; UNEP-GEAS, 2013; Watai, 2006).

3.6 DRONES ARE USEFUL IN CONDUCTING SOIL FERTILITY EXPERIMENTS IN FIELDS

A few primary reasons for adopting drones during agricultural experimentation are that, they are easy to handle and quick to obtain data using aerial imagery. They provide fairly accurate data without involving tedious soil/plant sampling. Drone technology does not involve destructive sampling and processing. Chemical analysis and extended laboratory procedures are totally absent. Rapid or instantaneous availability of data is a clear possibility, if appropriate drone machines, computers and software are utilized by researchers/farmers. Most importantly, drones are cheaper to operate and obtain data about field experiments. Drones can handle very large set of digitized data and store them too. Drone-derived data are much easier to over layer, compare and contrast, whereas data recorded by human scouts and skilled farm technicians need careful arrangement and processing. Drones allow us to monitor several key physiological parameters of crops. Aerial assessments can be done swiftly, accurately and repeatedly, based on which we can deduce useful inferences.

Satellite imagery with low resolution could not be used consistently to evaluate a field experiment. The pixel size caused inaccuracy in the estimation of NDVI. Some pixels were wide enough to cover two different plots and even several of them with different treatments. The resolution of satellite imagery was less than that required to decipher NDVI and leaf area within a small plot (Reyniers and Vrinsts, 2006; Aguera et al., 2011). In such cases, a drone, that flies close to crop canopy and avoids noise (haze), is needed. At 75 ft above crop canopy, a Tetracam aerial imager can reach a resolution of 2.8 cm. At this accuracy, individual plots can be assessed for crop growth, NDVI, leaf-N and chlorophyll, rather accurately (Aguera et al., 2011; Hunt et al., 2014).

Drone imagery is nondestructive unlike ground-based analysis. Aspects such as collecting soil/plant samples, chemical processing and estimating plant-N/soil-N are laborious. We have adopted such chemical assays since several decades (Brady, 1975; Jackson, 1973). However, drone technology uses spectral analysis. Drone-based techniques are quick and efficient, in terms of cost of analysis of field experiments. Forecasts suggest that drone-aided techniques are set to take over many of farm functions and analytical activity.

Agricultural field experimentation involves intensive monitoring, sampling, chemical analysis, data collection and the usual farm drudgery. Field experimentation is also costly. The idea here is to see how the drones and their sensors could reduce the burden of cost for experimental evaluation. Drones could also reduce farm drudgery by scientists and workers alike. Drones with a range of sensors in the payload could be highly efficient, in terms of accurate data collection and storage. Computers with appropriate software could then analyse such data. They offer tangible suggestions to farmers, swiftly and at low cost. Drones are excellent aerial vehicles to surveillance the experimental fields in a farm. They can measure crop growth parameters at short intervals. Drones offer data at a fast pace to researchers. Incidentally, sensors placed on balloons (Boike and Yoshikawa, 2003), blimps (Inoue et al., 2000) and remote-controlled aircrafts have also been used to collect regular data about crop's response to agronomic procedures (see Zhu et al., 2009; Krishna, 2016; Thamm 2011; Pudelko et al., 2012).

Drones could be adopted to study the influence of soil moisture and crusting on germination of seeds. As stated earlier, data collected using cameras that operate at visual band width offer details about wheat seed germination. They are highly correlated with ground reality data collected by farm scouts (Khot et al., 2014). Report by Sullivan et al. (2007) suggests that drones fitted with IR sensors to measure TIR emittance could be used to detect water stress in the cotton canopy. They used an UAV with TIR sensor to judge cotton grown in the Tennessee Valley Research and Experimental Station at Belleville, Alabama, USA. Cotton crop was exposed to treatments involving organic residues from winter wheat. The crop was given different levels of irrigation. Experimental variability was higher if ground-based TIR was used. However, coefficient of variability reduced if drone-based remote sensing was adopted. Over all, it is believed that UAV with TIR could be used to obtain data from experiments that involve water stress and organic matter supply to cotton crop.

Drone-aided field experimentation is feasible. A drone is perhaps a best bet, if we have to acquire field data without destructive sampling of crop plants. In fact, data from multilocation involving soil fertility trials could be acquired, collated and analysed rather rapidly, using drones. Over all, multilocation trials will cost much less if drones are adopted for periodic scouting and data collection.

Soil fertility is among the most important factors that affects rice production. Let us consider an example related to soil fertility. It involves assessing rice crop's response to fertilizer-N supply in a field location. Regarding soil-N, farmers adopt a range of different agronomic procedures such as application of inorganic fertilizer-N, organic manures, foliar sprays of urea-N, residue recycling, adopting suitable crop rotations that include legumes and so forth. Fertilizer-N supply has to be optimum. Matching soil-N supply with crops' need at various stages is the crux. Soil-N deficiency could be detrimental to crop growth and grain formation, whereas excessive supply of fertilizer-N results in its accumulation in the soil profile. Such accumulated soil-N could be vulnerable to loss via erosion, seepage, runoff, emissions as NO_2, N_2O or N_2 in arable soils and as NH_4 in inundated rice fields. To arrive at optimum levels of fertilizer-N dosage, experts usually conduct field trials by applying fertilizer-N and/ or other major nutrients such as P and K. Such experimentations involve tedious sampling of soils prior to seeding, at the time of planting, then in the growing season, to assess soil-N recovered into plants. This helps farmers to apply fertilizer-N in split dosages at various stages of the crop. This procedure helps in improving fertilizer-N use efficiency.

As stated earlier, several different procedures are adopted to assess plant-N status. Many of them are tedious, time consuming and costly to perform. A few, like, hand-held leaf chlorophyll meters could be swift to collect data. But, they are apt only at leaf or plant scale. It is difficult to adopt the results for the entire field of say 10,000 ha. On the other hand, satellite techniques lack in resolution and could be affected by haze and cloud. They are not apt for single field of 100–1000 ha. They say, if satellite data are used, a single frame at that resolution (pixels) may include two different treatments. Hence, it is preferable to adopt drones that fly closely above the crop canopy. Drones can effectively separate fields, plots and can even allow us to monitor a single seedling, using GPS tags.

I'nen et al. (2013) state that hyper-spectral imaging of cereal crop, using small and lightweight drones, is useful. It helps one to arrive at precise fertilizer dosages through field experimentation. Zhu et al. (2009)

have conducted a field trial with different fertilizer-N inputs on rice. They have made detailed assessment of fields for variability, with regard to soil fertility. They have estimated various traits such as NDVI, leaf chlorophyll, plant-N, moisture, growth pattern and grain productivity. They have used sensors (visual, NIR and IR) to obtain data about rice crops' response to fertilizer-N supply. Zhu et al. (2009) have concluded that drones with sensors to image the crop colour could be utilized to assess plant-N status. They can also detect crop's response to fertilizer-N supply. Crop's response to fertilizer-N could be predicted by using aerial imagery for leaf colour, chlorophyll content and moisture stress, if any. Drone-aided imagery could be cost efficient. Requirement for skilled farm workers to scout the crop is avoided. Fertilizer supply at variable rates too could be possible, depending on the drone model. The computer software in the payload that helps in variable-rate application of liquid fertilizer-N is important (see Yamaha Inc., 2015; RMAX, 2015).

Precision farming methods supposedly reduce need for fertilizer and improve crop yield (see Krishna, 2013; Lowenberg-Deboer, 2003, 2006; I'nen et al., 2013; Kaivosoja et al., 2013). Precision farming requires detailed knowledge about spatial variations of soil fertility (e.g., soil-N), moisture distribution in the surface and rooting zone. It also requires previous yield data (maps) derived, using GPS-guided combine harvester. Management zones are demarcated using such primary data. Zhang et al. (2014) state that variable-rate methods and GPS-guided distribution of fertilizers form the core of precision agriculture. During yester years, field/crop variability in productivity was estimated, using manual scouting. Previous years' yield maps from GPS-combine harvesters were also utilized. However, during recent few years, Low Altitude Airborne Remote Sensing, in other words, use of low flying drones, is gaining acceptance. Drones are getting popular in conducting field trials, to assess soil fertility and crop's response to fertilizer amendments (Drone Analyst, 2015).

'Agricultural drones' with multispectral sensors offer basic images and digital data required to practise variable techniques. Drone imagery also helps in monitoring and collecting data from experiments that assess effect of fertilizer on crops. Let us consider an example from Ontario region of Canada. Here, Zhang et al. (2014) have examined the influence of organic manures and chemical fertilizers on wheat, barley and soybean growth, using drone imagery. They have used an agricultural drone, a quadcopter (Ayeron Scout) produced by Aeyron Labs Inc., Ontario Canada, fitted with multispectral cameras (ADC Lite-Teracam). It collects images with GPS tag and covers an area of 25 ac per flight. They examined fields provided with organic manure only (9.37 gallons organic manure ha^{-1}), organic manure and chemical fertilizers (9.37 gallons organic manure plus 185 kg N ha^{-1}) and chemical fertilizers (371 kg N ha^{-1}) only. Later, they obtained ortho-mosaics and processed them. They have used computer software called *Pix4D mapper* to obtain colour images of crop's response to fertility treatments. The NDVI values for plots treated with organic manure only were weakest in vigour. It is represented by darkest zone in the IR images. Drone-derived images clearly depicted large difference in NDVI values for organic manure and chemical fertilizer treatments. Plots with organic manure only yielded 1.73 t grains ha^{-1}, those with both organic manure and chemical fertilizers gave 2.27 t grains ha^{-1}, whereas those with only chemical fertilizers gave 2.97 t ha^{-1}. Overall, drones seem efficient in collecting data about crop fields during precision farming. They cost much less compared with skilled farm worker. They aid regular monitoring of field experiments. Aerial imagery helps in deciphering differences in soil fertility. It helps to map the variations and record NDVI and crop growth trends throughout the season.

Tremblay et al. (2014) have tried to study the effect of interaction between soil-N fertility and irrigation on maize growth. They have also used ground-based vehicles fitted with sensors that operate, using proximal technology (SPAD chlorophyll meter). They have used satellite-based hyperspectral imagery. Importantly, drones fitted with a series four cameras that operate at R, G, B, NIR and IR band widths were also evaluated. They grew maize at two levels of fertilizer-N supply (0 and 200 kg N ha^{-1}) with and without irrigation. The experimental fields were located at the Agriculture-Canada Experimental Station, in the Montague region of Quebec, Canada. They have reported that soil-adjusted vegetation index (NDVI, GNDVI) obtained, using drones, was correlated to fresh biomass, better than readings from satellite imagery ($r^2 = 0.93$ for drones and 0.88 for satellite imagery). LAI from UAV cameras were highly correlated to biomass ($r^2 = 91$). Chlorophyll estimation using drone-based cameras provided accurate estimates of plant-N status, when the crop was at seedling stage. There are few suggestions from this study. First, agricultural experimentation could become easier, if drones are adopted. Accurate data about soil fertility and its influence on crop growth traits could be collected, using drones. Drones could be swifter and cost less to evaluate a series of fertilizer formulations, and their dosages. Even different geno-types of crops and their responses to soil fertility variations could be studied, using drone-aided aerial imagery. Tremblay et al. (2014) further state that, finer resolution and accuracy of UAV imagery helps farmers to manage irrigation channels and drainage.

Our basic aim in mending soils, irrigating and selecting proper agro-nomic procedures has been, to maximize grain/forage yield. At the same time, we have to reduce cost on inputs, farm labour and manage-ment. Usually, once genotype is decided, it is the soil fertility and water resources that have dictated farmers' yield goals. Soil fertility variations are to be removed. Consistently, optimum nutrient availability levels are to be maintained. *Drones are among the recent techniques to arrive into the farming world, with a promise to improve and impart uniformity to soil fertility.* They are apt to manage large farms and attain yield goals. Drones, no doubt, reduce cost on human labour. They are also known to reduce inputs, such as fertilizers, pesticides and herbicides, marginally or in some locations significantly. A recent report suggests that during recent years, farmers situated in established, high yielding agrarian region such as Great Plains of USA, who specialize in producing maximum tonnage. They have directed their efforts to reach really high-yield goals (e.g. 171 bu maize grains ac^{-1}). They have been striving to reach the maximum potential yield in that location. They seem to bestow less importance to reduction of cost on inputs and labour or production efficiency (Precision Farming Dealer, 2016). In the present context, we may have to note that drones have their role cut out, in all these locations. Drones could be of great value in conducting multilocation experimental trials. Drones seem to have bright future in commercial and experimental farms of all sizes and intensities of cropping. They suit farmers of different ambitions. Agri-cultural researchers may aim at using drones to reduce cost on inputs and at the same time enhance efficiency and accuracy of operations. Drones are apt if the aim is to reduce farm drudgery.

In the near future, we may encounter drones with multispectral sensors more frequently in the agricultural experimental fields. It means, we will depend more on data derived from spectral methods and remote sensing. Such a change will reduce our dependence on wet chemical analysis of soil/plant samples. We may have to realize that in addition to drones, there are several other farm robots and on-the-go farm instruments devised to measure soil EC, moisture, pH, soil NO_3-N, SOM and so forth. These vehicles and instruments are dependent on optical (spectral), electrical and magnetic principles and not on wet fluid-based chemical assays. Therefore, sensors based on proximal technology may become common, during measurement of a range of soil/crop parameters in agricultural experimental fields. Together, these procedures, and many more in the making will replace wet analysis of soil/plant in the laboratories by using test-tubes, aliquots of samples, reflex condensers, soxlet extractions, centrifugation of soil samples, chemical reactions and so forth. A big change in approach in the conduct and analysis of agricultural experiments is imminent, as drones begin to flourish in farm land.

3.7   AGRICULTURAL CONSULTANCY SERVICES FOR DRONE-AIDED FERTILIZER SUPPLY

Agricultural researchers and farmers alike have used satellite imagery, to show differences in landscape and their soil fertility. Satellite images have also helped farmers to monitor crops and trace changes in growth pattern and maturity, but with relatively low resolution. Such methods have ensured fairly accurate management of farms, particularly in choosing crops, cropping systems, seeding density, fertilizer application, watering and timing of grain harvests. However, in the present context, we are concerned more with aspects related to soil and how they affect crops, if they are managed, using drones. We are actually in a transition from satellite techniques to drones-based assessment of soils and crops (Crop Site, 2015).

Agricultural consultancy companies are switching rapidly into using drones. They use drones to scout the farmers' fields routinely for soil characteristics, plant stand, crop vigour, nutrient deficiency and while deciding need for fertilizer supply (see Table 3-1). They are using precision farming procedures to ascertain fertilizer-N demand, at each location in a field. For example, 'Wingscan', a private agency, detects variations in crop growth status, drought stress, if any, and nitrogen need of plants. Then, the data are analysed using appropriate computer decision-support programs. They aim at offering most profitable decisions, in terms of fertilizer supply and crop productivity. Farmers are also alerted about periodic needs for top-dressing fertilizer-N to the crop. The drone companies use high-quality digital aerial imagery (Farm Intelligence, 2014). Farmers can subscribe to consultancy agencies to get regular alerts and information about several aspects of soil and crop management. For example, farmers obtain information about plant stand, seedling population, soil compaction, loss of seedlings, soil erosion, top soil loss, nutrient deficiency, fertilizer needs at different locations in a field, grain maturity and yield traits. When farmers subscribe to such crop consultancy companies, they reduce costs on buying a drone and computer software. Farmer's need for computer technicians to help him in decoding the ortho-images and arriving at appropriate remedial measures and fertilizer dosages is also avoided. Drones, in fact, have an important role in soil/crop scouting, soil fertility management, irrigation and environmental mapping regularly and periodically (Kooistra et al., 2014; 2015; Bartholomeus et al., 2015; Anders et al., 2013). These functions of drones are needed irrespective of whether it is done by an individual farmer or by private agricultural consultancy agency.

Let us consider yet another example of agricultural consultancy agency that offers variety of services related to soil fertility mapping, fertilizer prescription and application by using variable-rate methods. For example, soil information is provided by Trimble Inc., located at Sunnyvale in California, USA. It offers drone imagery that helps in detecting soil variability with high precision (Trimble Inc. 2015a, 2015b, 2015c, 2015d). Images help in the formation of management blocks, managing fertilizer supply, regulating irrigation, assessing requirement of soil amendments such as gypsum, identifying nutrient runoff and seepage trends in the field. The key features of fields such as soil texture, colour, soil compaction trends in the field, moisture retention and soil fertility are included. Most importantly, the consultancies offer farmers with variable-rate maps and digital data (in chips) that could be incorporated into fertilizer inoculation vehicles. Such variable-rate application maps are of immense value to farmers. Fertilizer needs for a given yield goal get reduced. Therefore, cost of production could be reduced. Specifically, nutrient management advisory allows farmers to compare aerial maps and ground reality data, regarding soil fertility (nutrient) levels. Farmers can also reduce on labour cost by targeting soil/crop scouting to specific locations. The crop growth images help farmers in identifying locations suffering nutrient dearth. Such early warning through drone imagery helps in correcting fertility anomalies. Mid-way through the season, crop's response to fertilizer supply can be compared. In general, the consultancy companies, using drone imagery, can alert farmers about soil fertility status (Trimble Inc., 2015b).

Now, let us consider a specific drone model used by agricultural agencies and what it offers to farmers. Trimble's UX5 is an example of flat-winged field drone. It has a visual camera and NIR camera attached. It flies past a field and scouts the crop for mineral deficiency. Therefore, management blocks with nutrient deficiencies and those needing fertilizer could be marked. About 75 ha of field is imaged at 2.5 cm pixel^{-1} resolution in a single flight. The images (2D and 3D) could be processed instantaneously for farmers. They may take note of data and analyse on a computer screen (Trimble Inc., 2015e).

Drone-aided analysis of crop stand and nutrient needs is now in vogue, in different parts of the world. For example, in New Zealand, drone-aided visual and IR sensing is said to help farmers. Farmers are informed about variations in soil fertility, within in their fields. Regular crop health maps (NDVI, leaf chlorophyll) are also provided to farmers, through private drone firms (Aerial Imaging Services Ltd, 2015). These drone companies supply variable-rate maps (digital data) that could be used to supply fertilizer-N to crop at appropriate rates using variable-rate applicators.

Charlotte UAV Inc. is another example of a drone company offering soil/crop-related services to farmers. They offer drone (unmanned aerial) services, aerial photographs (2D and 3D) of farms, irrigation channels and crops on a daily basis. They conduct measurement of NDVI and crop stress, that is, thermal imagery to detect soil moisture level. Digital thermography is useful while examining crop and natural vegetation for moisture status. This drone company also specializes in imaging lakes and irrigation channels. They aim specifically at recording pollution, if any (CharlotteUAV, 2016).

In Israel, agricultural consultancy agencies (e.g., Sensilize) are trying to offer critical information required for precision farming, as quickly as possible to farmers. They have developed a composite system known as 'Robin eye System'. It integrates a series of high-resolution spectral sensors (Visual, NIR, IR and thermal range) and computer software that analyses the ortho-images (see Leichman, 2015). The farmers are provided with colour images of their crop fields. They also supply advisory about soil–moisture and crop–moisture status, NDVI and crop health within 24 h after the drone has imaged their fields. As the data are offered on web site, farmers can access and download digital data. Such data could be directly utilized in variable-rate applicators.

KEYWORDS

- **agricultural drone system**
- **multispectral sensors**
- **soil fertility mapping**
- **drone-derived imagery**
- **soil organic matter**
- **drone-aided fertilizer supply**

REFERENCES

Acevo-Herrara, R.; Aguasca, A.; Boscsh-Lluis, X.; Camps, A.; Martinus-Fernandez, J.; Sanchez-Martin, N.; Perez-Gutiarrez, C. Design and First Results of an UAV Borne L-band Radiometer for Multiple Monitoring Purposes. *Remote Sens.* **2010,** *2,* 1662–1669.

Aerial Imaging Services Ltd. *AIS Infra-Red Sensing.* AIS, Whangarai; New Zealand; 2015, pp 1–5. http://http://drones4crops.co.nz (accessed May 5, 2015).

AgEagle Aerial Systems Inc. *Aerial Technology.* AgEagle Aerial Systems: Kansas, USA, 2016; pp1–8 http://ageagle.com//technology/ (accessed February 16, 2016).

AggieAir. A Remote Sensing Unmanned Aerial System for Scientific Applications. Utah State University: Logan, UT, USA, 2015; pp 1–3. http://aggieair.usu.edu/node/292 (February 2 2016).

Aguera, F.; Carvajal, F.; Perez, M. Measuring Sunflower Nitrogen Status from an Unmanned Aerial Vehicle-Based System and on the Ground Device. *International Archives of the Photogrammetry, Remote Sensing and Spatial Information Sciences.* **2011,** *38,* 1–5.

Alvarez, R. Predicting Average Regional Yield and Production of Wheat in the Argentine Pampas by an Artificial Neural Network Approach. *Eur. J. Agron.* **2007,** *30,* 70–77.

Amaraj, L. R.; Molin, J. P.; Schepers, J. S. Algorithm for Variable-Rate Nitrogen Application in Sugarcane, Based on Active Canopy Sensor. *Agron. J.* **2015,** *107,* 1513–1523.

Anders, N.; Keestra, S. D.; Soumaleinen, J. M.; Bartholomeus, H.; Kooistra, L. Monitoring geomorphological change with Unmanned Aerial Vehicles. In: *Proceedings of the 8th IA International Conference on Geomorphology*, Paris, France, 2013, p 233; Abstract No S 472

Anderson, J. In: *Iowa city farmer uses drone to aide Precision Farming.* Demoines Register. 2014, pp1–6. http://www.desmoinesregister.com/histroy/money/agriculture/2014/10/27/iowa-city-farmer-uses-drone-aide-farming -precision. /18031077/ (accessed June 26, 2015).

Anikwe, M. A.; Ngwu, O. E., Mbah, C. N.; Onoh, C. E.; Ude, E. E. Effect of Groundnut Cover Crops On Soil Loss and Physico-Chemical Properties of an Ultisol in Southern Nigeria. *Niger J. Soil Sci.* **2007,** *17,* 94–97.

Antic, B.; Culibrk, D.; Crnojevic, V.; Minic, V. An Efficient UAV-Based Remote Sensing Solution for Precision Farming. An Efficient UAV Based Remote Sensing Solution for Precision Farming. In: *Bioscience- The First International workshop on ICT and Sensing Technologies in Agriculture, forestry and Environment Novi Sad, Serbia* 2010, pp 223–228.

Arregui, L. M.; Lasa, B.; Lafarga, A.; Iranata, I.; Baroja, E.; Quemada, M. Evaluation of Chlorophyll Meters as Tools for Nitrogen in Winter Wheat Grown Under Humid Mediterranean conditions. *Eur. J. Agron.* 24:140–148

AscTec. UAV Aerial Imaging. Ascending Technologies Inc. 2016, pp 1–8 http://www.asctech.de/en/uav-uas-drone-applications-uas-arial-imaging-hr-photos-skills/ (accessed Feb 11, 2016).

AutoProbe Technologies. Precision Agriculture Starts with a Precise Sample, 2016, pp 1–18. http://www.agobotics.com (accessed March 3, 2016).

Bartholomeus, H.; Suomalainen J.; Frank, J.; Kooistra, L. Design of an UAV-Based Hyperspectral Scanning System and Application in Agricultural and Environmental Research. Presentation Geospatial World Forum; Lisabon, Portugal; 2015, pp 1–5.

Bartholomeus, H.; Soumalainen, J.; Kooistra, L. Estimation of within Field Variation of SOM Using UAV Based RGB and Elevation Data. In: *Proceedings of the EGU general Assembly*, Vienna, Austria. Geophysical Research,, 2014; Abstracts No. 16 EGU2015-5660.

Bationo, A.; Kihara, J.; Vanlauwe, B.; Waswa, B.; Kimetu, J. Soil Organic Matter Dynamics, its Functions, and Management in West African Agro-ecosystems. *Agric. Syst.* **2006**, *94*, 13–25.

Bazzoffi, P. Measurement of Rill Erosion Through New UAV-GIS Methodology. CREA-ABP; Italy, 2015, pp 1–15. http://dx.doi.org/10.408/ija.2015.708 http://www.agronomy.it/index.php/agro/article/veiw/708 (accessed Feb 10, 2016).

Bendig, J.; Willkomm, M.; Tilly, N.; Gnyp, M. L.; Bennertz, S.; Qiang, C.; Miao, Y.; Lenz-Wiedemann, V. I. S.; Bareth, G. Very High Resolution Crop Surface Models (CSMs) from UAV-Based Stereo Images for Rice Growth Monitoring in North China. *International Archives of the Photogrammetry, Remote Sensing and Spatial Information Sciences* **2013**, *XL*, 45–50.

Beyer, L. Soil Geography and Sustainability of Cultivation. In: Soil fertility and Crop Production Krishna K. R. (Ed.). Science Publishers Inc.,: Enfield, New Hampshire, USA, 2003; pp 33–63.

Bielders, C.; Michels, K.; Rajot, J. On Farm Evaluation Of Wind Erosion Control Technologies. 2002a, pp 1–5. ICRISAT.Org CCER.htm (accessed Feb 20, 2016).

Bielders, C.; Rajot, J.; Amadou, M.; Skidmore, E. On Farm Quantification of Wind erosion Under Traditional Management Practices. 2002b, pp 1–6. ICRISAT.Org CCER.htm (accessed Feb 20, 2016).

Blackmer, T. M.; Schepers, J. S. Aerial Photography to Detect Nitrogen Deficiency in Corn. *J. Plant Physiol.* **1994**, *148*, 440–444.

Boike, A. K. M.; Yoshikawa, K. Mapping of Periglacial Geomorphology Using Kite/Balloon Aerial Photography. *Permafrost Periglacial Process* **2003**, *14*, 81–85.

Bousset, J. West Africa: Can Down to Earth Innovations Keep Hunger at Bay in the Sahel. Desertification. 2016, pp 1–5. https://desertification.wordpress.com/tag/sahel (accessed Feb 18, 2016).

Brady, N. C. *Nature and Properties of Soils*. Prentice Hall of India: New Delhi, 1975; p 576.

Buis, A.; Murphy, R. New Satellite Data will Help Farmers Facing Drought. 2014, pp 1–6. http://www.nasa.gov/jpl/smap/satellite-data-hep-farmers-facing-drought-201408#VJZHoF4AA (accessed Dec 22, 2014).

Burkert, A.; Mahler, F.; Marschner, H. Soil Productivity Management and Plant Growth in the Sahel: Potential of an aerial monitoring technique. *Plant Soil* **1996**, *80*, 29–38.

Bybordi, S.; Reggiani, L. Drones in Agriculture: Applications and Outlook. Departmento de Electronica Informzione e Bioingegneria, Politecnico; Milano, Italy. Internal Report 2015, pp 1–28.

Candiago, S.; Remonino, F.; DeGiglio, M.; Dubbini, M.; Gatellit, M. Evaluating Multispectral Images and Vegetation Indices for Precision Farming Applications from UAV Images. *Remote Sens.* **2015**, *7*, 4026–4047.

CharlotteUAV. Drone Integration Experts. 2016, pp 1–4. http://www.charlotteuav.com (accessed March 3, 2016).

Crawford, K.; Roach, Dhillon, R.; Rojo, F.; Upadhyaya, S. An Inexpensive Aerial Platform for Precise Remote Sensing of Almond and Walnut Canopy Temperature. *Proceedings of 12th International Conference on Precision Agriculture*, Sacramento, California, USA, 2014; p 27.

Crop Site. How Can Drones Make Farming Profits. 2015, pp 1–3. http://www.thecropsite.com/news/17406/how-can-drones-make-farming-profits/ (accessed June 28, 2015).

D'Oliere-Oltmans, S.; Marzolff, I.; Peter, K. D.; Ries, J. D. Unmanned Arial Vehicle (UAV) for Monitoring Soil Erosion in Morocco. *Remote Sens.* 2012, *4*, 3390–3416. DOI: 10.3390/rs4113390 (accessed Nov 18, 2015); 1–12.

Donald, B. Drones and AVs: What is Available Now and What is Possible in the Future. Grain Research Development Corporation, Barton, Australia. 2014, pp 1–5. http://www.grdc.com.au/research-and-development-in-the-future (accessed Feb 13, 2016).

Drone Analyst. Precision Agriculture. 2015, pp 1–3. http://droneanalyst.com/agenda/precision-agriculture/ pp 1–3 (accessed May 14, 2015).

Effren, D. Autocopter- The Precision Ag Solution. *PR News Wire*, 2014, pp 1–7. http://ireach.prnews.com (accessed Aug 25, 2014).

Eltner, A.; Mulsow, C.; Maas, H. G. Quantitative Measurement of Soil Erosion from TLS and UAVData. *International Archives of the Photogrammetry, Remote Sensing and Spatial Information Sciences.* 2013, *XL*, 119–123.

Esfahani, L.; Torres-Rua, A.; Jensen, A.; McKee, M. Fusion of High Resolution Multi-Spectral Imagery for Surface Soil Moisture Estimation Using Learning Machines. Utah State University: logan, USA, 2014a; pp 1. http://www.digitalcommons.usu.edu/do/search/?q=author_iname%3A22Hassan-Esfahani522%20author_fname%22Leila%22&start=0&context=656526 (accessed June 28, 2015).

Esfahani, L.; Torres-Rua, A. M.; Ticlavilca, A.; Jensen, M.; McKee, M. Topsoil Moisture Estimation for Precision Agriculture Using Unmanned Aerial Vehicle Multispectral Imagery. IEEE International Geoscience and Remote Sensing Symposium, 2014b, pp 1–4.

Esfahani, L.; Torres-Rua, A.; Jensen, A.; McKee, M. Assessment of Surface Soil Moisture Using High Resolution Multi-Spectral Imagery and Artificial Neural Network. *Remote Sens.* **2015**, *7*, 2627–2646. DOI:10.3390/rs70302627.

FAO. Soil Map of the World. *Technical paper20 ISRIC*, Wageningen, Netherlands, 1998, Revised Reprint p 140.

FAO. Use of Fertilizers by Crop and Region. Food and Agricultural Organization of United Nations, 2007; pp 1–7. http://fao.org/docrep/007/y5210e/y5210e09.htm pp 1–7 (accessed April 12, 2013).

Farm Intelligence. Transforming Precision Agriculture into Decision Agriculture. Wing-Scan, 2014; pp 1–5. http://fi2salesandleasing.com/wingscan.html (accessed May 5, 2015).

Farms.com. Precision Agriculture Economics. *Precision Agriculture Conference*, Ontario, Canada, 2015; pp 1–7. http://www.farms.com/precision-agriculture/economics/ (accessed May 5, 2016).

Farm World. Real Customers Real Results. 2014, pp 1–30. http://www.farmworld.ca/ drones/results.aspx pp 1–30 (accessed Feb 2, 2016).

Fox, R. H.; Walthall, C. L. Monitoring Technologies to Assess Nitrogen Status. In: *Nitrogen in Agricultural systems*; Schepers, J. S., Raun, W. S., Eds.; American Society of Agronomy: Madison Wisconsin, USA Agronomy Monograph, 2008; Vol. 49, pp 647–674.

Franzen, D. W.; Kitchen, N. R. Developing Management Zones to Target Nitrogen Applications. International Plant Nutrition Institute: Norcross, USA, 2011; pp 1–14. http:// www.ini.net. /ppiweb/ppibse.nsf/$webindex/article=772BE&3B85695A005A12E99F5 E03CD (accessed June 27, 2011).

Gallacher, D. Applications of Micro-UAVs (drones) for Desert Monitoring: Current capabilities and requirements. 2015, pp 1–24. http://www.acaddemia.edu/11338070/applications_of_micro-UAVs_drones_for_desertification (accessed May 5, 2016).

Gallmeyer, B. A.; Aoyagi, M. Imaging from an Unmanned Aerial Vehicle: Agricultural surveillance and Decision Support. *Comput. Electron. Agric.* 2004, *44*, 49–61.

Garcia, F. O. Nutrient Best Management Practices for Wheat Fertilization Practices for Intensive Wheat Production in Southern Latin America. 2013, pp 1–3. http://ipni.net/ ppiweb/itmes.nst/$webindex/48E840CAD00D2E590325752900710F96?opendocumen t&print=htm (accessed April 20, 2013).

GIS AG maps. Yield Map Based Management Zone Development. 2014, pp 1–16. http:// www.gisagmaps.com/yield-based-management-zones/ (accessed June, 30, 2015).

Goli, N. Researcher Uses Drones to Measure Soil Moisture. The University of Alabama: Huntsville, Alabama, USA, MS Thesis. 2015, p 1. http://www.chargetimes.com/2092/ science-and-technology/researcher-uses-drones-to-measure-soil-moisture.   Htm (accessed January, 26, 2016).

Grassi, M. 5 Actual Uses for Drones in Precision Agriculture Today. 2014, pp 1–3. Drone-Life.   http://dronelife.com/2014/12/30/5-actual-usees-drones-precision-agriculture-today/ (accessed May 13, 2015).

Gronau, I. Autonomous Tractor Corporation Lays Plan to Slowly Introduce Driverless Tractor. Precision Farming Dealer. 2016, pp 1–5. http://www.precisionfarmingdealer. com/articles/2013-autonomous-tractor-corporation-lays-plans-to   introduce-driverless-tractor (accessed Feb 28, 2016).

Guo, Y.; Shi, Z.; Li, H. Y.; Triantafilis, J. Application of Digital Soil Mapping Methods for Identification of Salinity Management Classes Based on a Study on Coastal Central China. *Soil Use Manage.* **2014**, *29*, 445–456.

Han, S.; Hendrickson, L.; Ni, B. *Comparison of Satellite Remote Sensing and Aerial Photography for Ability to Detect in-Season Nitrogen Status Stress in Corn.* An ASAE meeting presentation. Paper No 01-1142. ASAE: St Joseph, MI, USA, 2001; pp 1–3.

Hardin, P. J.; Hardin, T. J. Small Scale Remotely Piloted Vehicles in Environmental Research. *Geography Compass.* 2001, *4*, 1297–1311.

Hinkley, E. A.; Zajkowski, T. USDA Forest Service-NASA: Unmanned Aerial Systems Demonstrations-Pushing the Leading Edge in Fir Mapping. *Geocarto. Int.* 2011, *26*, 103–110.

Hunt, E. R.; Hively, W. D.; Daughtry, C. G. T.; McCarty, G. W.; Fujikawa, S. J.; Ng, T. L.; Tranchitella, M.; Linden, D. S.; Yoel, D. W. Remote Sensing of Crop Leaf are Index Using Unmanned Airborne Vehicles. *Proceedings of a Conference Pecora-The future of Land Imaging*, Denver, Colorado, USA, 2008; pp 1–12. http://www.asprs.org/a/publication/proceeding/pecora17/18.pdf.

Hunt, E. R.; Horneck, D. A.; Hamm, P. B.; Gadler, D. J.; Bruce, A. E.; Turner, R. W.; Spinelli, C.B.; Brungardt, J. J. Detection of Nitrogen deficiency of Potatoes Using Small Unmanned Aircraft Systems. *Proceedings of 12th International Conference on Precision Agriculture*: Sacramento, California, USA, 2014; p 16.

I'nen, I. P.; Saan, H.; Kaivosoja, J.; Honkavara, E.; Pesnonen, L. Hyperspectral Imaging Based Biomass and Nitrogen Content Estimations from Light-Weight UAV. *Remote Sens. Agric. Ecosyst. Hydrol.* 2013, *15*, 87–88. http://dx.doi.org/10.1117/12.2028624 (accessed May 23, 2015).

Inoue, Y.; Moringa, S.; Tomita, A. A Blimp-Based Remote Sensing System for Low Altitude Monitoring of Plant Variables: A preliminary Experiment for Agricultural and Ecological Explanations. Int. *J. Remote Sens..* 2000, *21*, 379–385.

Jackson, M. L. Soil Chemical Analysis. Prentice Hall Inc.: Englewood cliffs, New Jersey, USA, 1973; p 483.

John Deere Inc. Precision Manure Application with NIR Technology. Precision Farming Dealer. 2015, pp 1–2. https://www.precisionfarmingdealer.com/articles/1046-precision-manure-application-with-nir-technology (accessed August 12, 2016).

Kaleita, A.; Tian, L.; Hirschi, M. Relationship between Soil Moisture Content and Soil Surface Reflectance. *Trans. Am. Soc. Agric. Eng.* 2005, *48*, 1979–1986.

Kaivosoja, J.; Peonen, L.; Kleemola, J.; I'nen, L.; Salo, H.; Honkavaara, E.; Saari, H.; Makynen, J.; Rajala, A. A Case Study of a Precision Fertilizer Application Task Generation for Wheat Based on Classified Hyperspectral Data from UAV Combined with Farm History Data. *Remote Sens. Agric. Ecosyst. Hydrol.* 2013, *15*, 8887OH. DOI: 10.1117/12.2029165 (accessed May 23, 2015).

Karyotis, T.; Panagopoulos, A.; Alexiou, J.; Kalfountzos, D.; Pateras, D.; Argyropoulos, G.; Panoros, A. Nitrates Pollution in a Vulnerable Zone of Greece. *Commun. Biometry Crop Sci.* 2006, *1*, 72–75.

Khan, A. Low Power Greenhouse Gas Sensors for Unmanned Aerial Vehicles. *Remote Sens.* 2012, *4*, 1355–1368.

Khot, L.; Sankaran, S.; Cummings, T.; Johnson, D.; Carter, A.; Serra, S.; Musacchi, S. Unmanned Aerial Systems Applications in Washington State Agriculture. *Proceedings of 12th International Conference on Precision Agriculture*: Sacramento, California, USA, 2014; p 129.

Koh, L. P.; Wich, S. A. Dawn of Drone Ecology: Low Cost Autonomous Aerial Vehicles for Conservation. *Trop. Conserv. Sci.* 2012, *5*, 121–132.

Kooistra, L.; Suomalainen, J.; Anders, N.; Franke, J.; Bartholomeus, H.; Keestra, S.; Mucher, S. Opportunities for UAS as SMART Inspectors in Environmental Monitoring Applications. The Unmanned Systems Expo. The Hague: Netherlands, 2015; pp 1–6.

Kooistra, L.; Soumalainen, J.; Franke, J.; Bartholomeus, H.; Mucher, S. E.; Becker, R. Monitoring Agricultural Crops Using Hyperspectral Mapping Systems for Unmanned Aerial Vehicles. In: *Proceedings of the EGU General Assembly*, Vienna, Austria. Geophysical Research Abstracts 16 EGU2014-2790.

Krishna K. R. Soil Fertility and Crop Productivity. Science Publishers Inc.: New Hampshire, USA, 2002; pp 91–108.

Krishna K. R. 2003 Agrosphere: Nutrient Dynamics, Ecology and Productivity. Science Publishers Inc.: New Hampshire, USA; pp 241–279.

Krishna, K.R. 2008 Peanut Agroecosystem: Nutrient Dynamics and Crop Productivity. Alpha Science International Ltd. Oxford, England, pp 75–116.

Krishna, K.R. 2013 Precision Farming: Soil Fertility and Productivity aspects. Apple Academic Press Inc.: Waretown, New Jersey, USA; pp 91–108.

Krishna, K.R. 2015a The Cerrados of Brazil: Natural Resources, Environment and Crop Production. In: Agricultural Prairies: natural Resources and Crop Productivity. Apple Academic Press Inc.: Waretown, New jersey, USA; pp 90–142.

Krishna, K.R. 2015 b Savannahs of West Africa: Natural Resources, Environment and Crop Production. In: Agricultural Prairies: natural Resources and Crop Productivity. Apple Academic Press Inc.: Waretown, New Jersey, USA; pp 246–312.

Krishna, K. R 2015c Prairies of South Asia: Natural Resources, Environment, and crop production. In: Agricultural prairies: natural Resources and crop productivity. Apple Academic Press Inc.: Waretown, New jersey, USA; pp 313–399.

Krishna, K. R. 2016 Push Button Agriculture: Robotics, Drones, soil and Crop management. Apple Academic Press Inc.: Waretown, New Jersey, USA; pp 470.

Kyverga, P. M.; Blackmer, T. M.; Pearson, R. Normalization of Uncalibrated Late Season Digital Aerial Imagery for Evaluating Corn Nitrogen Status. *Precis. Agric.* **2012**, *13*, 2–16.

Lacava, T.; Brocca, L.; Irina, F.; Malone. F.; Moramarco, T.; Pergola, N.; Tramutoll, V. Soil Moisture Variability Estimation Through AMSU Radiometer. *Eur. J. Remote Sens. 2012*, *45*, 89.

Leichman, A. K. Drones-the Future of Precision Farming. 2015, pp 1–4. http://www.israel21c.orgheadlines-the-future-of-precision-agriculure (accessed June 21, 2015).

Lopes, A. S. Soils Under Cerrados. A Success Story in Soil Management. *Better Crops Int.* **1996**, *10*, 9–15.

Lowenberg-Deboer, J. Precision Farming in Europe. 2003, pp 1–3. http://agriculture.purdue.edu/ssmc/newsletters/june03_PrecisionAgEurope.htm (accessed March 23, 2011).

Lowenberg-Deboer, J. Economics of Variable Rate Planting for Corn. 2006, pp 1–8. http://agrarias.tripod.com/precision-agricuture.htm (accessed March 25, 2011).

Lund, E. Veris pH mapper. Veris Technologies Inc.: Kansas, USA, 2011; pp 1–3. http://www.veristech.com/products/soilpH.aspx (accessed Aug 10, 2014).

Marzolff, I.; Poesen, J. The Potential of 3D Gully Erosion Monitoring with GIS Using High Resolution Aerial Photography and Digital Photogrammetry System. *Geomorphology* **2009**, *111*, 48–60.

Marzolff, I.; Ries, J. B.; Albert, K. D. Aerial Photography for Gully Monitoring in Sahelain Landscapes. *Proceedings of Second Workshop of the EARSeL special interest group on Remote Sensing for Developing Countries*. Bonn, Germany, 2002; pp 18–20.

Marzolff, I.; Ries, J. D.; Poesen, J. Short Term vs Medium Term Monitoring for Detecting Gully-Erosion Variability in a Mediterranean Environment. *Earth Surf. Processes Landforms* **2011**, *36*, 1604–1623.

Michels, K.; Shivakumar, M. V. K.; Allison, B. E. Wind Erosion Control Using Crop Residue. Soil Flux and Soil particles. *Field Crops Res.* **1995**, *40*, 101–110.

Microdrones GMBH. Flying Farmers – Future Agriculture: Precision Farming with Microdrones. 2015, pp 1–5. http://www.microdrones.com/en/applicationsgrowth-markets/microdrones-in-agricudlture (accessed May 5, 2015).

NASA. NASA Flies Dragon Eye Unmanned Aircraft into Volcanic Plume. National Aeronautics and Space Agency.2013. http://climate.nasa.gov/news/891 (accessed Feb 20, 2016).

NASA. NASA to Launch Soil Moisture Active Satellite. 2014, pp 1–8. http://www.agprofessional.com/new/NASA-to-launch-soil-moisture-active-satellite-274361731.html (accessed Nov 16, 2014).

NDSU. Fertilizer Requirements of Corn. NDSU Extension Service. 2012a, http://ag.ndsu.edu/procrop//cm/cmflr04.htm (accessed April 10, 2013).

NDSU. Fertilizer for Winter Wheat. NDSU Extension Service. 2012b, pp 1–8. http://www.ag.ndsu.edu/procrep/wwh.flrwin04.htm (accessed April 10, 2013).

Neugerg, F.; Kaiseer, A.; Schmidt, J.; Becht, M.; Haas, F. Quantification, Analysis and Modelling of Soil Erosion on Steep Slopes Using LiDAR and UAV Photographs. *Proceedings of a Symposium on Sediment Dynamics.* New Orleans, Louisiana, USA, 2015, IAHS Publication No. 367: 51–76.

NOAA. NOAA invests US$ 3 million for Unmanned Aircraft System Testing Pilotless Craft to Gather Data for Hurricane Forests, Climate, West Coast Flood Warnings and Atmospheric Gases. National Oceanic and Atmospheric Administration. 2008, pp 1–12. http://www.esrl.noaa.gov/news/2008/uas.html (accessed March 19, 2016).

NOAA. NOAA Scientists Part of NASA-Led Mission to Study the Damaging Storms with Unmanned Aircraft. New Instruments. National Oceanic and atmospheric Administration, 2012, pp 1–7. http://uas.noaa.gov/news/target-hurrricane.html (accessed March 20, 2016).

Nowatzki, J.; Bajwa, S. G.; Schatz, B. G.; Anderson, V.; Harnisch, W. L. Verify the Effectiveness of UAS-Mounted Sensors in Field Crop and Livestock Production Management Issues. *Proceedings of 12thh International Conference on Precision Agriculture*, Sacramento, California, USA, 2014; p 41.

O'Leary, J. Iowa City Farmer Uses Drone to Aide Precision Farming. Demoines Register. 2014, pp 1–6. http://www.desmoinesregister.com/histroy/money/agriculture/2014/10/27/iowa-city-farmer-uses-drone-aide-farming-precision. /18031077/ (accessed June 26, 2015).

Oryza. Can Rice Farmers Use Drones for Fertilizer Management? Oryza.com. 2014, pp 1–2. http://www.oryza.com/news/rice-news/can-rice-farmer-use-drones-fertilizers-management (accessed Aug 3, 2014).

Papadopoulos, A.; Papadopoulos, F.; Tziachris, P.; Metaxa, I.; Iatrou, M. Site-Specific Agricultural Soil Management with the Use of New Technologies. *Globe NEST J. 16*, 59–67.

Peng, S.; Garcia, F. V.; Laza, R. C.; Sanico, A. L.; Visperas, R. M.; Cassman, K. G. Increased N-use Efficiency Using Chlorophyll Meter on High Yielding Irrigated Rice. *Field Crops Res.* **1996**, *47*, 243–252.

Peng, S.; Laza, R. C.; Garcia, F.V.; Cassman, K. G. Chlorophyll Estimates Leaf Area-Based N Concentration of Rice. *Commun. Soil Plant Anal.* **1995**, *26*, 927–935.

Precision Farming Dealer 2016 Survey: Farmers more interested in boosting yields than cost cutting. Industry News. 2016, pp 1–8. Http://www.precisionfarmingdealer.com/articles/1998-survey-farmers-more-interested-in-boosting-yields-than-cost-cutting (accessed Feb 16, 2016).

Precision Hawk LLC 2016 Lancaster Platforms in Agriculture. 2016, pp 1–4 http://www.precisionhawk.com/index.htm#industries (accessed Aug 5, 2016).

Pudelko, R.; Stuzynski, T.; Borzecka-Walker, M. The Suitability of Unmanned Aerial Vehicle (UAV) for the evaluation of experimental fields and crops. Agriculture 99: 431–436.

Quemada, M., Gabriel, J. L. and Zarco-Tejada, P. 2014 Airborne Hyperspectral images and ground level optical sensors as assessment tools for maize fertilization. Remote Sensing 6: 2940–2962 DOI: 10.3390/rs6042940 (accessed Sep 10, 2015).

Reed, J. J., Tarpley, L., McKinnon, J. M., Reddy, K. R. 2002 Narrow band reflectance ratios for remote estimation of nitrogen status in cotton. Journal of Environmental Quality 31: 1442–1452.

Ries, J. B.; Marzolff, I. 2003Monitoring of gully erosion in the Central Ebro Basin by large scale aerial photography taken from a remotely controlled blimp. CATENA 50:309–328 -- under experiments

Reyniers, M.; Vrinsts, E. 2006 Measuring wheat nitrogen status from space and ground based platform. International Journal of Remote Sensing 27: 549–567

Reyniers, M.; Vrinsts, E.; Baerdemacker, J. 2004 Fine-scaled optical detection of nitrogen stress in grain crops. Optical Engineering 43: 3119–3129

RMAX, 2015 RMAX specifications. Yamaha Motor Company, Japan. http://www.max.yamaha-motor.drone.au/specification pp 1–4 (accessed Sep 8, 2015).

Robota LLC. Complete Agricultural Systems Mapping. Robota LLC, USA, 2015; pp 1–17. http://www.robota.us/Complete-Systems/b/10743241011 (accessed May 16, 2015).

Roose, E.; Barthes, B. Organic Matter Management for Soil Conservation and Productivity Restoration in Africa: A Contribution from Francophone Africa Research. *Nutr. Cycling Agroecosyst.* **2001**, *61*, 159–171.

Rosas, F. World Fertilizer Model. The NPK world model. College of Agriculture and Rural Development. Iowa State University, Ames, Iowa. Working paper, No WP520. 2011, pp 1–23. http://www.ageconearch.umn.edu/bitream/103223/2/11-wp_520-NEW.pdf (accessed July 23, 2013).

Scharf, P. C.; Lory, J. A. Calibrating Corn Colour from Aerial Photographs to Predict Side Dress Nitrogen Need. *Agron. J.* **2002**, *94*, 397–404.

SenseFly Inc. eBee Sensefly: The Professional Mapping Drone. SenseFly Inc. A Parrot company, Chesseaux-Lousanne: Switzerland, 2016; pp 1–5. https://www.sensefly.com/drone/ebee.html (accessed Jan 29, 2016).

Shannon, K. Making Better Management Decisions with Precision Ag Data: Today and into the Future. University of Missouri Extension Service. 2014, pp 1–14. http://wwwextension.missouri.edu/boone/documents/PrecisionAg/MakingBetterMgtDecisionsUsing PrecAgData_2014.pdf (accessed May 23, 2015); pp 1–58.

Soil Survey Staff 1998 Keys to soil Taxonomy. USDA-NRCS:Washington, D.C. 8th Edition. 1998, p 326.

Sterk, G.; Herman, L.; Bationo, A. Wind Blown Nutrient Transport and Soil Productivity Changes in Southwest Niger. *Land Degrad. Dev.* **1996**, *7*, 325–335.

Sterk, G.; Stroosnijder, L.; Raats, P. A. C. Wind Erosion Process and Control Techniques in Sahelian Zone of Niger. 1998, pp 1–14. http://www.ksu.edu/symposium/proceedings/sterk.pdf (accessed May 25, 2014).

Suguira, R.; Noguchi, N.; Ishii, K. Remote Sensing Technology for Vegetation Monitoring Using Unmanned Helicopter. *Bioprocess Eng.* **2005**, *90*, 369–379.

Sullivan, D. G.; Fulton, J. P.; Shaw, J.; Bland, G. Evaluating the Sensitivity of an Unmanned Thermal Infrared Aerial System to Detect Water Stress in a Cotton Canopy. *Trans. ASABE.* **2007**, *50*, 1955–1962.

Tara de Landgrafit. Flying with Drone Technology in Agriculture. Rural. 2014, pp 1–2. http://www.abc.net.au/news/2014-07-04/wach-drones/5569770 (accessed Jan, 25, 2016).

Taylor, J. Crop Scouting with Drones-A Case Study in Precision Agriculture. Drone yard. 2014, pp 1–5. http://droneyard.com/2014/08/18/crop-scouting-agriculture/ (accessed Oct 10, 2015).

Thamm, H. P. SUSI-62 A Robust and Safe Parachute UAV with Long Flight Time and Good Payload. Conference on Unmanned Arial Vehicles in Geomatics, Zurich, Switzerland. *International Archives of the Photogrammetry, Remote Sensing and Spatial Information Sciences 2011*, *38*, 1–6.

Tigue, K. Universities of Minnesota Research Group Pushes for Ag Drones. Precision Farming Dealer. 2014, pp 1–5. http://www.precisionfarmingdealer.com/articles/781-university-of-Minnesota-Research-group-pushes for-ag-drones (accessed Feb 3, 2016).

Tremblay, N.; Wang, Z.; Cerovic, Z. G. Sensing Crop Nitrogen Status with Fluorescence Indicators: A Review. *Agron. Sustainable Dev.* **2011**, *32*, 451–464.

Tremblay, N.; Vigneault, P.; Belec, C.; Fallon, E.; Bouroubi, M. Y. A Comparison of Performance Between UAV and Satellite Imagery for N Status Assessments in Corn. *Proceedings of 12th International Conference on Precision Agriculture*, Sacramento, California, USA, 2014; p 19.

Trimble Inc. Soil Information System. Trimble Inc.: Sunnyvale, California, USA, 2015 a; pp 1–2. http://www.trimble.com/Agriclture/sis.aspx (accessed May 20, 2015).

Trimble Inc. Agronomic Services. Trimble Inc.: Sunnyvale, California, UAS, 2015 b; pp 1–2. http://www.trimble.com/Agriculture/agronomic-services.aspx (accessed May 20, 2015).

Trimble Inc. Nutrient Management VRA. Trimble Inc.: Sunnyvale, California, USA, 2015 c; pp 1–3 (accessed May 20, 2015).

Trimble Inc. PurePixel Precision Vegetation Health Solution. Trimble Inc.: Sunnyvale, California, USA, 2015 d; pp 1–3 (accessed May 20, 2015).

Trimble Inc. Trimble UX5 Aerial Imaging Solution for Agriculture. Trimble Inc.: Sunnyvale, California, USA, 2015e; pp 1–3 (accessed May 20, 2015).

Tsouvaltsidis, C.; Zaid Al Islam, N.; Benari, G.; Vreckalic, D.; Quine, B. Remote Spectral Imaging Using a Low Cost UAV System. Paper presented at international conference on Unmanned aerial vehicles in Geomatics. Toronto, Canada. *The International Archives of the Photogrammetry, Remote Sensing and Spatial Information Sciences.* **2015**, *XL*, pp 25–31.

UNEP-GEAS. An Eye in the Sky- Eco-Drones. United Nations Environment Program-Global Environmental Alert Service. 2013, pp 1–5. http://www.unep.org/gaes (accessed Feb 25, 2016).

UNCSAM. Background on Appropriate Precision Farming for Enhancing the Sustainability of Rice Production. United Nations Centre for Sustainable Agricultural Mechanization (UNCSAM) and Malaysian Agricultural Research and Development Institute (MARDI), Selangor, Malaysia, 2015 pp 1–9. http://unscam.org/publication/PreRice-Farm.pdf (accessed Feb 5, 2016).

USDA-NRCS. Precision Nutrient Management. Planning. United States Department of Agriculture: Beltsville, USA, 2010, Agronomy Technical Note No 3. pp 1–14.

USGS. UAS Raven Flight Operations to Monitor Bank Erosion on the Lower Brule Reservation in South Dakota. United States Geological Society, 2013. http://rmgsc.cr.usgs.gov/UAS/missouriRiverErosion.shtml (accessed Feb, 20, 2016).

Varvel, G. E.; Wilhelm, W. W.; Shanahan, J. F.; Schepers, J. S. An Algorithm for Corn Nitrogen Recommendations Using a Chlorophyll Meter Based Sufficiency Index. *Agron. J.* **2007**, *99*, 701–706.

Vellidis, G.; Tucker, M.; Perry, C.; Kevin, C.; Bednaz, C. A Real Time Wireless Smart Array for Scheduling Irrigation. *Comput. Electron. Agric.* **2008**, *61*, 44–50.

Watai, T, A. Light Weight Observation System for Atmospheric Carbon Dioxide Concentration Using Unmanned Aerial Vehicle. *J Atmos Oceanic Technol.* **2006**, *23*, 700–710.

Yamaha Inc Spray Equipment. Yamaha Company, Australia. 2015, pp 1–7. http://www.rmax.yamaha-motor-com.au/spray/equipment.html (accessed Feb 13, 2016).

Yin, Z.; Lei, T.; Yan, Q.; Chen, Z.; Dong, Y. A Near-Infrared Reflectance Sensor for Soil Surface Moisture Measurement. *Comput. Electron. Agric. 2013, 99,* 101–107.

Yintong Aviation Supplies Company Ltd. A Precision Agriculture UAV. 2012, pp 1–3. http://www.china-yintong.com/en/productshow.asp?sortid=7&id=57 (accessed July31, 2014).

Zarco-Tejada, P. J.; Gonzalez Dugo, V.; Berni, J. A. J. Fluorescence, Temperature and Narrow Band Indices Acquired from a UAV Platform for Water Stress Detection Using a Micro-Hyperspectral Imager and a Thermal Camera. Remote *Sens. Environ.* **2012**, *117*, 322–337.

Zarco-Tejada, P. J.; Morales, A.; Tsti, L.; Villalobos, R. J. Spatio-Temporal Patterns of Chlorophyll Fluorescence and Physiological and Structural Indices Acquired from Hyperspectral Imagery as Compared with Carbon Fluxes Measured with Eddy Covariance. *Remote Sens. Environ.* **2013**, *1333*, 102–115.

Zarco-Tejada, P. J.; Miller, J. R.; Mohammed, G. H.; Noland, T. L.; Sampson, P. H. Scaling Up and Model Inversion Methods with Narrow Band Optical Indices for Chlorophyll Content Estimation in Closed Forest Canopies with Hyperspectral Data. *IEEE Trans. Geosci. Remote Sens.* **2001**, *39*, 1491–1507.

Zaman, B.; McKee, M.; Neale, C. M. U. Fusion of Remotely Sensed Data for Soil Moisture Estimation Using Relevance Vector and Support Vector Machines. *Int. J. Remote Sens. 2012, 33,* 6516–6552.

Zhang, C.; Walters, D.; Kovacs, J. M. Applications of Low Altitude Remote Sensing in Agriculture Upon Farmer's Requests- A Case Study in Northeastern Ontario, Canada. *PLOS One* **2014**, pp 1–9. http://journals.plos.org/plosone/article?id=10.1371/journal-pone.0112894 (accessed June 25, 2015).

Zhu, J.; Wang, K.; Deng, J.; Harmon, T. Quantifying Nitrogen Status of Rice Using Low Altitude UAV-Mounted System and Object-Oriented Methodology. *Proceeding of the ASME 2009 International Design Engineering Technical Conferences and Computers and Information in Engineering Conference*, San Diego, California, USA, 2009; pp 1–5.

CHAPTER 4

DRONES IN PRODUCTION AGRONOMY

CONTENTS

4.1   INTRODUCTION

Drones are of great importance in agricultural and industrial settings. Current forecasts suggest that drones could find elaborate use in crop husbandry. A wide range of agronomic procedures may become efficient and accurate, if we adopt drone technology. Drones, with their multispectral and thermal sensors, would actually form the most crucial component of production agronomy in future. Agricultural scientists believe that drone technology can revolutionize crop production tactics. Drones could induce higher grain harvests at better production efficiency. In this chapter, let us discuss the knowledge accrued till date about drones and their role in crop monitoring, basal and in-season fertilizer supply, irrigation, and yield forecasts. Drones are versatile instruments, and they could be put to work in all agro-environments. Also, they could be utilized during production of a wide range of different crops. However, drones have yet to dominate the crop land. They have a lot of ground and airspace to cover, with regard to major crops of the world such as maize, wheat, rice, legumes, oil seeds and cash crops. A few plantation crops have only been exposed to drone technology. Drones may have to be tested and their performances standardized, on variety of crops grown in different agrarian regions. Right now, the information is relatively rudimentary even with major crops.

4.2   DRONES TO GUIDE AGRONOMIC PROCEDURES IN CROP FIELDS

Production agronomy envisages plentiful harvest of grains/fruits by farmers. Most recent of the methods to enter farm world with a promise to maximize grain/forage production is the 'Drone Technology'. It is based on multispectral images and variable-rate application of inputs, using drone acquired data (DMZ Aerial, 2013; DMZ Aerial Autonomous Scouting Robotics, 2013). There are indeed several aspects of drone technology that need consideration. They are terrain, soil type, its fertility/productivity, crop species cultivated, disease/pest pressure in the location, agronomic procedures that are mandatory and those that are optional but enhance crop yield. The agronomic procedures have to be timed and conducted efficiently, if farms have to reap better harvests. Agricultural drones are being touted as among the best and most recent techniques. They supposedly help individual farmer/farming companies in a wide range of ways, so that, agronomic procedures are accurately conducted. Let us consider an example. Following is a list drawn from SenseFly (2015) that depicts how drones could be adopted sequentially, while conducting various soil and crop management procedures:

a.   First and foremost, we have to choose the best sensors possible considering the drone machine, crops and purposes such as collecting data [normalized difference vegetative index (NDVI), green normalized difference vegetation index (GNDVI), leaf chlorophyll, crop-N status, crop water stress index (CWSI) etc.]. The sensors most commonly used to judge crops during crop production are (i) R, G, B (red, green and blue) in the visual range to obtain 3D and elevation of field, plant counting and assessing crop stand; (ii) near infrared (NIR) sensor is used to assess soil properties, moisture, crop health, soil erosion analysis and plant counting; (iii) Red-edge is used to assess crop health, plant count and water management; (iv) ThermoMap sensors are used to conduct crop physiological analysis, irrigation scheduling, yield forecasting; Multispec4 is used to conduct plant count, crop health, phenotyping and so forth.

b. Next step is to plan drone's flight path giving due consideration to the field, crop species and its spread in the field. We should define takeoff and drone landing location appropriately. Set the sensors on the drone for appropriate resolution and timing of photographing events or videographing and so forth.

c. Drones have to be launched correctly to reach the correct height, above the crop canopy. We have to monitor the drone flight throughout, until it is back.

d. Download the data, images and transfer it to processing unit on the ground; perform initial processing using appropriate computer and software; generate vegetation indices data for the entire field and show its variations; identify insect/disease affected locations on the digitized maps; select and pick a few soil and crop samples from fields, if ground reality data are to be obtained.

e. Match drone's observation with ground reality data on hand; export data if further analysis and agronomic prescriptions are required. Most importantly, collect data on vegetation indices, leaf area index (LAI), leaf chlorophyll and thermal infrared (IR) data.

f. Utilizing the data and maps depicting vegetation indices, also noting crop's needs, we have to arrive at appropriate recommendations.

g. In general, use drones to assess and map variations in vegetation indices, depict patterns of progress in canopy height, vigour, colour, density and so forth; develop drainage channels and avoid soil erosion, gauge severity of insect/pest damage; forecast grain/forage yield using indices.

In more simplified terms, Taylor (2015) states that a case study of drones and their utility in precision farming would include following items. They are flying drones with a full complement of sensors (R, G, B, IR, NIR) above the crop canopy, picturizing the crop and then analysing the data. First, spot light the areas of interest by viewing the planting area, mark the soil patches that may need special attention, by making a close-up scouting of bare soil. After seeding and germination, mark the areas with growth depression and those with good vigorous seedling. Mark areas with loss of seedlings. Make a detailed view of the field for crop versus weed growth in the rows and in-between the rows. Determine crop height using drone imagery of entire field. Also, note the elevation properly. Note down vegetation indices and calculate crop health and nutrient status. Note down crop patches with drought or flooded conditions. Map the crop to decide on harvest schedules.

Drones are of great utility to farmers in all the different seasons, when the crop is in the field, growing rapidly and reaching maturity, also when it is not there during postharvest period. Overall, in a field setting where drone technology is adopted, following are the drone-related activities performed through different seasons:

a. In spring, drones could be utilized to image the soils in unplanted fallow fields. Tillage sequence and intensity could be decided using, drone images. Drainage lines could be laid based on 3D images from drone's sensors.

b. In summer, drones could be used to assess crop stand, count plants and assess growth variability of seedlings. They could be used to

PLATE 4.1 A flat winged drone scouting and imaging cereal fields in Kansas, Central Great Plains of United States of America.

Note: A drone carries a full complement of sensors for visual (red, green, blue), near infrared, red edge and thermal imagery. Farmers are offered images periodically after scouting the fields. They are also provided with prescription maps that is variable-rate maps for fertilizer and pesticide applicators. Periodic monitoring of crop disease is also a clear possibility, using such flat drones.

Source: Dr Tom Nicholson, AgEagle Aerial Systems Inc. Neodesha, Kansas, United States of America.

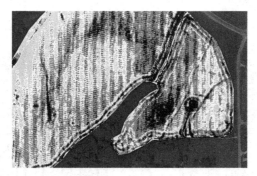

PLATE 4.2 In-flight instant processing of crop images to accrue data about NDVI, leaf chlorophyll, canopy temperature and disease/pest incidence.

Note: Drones are equipped with computers that process ortho-images instantaneously and offer a preview to farmers about the status of crops. Detailed analysis is conducted later at the ground station.

Source: Dr. Tom Nicholson, AgEagle Aerial Systems Inc., Neodesha, Kansas.

monitor crop stages and apply nitrogen split dosage of fertilizer-N appropriately (Plate 4.5).

c. In fall, drones could be used to conduct preharvest scouting for maturity of crops, grain-fill and grain maturity. Drones could also be used to conduct postharvest soil imagery, mainly to depict variations due to cropping and prepare for next sowing.

d. In winter, drones' imagery could be used to guide tractors and sprayers. The aim is to supply inputs and spray pesticides accurately. Drones could be monitoring progress of field work, by semi-autonomous and robotic vehicles (see SenseFly, 2015; Plate 4.1).

The above list is a generalized version of drone technology, with its sequential application in fields. Clearly, drones could be of great utility to agronomists. Yet, drone technology may need minor modifications based on specific crop species and geographic location. We may need development of few more amenities or accessories to drones (platforms), so that, they are efficient in a farm setting. Let us consider a few of them. Seeding field crops and establishing an optimum plant density commensurate with soil fertility status and yield goals is important. It is an essential aspect of production agronomy. Aerial images that drones offer at short intervals, particularly, during early stages of the crop could be of great value to farmers. In general, we know that farmers try to adjust seeding rate based on the location (management block), based on its fertility status and knowledge about previous years' data of the given block. Low-productivity

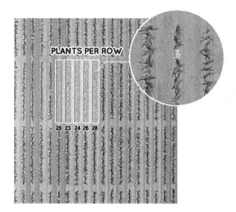 

PLATE 4.3   Drone imagery of sprouted maize seedlings and 'seed planter skips' (gaps).

Note: The gaps in seedling rows are usually identifiable a few days after planting. Drone image with its GPS tags can clearly identify locations where planters have skipped planting seeds or seeds have not germinated leaving gaps to be filled. As each planting hill could be accurately located, using GPS coordinates, repeat sowing to fill gaps can be restricted, to only such locations. It saves on time of scouting. Sowing entire patch is not needed, if tractors/planters are fitted with GPS RTK facility and data from drone are accurate.

Source: Ms. Lea Reich, Precision Hawk Inc., Raleigh, North Carolina.

blocks are sown to support lowered levels of planting density compared with medium or high-fertility areas. Seeding rate also depends on fertilizer and irrigation supply envisaged in a particular management block. Let us consider an example involving precision techniques for seeding. Wheat grain yield in low-fertility area was 110 bushels ac^{-1}, if fixed rate seeding was adopted and 123 bushels ac^{-1}, if variable-rate methods were used. In case of high-fertility locations, fixed rate provided 144 bushels ac^{-1}, whereas variable-rate methods gave 149 bushels ac^{-1}. No doubt variable-rate methods are superior to traditional methods of uniform fixed rate sowing (Shannon, 2014). However, during the seeding, then emergence and establishment of seedlings, a close observation is needed. Such surveillance could be very effectively provided by drones. Drones could also be used to help farmers in fixing management blocks for seeding. It is based on previous years' data and overlayering it with current situation. Drones can help farmers in high-density planting, particularly in forming management blocks that specifically cover high soil fertility areas. We can also mark low-fertility zones and reduce seeding rate, accordingly. They say, it is possible to monitor low and high-density planting of wheat carefully, using drones and ground stations that possess computer controls and ortho-mosaic processing facility (Plate 4.2). Even a single seedling that has emerged or a planting hill that has not emerged can be viewed, using drone cameras. They can be then marked, using Global Positioning System (GPS) tags (Plate 4.3). Therefore, drones offer a close watch of even single seedling. They could be adopted to watch the large fields for emergence and seedling establishment (e.g. gaps in seedling rows; see Plate 4.3).

4.2.1 DRONES AND RICE PRODUCTION PROCEDURES

Drones are getting ever more useful in accomplishing agronomic procedures during rice production. Agronomic procedures that involve use of drones, either directly or indirectly, are as follows:

a. Developing 3D images of rice fields
b. Field scouting
c. Land levelling
d. Variable-rate seeding
e. Monitoring seedling emergence
f. Obtaining values for NDVI, greenness area Index, leaf chlorophyll, leaf-N status to apply fertilizer-N and so forth
g. Spraying or dusting fertilizer-N formulation
h. Application of soil amendments such as lime or post-emergent herbicides (see Ishii et al., 2006; Krishna, 2016; UNSCAM, 2015; RMAX, 2015; Ministry of Agriculture, 2013; Schultz 2013).

Rice farmers in Southeast Asia are being guided to adopt aerial imagery of crop canopy, measure green area index (GAI) and calculate fertilizer requirements, using drones. They are also encouraged to adopt site-specific methods and variable-rate applicators (VRAs), particularly during supply of fertilizers and other amendments to soil/crop. Drone technology has been touted to supply fertilizer-N, using VRA. In fact, drones are used right at the first step to obtain aerial photos of field topography, its GPS location, soil type, softness of puddled soil and during water pounding. During in-season, drone imagery is used to collect data on GAI and plant moisture status. A simple, overall set of agronomic procedures adopted during rice farming in Malaysia that includes drone technology is as follows:

a. Grid sampling using drone imagery: actually, sampling spots are decided after consulting drone images
b. Manually collecting single-photon avalanche diode (SPAD) readings or Greenness index (GAI) using aerial images
c. Development of GAI maps and collecting digital data
d. Calculation of fertilizer-N requirement based on GAI maps
e. Then, calculation of fertilizer formulation needed to supply variable rates of Nitrogen, Phosphorus and Potassium
f. Use tractors to supply fertilizers uniformly or adopt VRAs;
g. A treatment map is installed into field computer (ground station); fertilizer quantity at each spot is calculated, using VRA monitor
h. Variable-rate spreader utilizes signals from computer, to control and alter outlet size for fertilizer discharge and placement in the field

4.3 VEGETATIVE INDICES: SOME USEFUL DEFINITIONS AND EXPLANATIONS

Believe it or not, it is the same type of sensors fixed on drones that serve both military and peaceful agricultural pursuits. The sensors, namely, high-resolution multispectral cameras, form the center piece of drone technology for most of the purposes that they serve. We have to be careful and selective in the use of sensors and computer software attached to drones or those placed at ground stations. They are used to guide drone's flight paths, conduct imagery, collect data and apply chemicals correctly to crops. Drones collect some of the most useful data, at an unimaginably rapid pace, for farmers to rely on during a crop season. Vegetative indices (VIs) are the most crucial data that drones gather and provide to the decision-making computers. Let us consider basic facts, definitions and some explanations about VIs of crops, and their relevance to agronomic procedures.

4.3.1 NORMALIZED DIFFERENCE VEGETATIVE INDEX

NDVI is the earliest most frequently observed parameter using sensors placed on platforms such as satellites (QuickBird, Landsat-8), drones (Trimble Inc., 2015a, 2015b, 2015c; SenseFly, 2015; Candiago et al., 2015), low-flying parachutes (Antic et al., 2010; Pudelko et al., 2012;

Thamm, 2011; Thamm and Judex, 2006; Weir and Herring, 1999; Agri-botix, 2014) and ground-based pedestal. It is indicative of crop biomass accumulation, its greenness (chlorophyll content) and water status. NDVI is perhaps single most important parameter of relevance, to agronomists who rely on remote sensing as a method, to control crop growth and productivity in their fields.

$$NDVI = \frac{(NIR - R)}{(NIR + R)}$$

Where, NIR refers to near infrared reflectance, R refers to reflectance at red bandwidth.

NDVI is generally easy to calculate and it is a good parameter that depicts vegetation cover on the surface of earth. It is not a good measure of crops, if the fields are sparsely covered with crops/weeds or other species. Forests with dense and uninterrupted vegetation too may not be assessed properly, using NDVI. The unit of measurement for NDVI ranges from -1.0 to $+1.0$. Positive values suggest that greenness in a crop field or natural vegetation zone or forested zone is increasing. A negative value for NDVI indicates that aspects other than green vegetation are relatively more. Nonvegetative features such as water, soil, barren areas, desert sand, snow or clouds are increasing. The most common NDVI value encountered for a well-vegetated crop field or natural expanse is 0.2–0.9. NDVI values at 0.2–0.3 may be indicative of shrub or grass land not so thickly vege-tated. Thickly forested areas and intensively cultivated crop fields show up NDVI values in the range of 0.4–0.9 (Candiago et al., 2015; Huang, et al., 2010a, 2010b). Highly productive crops with profuse foliage and ability for high grain output, due to intensively supply of inputs, will show up high values of NDVI. There are also low-cost analog-to-digital converter (ADC) cameras (sensors) that *directly* provide values for different VIs (Huang et al., 2010a, 2010b).

4.3.2   GREEN NORMALIZED DIFFERENCE VEGETATION INDEX

This is mathematically a modified value derived much similar to NDVI, except that sensors used operate at green band width range, instead of red band width.

$$GNDVI = \frac{(NIR - G)}{(NIR + G)}$$

GNDVI is related more to photosynthetic radiation absorption. It is linearly related to LAI of the crops (Gitelson et al., 1996; Antic et al., 2010; Candiago et al., 20 1 5; Weir and Herring, 1999; Drone Remote Sensing, 2015). The GNDVI is actually indicative of chlorophyll content on leaf material in the vegetation. Again, GNDVI value too that ranges from 0 to 1.0 is indicative of higher quantity of leafy vegetation and chlorophyll content. GNDVI is also linearly related to LAI up to 2.5 $m^2{\cdot}m^{-2}$. Above this value, GNDVI is not responsive to changes in LAI. The saturation of VIs depends on several factors related to crop species, it foliage, chlorophyll content and so forth (Hunt et al., 2008). Hunt et al. (2008) report that for a specific case, such as soybean, the regression equation for values up to a LAI of 2.5 $m^2{\cdot}m^{-2}$, it is GNDVI $= 0.5 + 0.16 \times$ LAI and correlation is high at $R^2 = 0.85$. We may have to identify the saturation point and apply the equations judiciously for each crop field.

4.3.3   SOIL ADJUSTED VEGETATION INDEX

$$SAVI = \frac{NIR - R}{NIR + R + L} + (1 + L) \text{ (Haute, 1988)}$$

$$SAVI = (1 - L)(NIR - R)/(NIR + R + L) \text{(Haute, 1988; Antic et al., 2010)}$$

Where, NIR refers to NIR reflectance, R refers to reflectance at red band width. L is a parameter that takes a value between -1.0 to 1.0. When L$=0$ then Soil Adjusted Vegetation Index (SAVI)$=$NDVI that is interference from soil is nil. Low 'L' value is indicative high interference by soil or similar background. If 'L' value is 1.0 or nearer 1, then vegetation is uniformly dense and spread across crop field and interference from soil is least or nil.

4.3.4   LEAF AREA INDEX

A basic assumption is that NDVI is linearly related to LAI. Also, that maximum NDVI corresponds to maximum LAI.

$$\text{Leaf Area Index} = \text{LAI}_{max} \frac{\text{NDVI}_i - \text{NDVI}_{min}}{\text{NDVI}_{max} - \text{NDVI}_{min}}$$

For LAI values ranging from 2 to 6 $m^2 \cdot m^{-2}$, NDVI becomes saturated. Therefore, if nonlinear relationship is considered, then LAI estimates depend on factors such as morphological characteristics of crop, optical features of leaf and ground surface, cloud cover, haze and so forth (Antic et al., 2010). We may note that standard error for measurements about LAI may range 12–17%.

4.3.5 LEAF CHLOROPHYLL INDEX

Chlorophyll content = Chlorophyll concentration$_{(mg/kg\,biomass)}$ × Plant biomass$_{(kg)}$
    Canopy Chlorophyll Content Index = [Normalized Difference Red Edge (NDRE) $-$ NDRE$_{min}$]/(NDRE$_{max}$ $-$ NDRE$_{min}$) for chlorophyll and Nitrogen content (see Cammarano, 2010; Fitzgerald et al., 2006; Basso et al., 2015)
    NDRE = $(R_{790} - R_{720})/(R_{790} + R_{720})$ for chlorophyll and nitrogen (Barnes et al., 2000; Basso et al., 2015)
    Leaf chlorophyll content has direct relevance to agronomic procedures such as fertilizer-N supply. It indirectly influences crop management aspects related to biomass accumulation and grain production. Basically, leaf chlorophyll index is related to photosynthetic ability and carbon fixation in a crop. Chlorophyll a and b and carotenoid pigments actually indicate crop vigour and productivity status. Leaf chlorophyll data are obtained using reflectance at 435 and 735–750 nm wavelength band. Leaf chlorophyll index is derived using sensors that measure the reflectance of leaves and canopy. Leaf chlorophyll content is an indirect indicator of nitrogen status of leafs/plant (Jones et al., 2004). There is usually a strong statistical correlation between leaf chlorophyll and nitrogen content. Measurement of spectral reflectance at 705–735 and at 505–545 is useful, while predicting nitrogen and phosphorus concentration of leaf (e.g. wheat leaf), respectively (see Jones et al., 2004). Weckler et al. (2003) have reported that NDVI was strongly correlated to plant biomass, vegetative cover and chlorophyll content per unit area. For example, the chlorophyll content (mg plant^{-1}) is related to NDVI measured at 670 nm in spinach by an equation (Chlorophyll content = 1364 × NDVI − 370.41) and correlation coefficient is $R^2 = 0.75$. Each crop and location may need a certain standardization,

prior to adopting spectral value derived, using sensors on drones. $NDVI_{670}$ multiplied by biomass provides best estimate of chlorophyll content. If we use reflectance ratios, then, accuracy of chlorophyll estimates increases (Jones et al., 2004; Daughtry et al., 2000). Leaf chlorophyll estimation using sensors on drones is a rapid and nondestructive method. It is instantaneous. It does not need elaborate chemical extraction and estimations.

4.3.6 NITROGEN UPTAKE

The total N uptake by a certain area of crop located on the ground can be estimated, using modified NDVI that is GNDVI. Again, assumption is that GNDVI is directly linearly dependent on nitrogen uptake (QN). Moreover, we assume that maximum value for QN corresponds to maximum GNDVI. Total N uptake could be computed using the following formula:

$$QN_i = \frac{GNDVI_i - GNDVI_{min}}{GNDVI_{max} - GNDVI_{min}} \text{ see Antic et al., 2010; Lelong et al., 2008}$$

Standard error for measurements about GNDVI hovers around 13%.

The nitrogen balance index (NBI) is a useful measurement during crop-N management. Li et al. (2015) reported that during rice production, fluorescence emissions and polyphenol content could be measured nondestructively, using sensors on drones. The dark green colour index derived through aerial imagery helps in assessing canopy nitrogen (canopy-N) concentration in rice fields. It predicted N concentration in leaves and NBI with correlations $R^2 = 0.672$ and 0.713, respectively. Aerial monitoring and imagery using drones with sensors could therefore help in fertilizer-N management in rice fields.

4.3.7 THERMAL INDEX

The thermal cameras measure energy emitted from plants to produce IR images. The ortho-images are then processed using appropriate computer software. Such thermal sensors can decipher difference in crop canopy temperature up to 300°C (Eckelkamp, 2014). The thermal images of crop canopy provide 1.0-m resolution. The thermal imaging systems using IR cameras are highly pertinent, particularly when farmers intend to understand the crop's response, to soil moisture and other conditions. It allows farmers to obtain data about canopy temperature. It provides an idea about crop stress in response to environmental parameters such as rainfall pattern, irrigation and drought tolerance of the crop genotype and so forth. There are reports that thermal IR data are slightly more accurate than VIs derived, using visual and NIR reflectance. Thermal imagery is sensitive to changes in crop canopy temperature, and if the change is greater than 15% of original (Thomson and Sullivan, 2006). Thermal imagery could be adopted for regular monitoring of crops' response to drought stress. For example, Thomson and Sullivan (2006) believe that thermal imagery could be effective in monitoring rain-fed, dry land crops such as peanut, cowpea, millets and so forth. The utility of thermal imagery for regular scheduling of irrigation to crops has also been examined (Alchanatis et al., 2006). Such thermal IR sensors can provide CWSI values, when placed on unmanned aerial vehicles (UAVs) and used, to scout the crops. Influence of tillage treatments such as no-tillage, no-tillage with subsoiling, strip tillage and conventional tillage on soybean crop growth has been monitored, using thermal IR cameras (Thomson and Sullivan, 2006; Abuzar et al., 2009). Tattaris and Reynolds (2015) have utilized, thermal indices obtained using low-flying drones, to assess wheat genotypes grown in experimental plots. They have standardized drone-derived canopy temperature and NDVI values, by comparing and correlating them with those, derived from ground-based instruments. They further state that drone-derived data about canopy temperature and NDVI could be compared with final biomass and grain yield.

4.4   CROP PHENOTYPING USING DRONES: FIELD PHENOTYPING

We may note that monitoring crops and recording their phenotypic expression throughout the season, is almost essential. It is a useful aspect of any kind of crop production tactics that farmers or farming companies adopt. Crop phenotype is actually an end result of interaction between genetic constitution of the crop variety sown by the farmer and environment, and the agronomic procedures adopted by him. Crop phenotyping helps farmers to keep tract of progress of crop during a season. A farmer can at any time refer the phenotype that he observes in the field, with known data about crop genotype. Then modify agronomic procedures accordingly, only if necessary. Phenotyping involves tedious recording of

PLATE 4.4a A multispectral imagery of crop field showing variations in reflectance attributable to plant health.
Note: Such multispectral images of crop stand allow agricultural agencies to pinpoint locations in a field that need irrigation, fertilizers or pesticide sprays. Crop Health Indices supplied to farmers, by the drone companies, usually identify crop stress, track rogue plants and detect pest infestations.
Source: Ms. Lea Reich, Precision Hawk Inc., Raleigh, North Carolina.

several parameters related to crop. It begins from seed germination, seedling growth, tillering/branching, vegetative growth, anthesis, grain fill and maturity. This aspect is indeed time and cost consuming for both grain producing companies and experimental crop breeding research stations. Plant breeders routinely procure elaborate data sets before confirming the performance of parent lines, elite lines and commercial varieties. Drones with their ability for capture of data using sensors are perhaps, most useful farm instruments, to plant breeders and production unit managers (see Perry et al., 2012a, 2012b; Tilling et al., 2007; French, 2013; Hungry et al., 2010; Knoth and Prinz, 2013; Plate 4.4a,b, 4.5–4.7). Drones can swiftly fly over experimental fields with a large set of crop genotypes that are to be evaluated. Similarly, they rapidly cover a farm with large acreage of cereal/legume crop. Then collect aerial images and spectral data for each genotype. Currently, even plant height could be studied, using oblique shots obtained by cameras on drones. Baret et al. (2014) believe that drones

PLATE 4.4b   A multispectral imagery of maize field derived using a low-flying drone.
Note: Such images can be procured from farm agencies, at frequent intervals during a crop season, to assess growth, also to monitor nutrient and water status of crops. Images from thermal infrared cameras could be overlayered, to know the effect of variations in water status on crops.
Source: Dr. Tom Nicholson, AgEagle Aerial Systems Inc., Neodesha, Kansas.

are apt to conduct crop phenotype analysis, at a rapid pace. It allows us to revise decisions related to crop husbandry. There are also specific traits such as plant height, canopy and water stress index of crop that are perhaps best done, using a low-flying UAV fitted with visual and thermal sensors. Other methods may be inefficient, less accurate and may not be amenable for repeated deployment and collection of data.

One interesting fact about drone-aided phenotyping of crops in fields is that we observe and collect data about crop genotypes, analyse their genetic/phenotypic aspects and conclude based on crop's (genotypes) performance, totally in the open field. Single plant data and green house performances, which may not extrapolate accurately, are entirely avoided.

There is a range of field phenotyping equipment that are in vogue. Let us consider an example. Field phenotyping could be done by mounting the cameras, data collection equipment and analytical computers on tractors, field vehicles or drones. Drones are rapid and less costly. Farm scientists at the University of Copenhagen, Denmark have developed a field phenotyping system known as 'Phenofield' (Christensen, 2015). It comprises a 5-megapixel camera, computer vision system and advanced image analysis software. These are all mounted on a hexacopter (drone). The digital cameras procure data about a series of VIs, plant height, foliage,

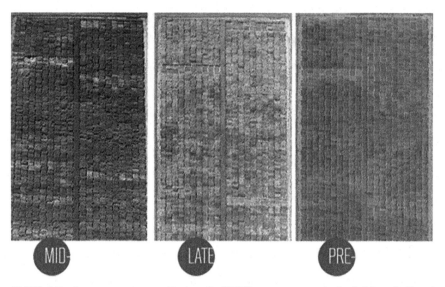

PLATE 4.5 In-season crop monitoring for NDVI, canopy growth, leaf chlorophyll and crop maturity.

Note: Periodic evaluation of the crop using multispectral sensors helps farmers in timing and managing several agronomic procedures. It also helps in forecasting grain/forage yield. *MID:* mid-season, *LATE:* late-season and *PRE:* preharvest mature crop stage.

Source: Ms. Lea Reich, Precision Hawk Inc., Raleigh, North Carolina.

leaf chlorophyll, crop growth rate and CWSI using thermal IR cameras and nutrient status. Whatever is the model of drones used, sensors and computer software, the bottom-line advantage with drone technology is that remote sensing allows analysis of crop phenotype swiftly. It is nondestructive and offers accurate data.

Let us consider an example dealing with a widely grown cereal, maize. Maize may often encounter soils with N deficiency. Sometimes fertilizer-N supply may not be adequate. In such cases, crop genotypes that survive better and still yield optimum grain yield are preferred. We have to make well-directed phenotypic analysis of crops, spatially and temporally, particularly to identify low-N or low-P tolerant varieties or even those resistant to diseases, pests and drought. In agricultural experimental stations, such phenotyping of all genotypes that are being evaluated for various traits may be costly. It may not be feasible, at times, due to logistics and lack of skilled farm workers. Ground-based imaging and analysis too could be tedious. It may still not offer best data at a rapid pace. It

PLATE 4.6   A drone scouting and measuring phenotypic traits of maize seedlings.
Note: Phenotypic data such as seedling height, NDVI, GNDVI, LAI, leaf chlorophyll status, plant -N and plant water status are recorded by drones.
Source: Dr. Mitchell Feine, Director, DMZ AERIAL Inc., Wisconsin.

could be time consuming, as fields are large and number of genotypes to be assessed is many. Zaman-Allah et al. (2015) have reported that experimental fields with maize could be assessed for a range of phenotypic traits such as NDVI, GNDVI, leaf/canopy chlorophyll and water stress index, using drones. VIs for low-N tolerance could be studied rapidly and best genotypes of maize identified. Multispectral analysis of plots with low-N input and check plots with optimum soil-N is needed. It could be done effectively using low-flying drones (Plate 4.4a, b). Clearly, experimental evaluation of most field and horticultural crops could be done, using drones with appropriate sensors on them. They further state that UAVs with their ability for phenotyping at high rapidity and repeatability are almost, critical for crop production (Plate 4.6). At present, aerial platforms and sensors are being utilized, mostly, to obtain phenotypic data relevant to crop growth, yield formation, disease and pest tolerance, drought tolerance and earliness. Monitoring and collecting periodic data about senescence process and grain maturity is an important procedure. Such a procedure too could be accomplished efficiently, using aerial imagery through drones.

Dreiling (2012) reported that phenotyping is among the best advantages that drones offer to plant breeders and crop production specialists.

Several phenotypic traits could be monitored, simultaneously, for several genotypes being tested in a field. Drones offer side-by-side comparison of crop genotypes, parent lines and crosses. Therefore, so that plant breeders could make choices, after being well-informed about the genotypes, agronomists, who utilized drones, could monitor progress and differences with regard to a range of phenotypic traits (Plate 4.6). Drone technology that concentrates on production aspects of crops is being popularized, in the wheat belt of Kansas State, United States of America (Dreiling, 2012). Phenotypic comparison and monitoring changes of wheat genotypes cost much less, if drone images or videos are utilized. Comparisons of genotypes are made using computer software that selects the best ones. Actually, experimental evaluation itself becomes easier, accurate and cost efficient, if drone-aided phenotyping is adopted. They say, in near future itself, many of the agricultural experimental stations will adopt, drone-aided phenotyping of field crops and their genotypes. Drones are also recommended for multilocation trials, wherein phenomics data collected by drones could be radioed or relayed to coordinating centers. It could be done rapidly and accurately using Internet facilities. Pooling data about genetics and phenotypic expression of crop varieties and elite lines could become easier (Krishna, 2016).

4.4.1   CROP SURFACE MODELS (CSM) DEVELOPED USING DRONE IMAGERY AND THEIR UTILITY

Anderson et al. (2014) point out that the era of drones and satellite-mediated crop husbandry is at the door step of getting popularized. Yet, we have not been able to obtain crop yield data of an individual plot or to note plot-to-plot variations rapidly. Our knowledge about crops and yield variations in different locations is still insufficient. Establishing crop surface and yield models, using drone/satellite imagery, is perhaps urgently needed. It helps farm agencies to analyse crops, their growth and productivity. Then forecast final yield more appropriately. Even on a larger scale, crop productivity models can be highly useful, to policy makers dealing at county or national levels. Crop surface models (CSM) that apply for the entire region of a particular crop species/genotype too could be prepared. They could be consulted periodically by governmental agencies, prior to supplying inputs or while fixing marketing trends. Researchers at IFPRI

(International Food Policy Research Institute, Washington, D.C.) state that knowledge about spatial variation in grain yield both at micro (field/plot) and macro (agrarian belt) is essential. It allows us to judge yield gaps and in devising input schedules appropriately, at variable-rates.

The basic idea is to process the drone images and develop CSM, using appropriate software. For example, *Agisoft Photoscan 0.90* and *ArcGIS* allow comparison of quantitative data. The results can be further combined with additional field data. It is said that measuring plant height, using drone imagery that is noninvasive and one that offers good correlation to plant biomass is the crux (see Hansen and Schjoerring, 2003; Thenkabail et al., 2000; Bendig et al., 2013a, 2013b). CSM can be touted to farmers through farm agencies. A really wide range of possibilities known from previous data could be shown to farmers and farming companies, as CSMs. They can select most appropriate CSM and follow crop production packages as suggested by drone agencies.

CSM have been utilized effectively in judging a range of physiological and morphological traits of crops. Plant height is among the easiest to estimate, using ground-based hyper-spectral sensors and those on UAVs. Plant height and VIs derived from both ground-based pedestals and UAVs were used, to forecast biomass accumulation pattern (Bendig et al., 2015). The CSM showing plant height data was a robust estimate of biomass, with a correlation coefficient of $R^2 = 0.83$. They say, combining VIs derived using R, G, B wavelength bands and plant height data, from CSM were excellent, in judging the crop biomass. Forecasts using such CSM and normalized ratio index (R, G, B, VIs) could be useful, in the prediction of biomass accumulation, by cereal crops.

Researchers at Michigan State University's Geology Department have aimed at ascertaining the value of drone-derived VIs, crop models and ground data, if any. They aimed to forecast crop biomass and grain yield (Azorobotics.com, 2014). They say, integrating crop models and drone imagery at an instance, then, comparing it with known crop growth/production trends could help, in accurate forecasting. Inputs to crops could be channeled more accurately to derive better efficiency. It improves farmer's profitability.

As stated earlier, forecasting crop growth, biomass accumulation trends and grain formation is an essential aspect of crop husbandry. Several of the agronomic procedures adopted during crop production, depend entirely on data at hand about the crop and accurate forecasts. Yield goals are often revised and inputs are rescheduled appropriately. Geipel et al. (2014) have stated that during precision farming, within field variation of crop productivity has been forecasted, based on drone or satellite-derived VIs. Actually, combinations of VI maps and CSM have enabled farm agencies, to forecast grain yield. They have utilized a series of data about VIs derived from drones and plant height from CSM, to forecast grain productivity. They have also mapped grain yield variations for the entire field. Statistical evaluations suggest that, yield forecasts based on CSM, plant height and VIs correlate excellently with a factor $R^2 = 0.74$. Crop grain yield forecasts made using CSM are accurate, to a greater extent, if done at mid-season of corn crop. They further suggest that, right now, there is less information about 3D CSM and their utility, during cereal crop production. Drones need to be utilized to develop suitable data, so that, farmers and farming agencies could predict yield, at various stages during the season. Later, revise decisions, if needed (Bendig et al., 2013a, 2013b).

Plant height is among the most important phenotypic traits that farmers and crop experts, consistently monitor and record. Plant height is related to crops' ability to accumulate biomass and nutrients. In the general course, recording crop height is tedious. It needs several skilled farm workers, if the field is large. However, during recent years, remote sensing and drone-aided aerial imagery are getting popular. Drones are being utilized increasingly, during collection of data about crop phenotype (Plate 4.6). Anthony et al. (2014) state that drones have a great potential in periodically estimating crop height, using oblique shots of ground and top of the plant. They say, drones with laser scanners and low-cost sensors capable of imagery at visual range suffices. The procedure can be adopted swiftly and as many times, in a crop season. Tree canopy height and volume too could be estimated by carefully maneuvering drone's path. Then by collecting images of ground and the canopy height. Usually, crop height estimates at several spots are pooled and average height is computed.

4.5   DRONES AND PRECISION FARMING OF CROPS

Foremost, drones and their adoption during farming has to be efficient and less costly compared with techniques that we already know. Drones should be useful to farmers, if they have to be accepted as a regular farm instrument. A bottom-line question is 'in how many ways drones seem better than methods we already know'? However, we already know that drones are now gaining ground during precision farming (Moorehead, 2015).

According to reports by AUVSI (Association for Unmanned Arial Vehicle System International), precision farming and drones are best fits. Drones aid farmers with most needed data about soil and crop growth variability (Ehmke, 2013). To a certain extent, drones also help in variable-rate supply of inputs at a relatively rapid pace. There are indeed several drone companies and established agricultural consultancy agencies that help farmers. They offer timely decisions and prescriptions about precision farming. But, many of them are currently adopting drone-aided imagery. Drone imagery has been utilized, to first scout the natural vegetation, the soil and crop stand, then trace and map variations in nutrient status of crops. Some of the most common agronomic services offered, so that, farmers could easily practice precision farming are nutrient management, mainly fertilizer-N. Split applications of fertilizer-N at variable-rates could be best accomplished, using drones (Trimble, 2015a, 2015b, 2015c; Misopolinos et al., 2015).

Precision farming could also be construed as conventional farming, but at a different scale. Instead of adopting procedures for entire 200 ha, in a farm, we try to split it into small plots of 0.20–2.5 ha. Then, perform all agronomic tasks (Carlson, 2015). This procedure involves careful planning and formation of 'management blocks'. Actually, precision farming involves at least three basic steps. They are (a) gathering field data using ground-based or airborne (piloted aircraft) or drone-aided or satellite-based methods; (b) The data has to be suitably formatted and processed such as a field map, digital data sets or series of computer directions; and (c) Such data and computer-guided decisions have to be utilized by VRAs, in order that, soil fertility or crop growth is uniform throughout the field (Modern Agriculture, 2015).

Precision technology aided by GPS has driven a change in the farming procedures followed by agrarian regions of developed world. Mechatronics and autonomous vehicles should be adopted and encouraged in the crop fields. They can further improve precision farming. There are opinions that next stage of revolution in farm world will be aided partly, by agricultural drones. Many of the agronomic procedures will become easy to perform. We may employ very few, if not negligible, levels of farm labor. Drones impart a greater degree of efficiency. They could enhance total grain/fruit yield (Ag Business and Crop Inc., 2015).

Over all, formation of management zones and adopting computer-generated decisions for variable-rate inputs using appropriate software forms the center-piece of precision farming (see Zhang et al., 2009). As such, drones and precision techniques together offer great advantages to farmers, particularly in terms of reduction in labor, ease of farm operations and profitability.

In the North American plains, drones are gaining in popularity among farm companies, particularly with those already dealing with precision farming methods. Reports suggest that in Prairie States of United States of America, drones are being introduced into the routine practices that form the core of precision techniques. Mainly, they are used in imaging crops, tracing growth variations and nutrient status (Patas, 2014). Precision methods that involve mapping variability of soil fertility, disease spread and drought incidence are gaining ground, in many agrarian regions. The aim is to apply remedial measures at appropriately matching variable-rates. Drones, no doubt, find a new role during precision farming. They fit best as versatile instruments to obtain maps that depict variation in crop stand. At the same time, drone technology is much cheaper than human scouts. In the Southern Plains, for example, farmers are using drones to accomplish a series of tasks relevant to precision farming (Cowan, 2015). Specifically, drones with sensors are used to scout winter wheat crop. They assess NDVI and crop stand, monitor and identify insect attack and its spread, if any. Wheat crop is also imaged by drones to assess leaf chlorophyll content. In other words, crop-N status and its variability in the entire field are assessed by drones. This step helps farmers to supply fertilizer-N rather efficiently, particularly, the split dosages at variable-rates to wheat seedlings (Cowan, 2015). Sugar beet crops too have been assessed for N uptake and status. Fertilizer-N is supplied at variable-rates, using digital data pertaining to crop canopy-N status. Drone imagery has also been utilized to assess CWSI. It helps farmers to apply irrigation, accordingly using variable-rate technology. One other aspect of importance is that drones are used in Southern Plains to assess disease damage to crops such as wheat, sorghum, legumes and soybean. The digital data and maps that drone images provide are used to apply fungicides, only to locations affected. Fungicides are applied and at variable-rates based on intensity of disease (see Chapter 6).

Lan et al. (2010) opine that aerial application of liquid/granular fertilizers, defoliators, growth regulators or pesticides has been accomplished, using agricultural flat-winged airplanes or helicopters. The current trend is to adopt precision farming, wherein inputs are applied at variable-rates. Drones with containers attached and VRA need to be manufactured in greater number, popularized and used. They could be handy to farmers, equally so, to farm companies with large acreage. Farm companies may have to opt for swarms of drones with aerial application facility. Drones that are less costly, efficient and easier to handle than airplanes could serve farmers in future.

Reports suggest that drones are amenable to collect data about sugarcane (*Saccharum officinarum*) crop exposed to different levels of fertilizer-N (0, N, 65 and 130 kg N ha^{-1}). Drones fitted with R, G, B and NIR sensors were used to collect data on NDVI, GNDVI and simple ratio pigment index (SPRI). The indices correlated with nitrogen status of sugarcane leaf and canopy-N. For example, SPRI correlated with leaf-N ($R^2 = 0.70$). Similarly, NDVI and GNDVI were correlated with crop canopy-N ($R^2 = 0.70$ and 0.64). These parameters were later used by computer software to arrive at appropriate fertilizer-N dosage to sugarcane crop (Lebourgeois et al., 2012). In due course, drone technology could be effectively used to reduce farm labour need and the fertilizer-N requirements. Drone-based techniques may also be used to avoid contamination of irrigation channels with excess fertilizer-N supply. Precision techniques and drones could serve the farmer in delaying or avoiding soil deterioration. This could be done by regulating fertilizer and chemical inputs to crop fields.

Pauly (2014) opines that drones have gained in popularity in precision agriculture. It is attributable to rapidity with which drones can offer data, about a series of crop traits, through spectral analysis. Sensors operative at NIR, IR and thermal band width along with visual band width (R, G and B) cameras have offered excellent data about crop vigour and a series of other related traits. Yet, there are problems such as high degree of soil and foliage shadow effects. Such interference could occur when drones fly close to the crop canopy and pick images at relatively higher resolution. Pauly (2014) says that noise encountered during imagery by drones could at times be more. It may have consequences on data accrued. Hence, multiple band spectral data are preferred to obtain values for NDVI and other plant growth indices, such as GVI, VCI and so forth. The suggestion is to rely on several parameters while assessing crop growth and vegetative vigour maps. Crop scouting strategies using low-flying drones could be tailored to reduce noises from soil and shadow.

Calibration against known standard readings from ground-based green seekers is also recommended.

Melchiori et al. (2014) are among the earliest to study maize grown at Entre Rios, in Argentina, utilizing drones. They have used Unmanned Aerial Vehicle Systems with hyperspectral cameras. They have recorded changes in LAI, NDVI and crop biomass. They have actually compared LAI data obtained using SPAD, handheld leaf meters and hyper-spectral cameras on drones. They state that aerial photographs obtained using R, G and B band width were sensitive and detected biomass and LAI changes accurately. Farmers may eventually use such data obtained using drones to apply precision techniques.

It seems Canola cultivation is expanding in the Central Plains region of United States of America. Farmers are trying to optimize the fertilizer-rate for this crop. Hence, experiments that examine the effect of fertilizer on biomass and productivity are sought. Farmers are particularly interested in deciding the in-season fertilizer-N supply rates. This necessitates rapid scouting of entire fields for LAI, NDVI and other traits. Drone usage to detect crop-N status could be an appropriate idea.

Ciampitti et al. (2014), for example, have used a small drone to study canola crop grown at the agricultural experiment station, in Manhattan, Kansas, United States of America. They found that NDVI and canopy temperature measured using NIR were highly indicative of crop growth and biomass accumulation by canola. The correlation between NDVI and biomass were best at flowering stage of the crop.

During the recent past, there has been greater emphasis on adoption of precision techniques. Aspects such as soil fertility mapping, variable-rate application of fertilizers and water, also grain yield maps, are getting popular. Precision farming methods were adopted in many agrarian regions, particularly in North America and Europe, to induce higher input efficiency and grain productivity. However, in future, they say a combination of robotics, drones and satellite-guided methods may serve farmers to accomplish most of the agronomic procedures. Drones may receive special attention during precision farming (Krishna, 2016). Drones help farmers to conduct scouting in a matter minutes that otherwise would have taken weeks. Drones are highly cost-effective and time efficient. They offer excellent accuracy through their digital data and imagery thus reducing the need for skilled farm labor. Drones are expected to revolutionize the way agronomic production techniques are handled and accomplished by farmers worldwide (SenseFly, 2015).

PLATE 4.7 Experimental plots of different cultivars of cereal that are maintained, to be evaluated and ranked using drone technology.

Note: The above cultivars are evaluated for their forage/grain yield productivity, and their ability to negotiate drought or disease/pest pressure in the location. Researchers in Wisconsin, United States of America collect data relevant to phenomics, using drone imagery. Researchers could offer yield forecasts to farmers using smart phones and ipads. Literally, farmers can compare agronomic performance (data) of different genotypes, even as they walk through the plots (in field). Farmers may also monitor and note performance of genotypes, periodically, on computer screens via Internet.

Source: Dr. Mitchell Feine, Director, DMZ AERIAL Inc., Wisconsin

4.6 EXPERIMENTAL EVALUATION OF CROP GENOTYPES USING DRONES

Periodic scouting of crops and their genotypes through rapid aerial imagery is among the greatest advantages that drones offer to crop scientists and farmers. There are indeed innumerable drone companies that produce flat-winged and copter drones. They have the capability to scout crops and acquire data to study the phenomics. Evaluation of crops becomes easier using drone imagery. For example, Taylor (2015) states that field workers noting crop performance spend more than an hour to just cover perhaps an acre. On the contrary, a drone with sensors just takes a 10-min flight to record data on phenomics. Drones accomplish imaging at a rate of 80 ac·h^{-1}. In Ontario, Canada, farmers can scout over 1000 ha of crops (corn, canola, wheat and soybean) in a day. Drones can collect data about crop growth, NDVI, leaf chlorophyll and water status of several different agronomically elite genetic lines. Drones cover genotypes planted in over 100 ac in just 20-min flight. Such rapidity and accuracy with which information is collected through aerial photography is perhaps impossible, if farm workers are employed (Dietz, 2015; Plate 4.7).

Reports from Rothamsted Agricultural Experimental Station in the United Kingdom state that drones have a great role to play in scouting, assessing phenotypes and recording forage/grain yield data aerially. At the above station, drones are being tested to see if they could be employed routinely, particularly to scout and offer useful data to evaluate and rank genotypes of wheat, barley, maize, legumes and so forth (Case, 2013). Reports suggest that parent lines and crosses could be evaluated in greater detail, rapidly and repeatedly during a crop season. This is something not possible too often, if skilled farm workers are employed. Annually, a very large number of elite entries, numbering a few thousands, from each of the major crops such as wheat, barley, maize, lentil and many other species are evaluated for agronomic traits. Their growth and yield expression is evaluated in different locations. Drones could serve the plant breeders/geneticists ably, by covering all genotypes at a rapid pace, and with greater accuracy. Drones that scout also allow accurate data storage and retrieval. This aspect is something not easily possible. Inaccuracies due to human fatigue and different causes could creep in large numbers. At the Rothamsted Experimental Station, Harpenden, United Kingdom, they expect to screen, evaluate and disseminate data of huge number of genotypes of each crop (Farming Online, 2013). This information holds true for several other agricultural centres worldwide, particularly those that are involved in crop evaluation and genetic improvement. In fact, if drone flights, computers and data analysis systems may get interconnected and networked precisely, several genotypes of a crop sown in multilocations across different agrarian regions could be monitored simultaneously. They could be compared instantaneously on a computer screen. Growth patterns of several genotypes observed aerially could be compared, side-by-side, in an instant on the computer screen (Plate 4.7).

Hunt et al. (2010) state that, at present, drones with sensors are able to collect useful data for conducting field trials and evaluating performance of maize genotypes. Drones are utilized to collect data on each of the maize genotype sown in the experimental plots of International Maize and Wheat Centre (CIMMYT), Mexico. They collect data such as NDVI, LAI, leaf chlorophyll, plant-N status and water stress index rather rapidly and repeatedly. They collect data once in every 3–7 days. Drone-aided collection of experimental data and assessment of maize genotypes' performance is efficient, both in terms of time and cost. Similar trials on maize genotypes have been conducted in Zimbabwe, again, by scientists of CIMMYT. Their aim was to select crop genotypes for growth and yield stability under semiarid conditions (Mortimer, 2013). Researchers at CIMMYT are also testing the utility of drone-derived data for yield forecasting. Lumpkin (2012) states that drones are nondestructive, rapid and could be flown repeatedly over the canopies of maize genotypes. Hence, they are being preferred by both maize researchers and commercial farms alike. With regard to large-scale experimental evaluation of wheat genotypes, researchers at CIMMYT, Mexico have clearly shown that drones could be adopted to estimate canopy temperature, NDVI and other physiological parameters relevant to biomass/grain formation (Tattaris and Reynolds, 2015). Drone-derived spectral data could help researchers to evaluate genotypes for several physiological traits in one go and rank them. In fact, drone-derived data may be effectively used to screen and identify new genes in wheat germplasm lines (Reynolds, 2009; Babar, 2006; Prasad, 2007; Pinto, 2010).

Fertilizer trials to standardize basal and split-N inputs are conducted periodically during sugarcane cultivation. Lebourgeois et al. (2012) have clearly shown that drone imagery of sugarcane exposed to at least three different levels of fertilizer-N could be experimentally evaluated. Drones avoid regular use of human scouts and skilled technicians to collect data about crop's response to fertilizer-N. Drones fly swiftly over sugarcane crop and collect data on indices such as NDVI, GNDVI and chlorophyll. Drones offer accurate data and at relatively lowered cost to experimental stations.

Field experiments involving several crop species have been regularly monitored and data have been procured, using drones. Crop's response to factors such as fertilizer-N supply and water has been evaluated, using drone imagery. For example, Hunt et al. (2014) have examined a potato crop grown at four different levels of fertilizer-N supply (112, 224, 337 and 449 kg N ha^{-1}). They have used drones fitted with Tetracam ADC lite cameras operative at NIR, red and green band width. They found that measurements of NDVI and GNDVI of potato crop was related directly to LAI and canopy size. Such experiments could help in optimizing fertilizer-N supply to potato crop, of course, by adopting drone technology.

Zhang et al. (2014) suggest that drone-derived images give detailed data about cereal genotypes, particularly genotypes that have a tendency to lodge in the field in response to fertilizer-N application. They say lodging and stem breakage happens when cereal genotypes are supplied with fertilizer-N. However, genotypes with stronger and thicker stem tolerate storms and do not lodge. Drone imagery at IR band width easily detects genotypes and plots where lodging is severe. Hence, drones could be very useful in imaging and identifying genotypes prone to lodging with high accuracy.

4.7   CROP PEST, DISEASE AND WEED CONTROL USING AGRICULTURAL DRONES

As stated earlier, in the general course, production agronomy involves series of well-weighed-out procedures based on previous data, current crop status and yield goals. Farmers adopt several different soil and crop management procedures to enhance nutrient recovery, growth and yield formation. At the same time, a field in the open is exposed to a few diseases and insect pests. Farmers have to negotiate such detriments through inherent (genetic resistance) ability of crop species and its genotype (cultivar). They also adopt escape mechanisms by adjusting planting date to avoid disease/pest peaks. Farmers frequently adopt regular and timely sprays of plant protection chemicals. Production agronomy does include these aspects. However, within this chapter only few examples have been listed. Chapters 6 and 7 include detailed discussions on drones and their role in disease, pest and weed control.

They say, drones could play a vital role in early and rapid mapping of weeds in the field, including volunteers from previous crop. This can lead to reduced use of herbicides. Drones are being evaluated for accuracy and efficiency for weed detection. They are adopted to judge the intensity of weed infestation and to apply control measures. Farmers could use computers endowed with detailed spectral data of several or all the weed species that localize, in a region or county. Weed detection could be easier and accurate, if sensors on drones are used. Weeds could also be identified using object-based image analysis (OBIA) (Pefia et al., 2013; Lopez-Granados, 2011). Farmers could then direct workers or use drones fitted with herbicide tanks and variable-rate sprayers, to apply herbicides only at spots that require it. The same digital maps could be used by robotic weeders to physically destroy weeds, if such a facility is available (Dobberstein, 2014; Dobberstein et al., 2014; Paul, 2014). Drones could also be effectively utilized to evaluate potency and dosage of herbicides in experimental fields. Basically, spectral data about weeds common to the region, identification of weed diversity and their intensity, selection of herbicide formulation and aerial spot or variable-rate spray using drones form the core of the procedure (Genik, 2015).

Drones are listed as useful gadgets to control diseases and pests that damage crops. Grain yield of major cereals cultivated across different agrarian regions gets reduced due to pests and diseases. The extent of disease/pest attack in a field could be assessed using multispectral sensors on drones. For example, in United States of America, Russian wheat aphid infestation is being detected using drone imagery. Crop scouting for disease/pest done using drones is highly cost and time effective, compared with skilled farm scouts. Drone surveys are also accurate and rapid.

Low-altitude remote sensing (LARS) done using drones has been effective, in identifying diseases/pests that occur on a range of plantation crops. A few examples are citrus greening disease on citrus plantations in Florida and the Laurel wilt on avocados (Ehsani et al., 2012). It helps farmers to take appropriate control measures at an early stage. Giles (2011) has reported that drones are getting popular with farmers, particularly those who aim at scouting crops rapidly and spraying them with plant protection chemicals.

4.8 DRONES USAGE DURING CROP PRODUCTION IN DIFFERENT AGRARIAN REGIONS

We can group the salient information regarding introduction and usage of drone during crop production. Here, in the following paragraphs, a few examples that prove introduction of drones across agrarian zones of different continents has been listed (not exhaustive). Drones have actually been adopted by farmers to conduct different agronomic procedures. They are primarily used to collect crucial data about soil, its fertility variations, crop growth pattern, nutrient status, water status and irrigation requirements, grain filling, its maturity and grain yield trends. In the second portion of this section, information about drones is grouped based on each crop species. Each crop needs specific agronomic procedures. Many of these could be answered by farmers ably, using a drone and the imagery they provide to him. Some of these aspects are delineated in slightly greater detail in the following paragraphs.

4.8.1 DRONES IN AGRICULTURAL BELTS OF DIFFERENT CONTINENTS

4.8.1.1 AMERICAS

Soon, drones could be serving farmers overwhelmingly in the Great Plains and other regions of North America. They could be accomplishing tasks such as crop monitoring, collection of data on phenomics (plant height, NDVI, GNDVI, CWSI), disease/pest affliction and weed infestation. In near future, several of the agronomic procedures adopted by farmers may actually be decided, using data derived by sensors on drones. Currently, farmers are being helped by several drone-producing companies and agricultural consultancy agencies, regarding regular surveillance of crop. They also help him in deciding suitable agronomic procedures during wheat, maize and soybean production (Rush, 2014). Plantation crops such as grapevines and citrus groves are also monitored regularly, using drone technology. Fertilizer and pesticide sprays are also conducted, using drones. Yet, there are many aspects of drone technology that needs standardization. Expertise about drones needs to be developed and disseminated to farmers (Ehsani et al., 2012). For example, use of 3D maps helps in directing drones to spray pesticides or disease control chemicals more accurately, considering the terrain (Rovira-Mas et al., 2005).

In Northeastern Ontario, Canada, low-flying drones such as LARS are being introduced into farming sector. Drones are being examined for feasibility for regular use. The drones fitted with both visual and IR sensors were examined for efficiency in scouting crops, mapping field drainage and monitoring fertilizer response trials. Such drones were also verified for their performance in the fields kept under precision farming (Zhang et al., 2014). It is said that farmers need access to processing software or companies that process the imagery and offer prescriptions regarding agronomic procedures. Prescriptions about fertilizer supply at variable-rates and irrigation timing is useful to farmers.

In the Pacific Northwest, potato cultivation is an important agrarian enterprise, particularly in the states of Oregon, Washington and Idaho. Drone companies have indeed introduced several of their models into scouting, imaging and even spraying activities in the potato fields. Mainly, drones with their full complement of sensors (visual, IR and NIR) are used to assess crop health, water stress index and nutrient deficiencies, if any. Here, drones reduce cost on scouting. They also reduce fertilizer requirements, since precision methods are adopted (Stevenson, 2015; Oregon State University, 2014). Reports by USDA (2010) have already made it clear that drones are expected to hover above the potato growing regions. Further, they would conduct scouting and obtain multispectral images. Drones would offer data about crop growth, biomass accumulation and yield. They may help nutrient management in potato fields by providing digital data to VRAs. It is said that in a mixed farm situated in Oregon, United States of America, drones are also useful in monitoring cattle herds, using the same sensors. The mixed pasture species, cattle health (body temperature) and herd movement could all be tracked, using low-flying drones. Agronomists at Oregon State University have also explored the possibility of enhancing production efficiency of Brassica oilseed crop, using drone technology. They are initially aiming at identifying weeds in the brassica fields, using drone imagery, then to appropriately spray fields with herbicides (Plaven, 2016).

Wild oats are a major weed species in the plains region that supports crops such as wheat, barley, canola and flax. Wild oats are robust weeds. They grow rapidly utilizing soil nutrients and water. As a consequence, they outgrow crop canopy and reduce photosynthetic light interception. Wild oats also compete for soil nutrients and moisture rather effectively. To control these and few other weed species, farmers in the wheat belt spend about 12–16 US$ extra ha^{-1}. Currently, drones with ability for accurate detection of wild oats in the fields, using their spectral signatures, are being developed. Drones with ability to spray herbicides are also being tested and adopted. Interestingly, cost/benefit analysis of drone usage in wild oat affected plots has been conducted. It indicated that for a 52-ac plot, farmers could save a net 900 US$ if drones were used instead of the regular scouting by farm workers. Manual spraying using human labour was costlier (Genik, 2015). In addition, there are several other species of weeds that infest crop fields in the Northern Plains. They could be treated and eradicated using drones. Basically, spectral signatures of weeds and crops have to be deciphered as significantly different. The data for weed spread and intensity has to be collected using drones, so that, variable-rate herbicide applicators could be used to control weeds.

PLATE 4.8   A flat-winged drone being readied to take off and scout the soybean field.
Note: Such low-flying drones are used to scout, for disease/pest patches and drought symptoms on the crop. They are also used to supply digital information to ground vehicles that spray pesticides. Drones accomplish the scouting and reporting with data about an entire field of 50 ac in a matter of minutes or at best under an hour. Otherwise, it takes a couple of days and tedious work involving walking, noting and mapping by farm workers.
Source: Dr. Tom Nicholson, AgEagle Aerial Systems Inc., Neodesha, Kansas.

Drones have been utilized in the wheat belt of Northeastern United States of America to scout the crop and to collect data from fields exposed to fertilizer supply (Hunt et al., 2010; Croft, 2013). Farmers could be using drones to scout, monitor and supply fertilizers and pesticides at variable-rates. Particularly, wheat and other crops such as corn and soybean grown in North Plains of United States of America (see Plate 4.8). In Michigan, drones are deployed to maximize grain yields of maize and soybean. But at the same time, farmers aim to reduce on use of excessive fertilizer-based nutrients. Drones are also used to monitor crop's need for water, and then manage irrigation schedules efficiently through drone-aided aerial imagery (Cameron and Basso, 2013). Farms are prone to high fertilizer usage. It may not be long before drones are also regularly used to monitor lakes and regions adjacent to farming belts that adopt intensive cropping techniques. For example, in Ohio, lakes and slush terrain are regularly monitored, using drone imagery. They are monitoring algal blooms in locations prone to fertilizer runoff. Rivers and streams close by to farms are also surveyed for excessive contaminations with weeds, toxic wastes and seepage. Drones are getting popularized in the cropping zones of Pennsylvania State, United States of America. Small and sleek drones with multispectral sensors are utilized to oversee large acreage of crops such as maize, soybean and wheat. In this state, drones are also adopted to conduct aerial sprays of pesticides (Eble, 2014).

Reports emanating from soybean belt of United States of America suggest that, drones have literally found a new purpose in crop management. Drones have been adopted to map soils that would be sown to soybean. Drones are used to assess soybean seed germination and seedling establishment, rather accurately. Drones are used to scout the entire soybean field, using multispectral sensors. Predetermined flight paths and software to process the ortho-mosaics are also utilized. Drones could surveil soybean fields for crop diseases and pests, and inform the farmers at an early stage of the problem. Drones could also be used to spray pesticides and herbicides. Drones cover large areas of crop field in a matter of hours. Drones offer greater accuracy with aerial images compared with maps produced by human scouts. Farm workers may need several days to accomplish the same task and at higher costs to farmer (United Soybean Board, 2015; Plate 4.8). Drones also help in creation of 3D maps of the terrain and soybean fields. Such aerial images are of great utility when GPS-tagged robotic or semiautomatic vehicles have to move in the field (Rovira-Mas et al., 2005).

In the Southwest Texas, drone specialists are trying to standardize the instruments (sensors). They intend to obtain information on crop growth via vegetative indices (VIs). They are also aiming at surveillance of crops for insect pests, diseases and monitoring water status (Starek, 2015). Drones are able to obtain 3D images of crops. They can also image individual plants, if needed, and then analyse for pest/disease pustules (fungal/ bacterial infections). Ultimately, they could be conducting crop surveillance and gaining data for computer-based decision-making about nutrients and water. Drones-aided agronomic procedures could be less costly, efficient and may offer similar yield levels to farmers. Project reports about drones and their role in crop production in the Texas Plains, clearly indicate that drones could play vital role in integrating several aspects of crop agronomy. Hoffmann (2008) states that drones could integrate and help farmers in conducting operations such as crop scouting, sensing the vegetation indices of individual crops in fields, obtaining data for variable-rate supply of nutrients and pesticides, identifying nutrient deficits and in regulating irrigation. There are indeed several ways that drones could help agronomists in performing tasks swiftly and accurately. It is just a matter of time when drones become more common in the Texan sky over crop fields.

Drones have been applied to identify avocado trees afflicted by diseases such as *Verticillium* and *Phytophthora*. More importantly, during recent years, drones are used to identify Laurel wilt's spread via beetles. Drone-aided imagery seems essential because the disease spreads rapidly. This disease is now prominent in the states such as Georgia and Florida, where it is causing devastation of avocados (Buck, 2015).

Drones are in vogue in the grape vineyards of California, United States of America. In some farms, drones are regularly utilized to assess crop vigour, fruit ripeness and detect disease/pest infestation, if any. They have also been used to spray plant protection chemicals. Drones are expected to make headway into vegetable and grapevine yards in California (Dobbs, 2013; Paskulin, 2013; Mortenson, 2013). Drones and a posse of robots with ability to remove weeds need to be integrated. We can also trace ripe grape bunches and harvest. It will then make agricultural operations, in general, more effective and easier.

Chilean crop production occurs mostly in the hilly and undulated terrain. Drones could therefore be an apt method to scout, assess crop growth and conduct a few agronomic operations. Drones could aid procedures such as fertilizer and pesticide sprays on crops. Drones are excellent bets, when it comes to treating crops in inaccessible locations. Wulfsohn and Lagos (2014) have reported that drones are being evaluated in the horticultural farms of Chile. Small drones fitted with multispectral cameras could be used to obtain ortho-mosaics. They are then processed to obtain colour images of orchards. They say, data about orchards are made available on laptops. Such data could also be utilized to assess and decide on fertilizer and irrigation application.

4.8.1.2   EUROPEAN PLAINS

Agricultural extension experts at the University of Munich, Germany have reported that, currently, more than 80% of farmers in Germany use electronic methods and documentation about field conditions, the crop growth and yield formation stages. They regularly use grain yield maps to ascertain soil fertility and crop productivity trends in their fields. Such digital data are extensively used prior to adoption of several agronomic procedures (Microdrones GMBH, 2014). They say, farming, which is among the oldest of human endeavours, is literally swarmed by methods such as site-specific or precision farming. In addition, they state that drones could be the major factor enhancing crop yield. Drones offer aerial images of the fields and crops at various stages of growth till maturity. The digital data obtained can be fed to computers on the tractors with VRAs. Precise management of crops reduces accumulation of soil nutrients and pesticides. Therefore, it delays or avoids onset of deterioration of agro-environment. Regarding soil fertility management, German drone companies visualize a massive reduction of fertilizer usage due to drone imagery and precision farming techniques (Microdrones GMBH, 2014).

Drones fitted with cameras that operate at visible and NIR range have been used to study the crop fields. Drones are used to assess crop growth and maturity status. They also detect plant water status and leaf chlorophyll. Drones, with ability for 3D imagery, offer excellent details regarding leaf colour, its health and grain formation. Such 3D imagery provides details on soil surface, erosion patterns, if any, humus content and so forth. Again, in Germany, drones are currently being evaluated for monitoring crops, obtaining crop/field maps and monitoring crop yield formation. Data about panicle initiation and grain maturity are also obtained using drone (Microdrones GMBH, 2014). Indeed, in a short span of time, we may find that most farmers in German agrarian regions use drones to accomplish a series of agronomic procedures. Drones could actually be a very useful gadget in their farms. Drones may help farmers and their families in a variety of ways. In addition, they could help in devising crop production systems.

Barley crop grown in the German plains have been assessed, using both drone and ground-based hyper-spectral analysis for plant height and other traits. They have used drone-derived imagery and CSM. Plant height data from CSM and VIs were utilized to forecast biomass accumulation by barley accurately. Predictions about biomass using CSM was more accurate, if done at the seedling stage (Bendig et al., 2015).

Drone-aided precision farming and smart farming is making it to airspace above the crop land in the German plains (Ascending Technologies GMBH, 2016). They say, high-performance precision guidance and multispectral sensors are aimed at revolutionizing crop production methods in Germany. Drones are currently in operation in only few locations, wherein they are helping agencies in soil fertility management, irrigation and vegetation monitoring.

In France, drones have passed through initial testing. They are now gaining ground in the agrarian regions. They are used mainly to scout the crop for germination, seedling health, LAI and grain formation. Drones have been used to assess nitrogen needs of field crops such as wheat, using vegetation indices (Lelong et al., 2008). In France (INRA, Auzeville), drones have been utilized to collect data about phenotypic traits and grain yield levels of crops, such as wheat. Drones have also been adopted in experimental stations that evaluate several genotypes of durum and bread wheat. Farm experts have examined the crops for growth and yield traits, using drone images. For example, Lelong et al. (2008) evaluated a set of wheat genotypes for performance under rain-fed conditions. They utilized parameters such as NDVI, GNDVI, LAI and leaf-N content to assess the performance of wheat genotypes.

Regarding corn, Geipel et al. (2014) state that maize grown in the European plains could be assessed at various stages of growth from seedling till harvest. Drones could provide high-resolution images and VIs data to develop CSM. Such data about plant height and CSM have excellent correlation with grain yield. Forecasts done using drone imagery about plant height and CSM correlate highly at $R^2 = 0.74$ (Bendig et al., 2013a, 2013b).

In Spain, drones are used to collect data about nitrogen status of sunflower. They are utilized to assess crop's water requirements and monitor for diseases (Aguera et al., 2011). Drones are used regularly to collect information on vegetation indices and leaf chlorophyll. They are also used to obtain thermal images of crop canopy. Drone-aided precision farming could be useful and less costly. In the Spanish agrarian belts, drones are also being evaluated to monitor crops such as maize and wheat for occurrence of weeds. They are evaluating a technique known as OBIA. Data from drone imagery could then be used to conduct agronomic procedures such as weeding, thinning and interculture (Torres-Sanchez et al., 2015).

Drones have made a strong bid into grape vine yards of Europe. Chevigny (2014) has shown that drones could be of great utility in the vineyards. Drones are used right from the stage fields and soils types are chosen for establishment of grape vine yards. Drones can help farmers in surveying soils, mapping the variations in soil characteristics (top soil), providing a 3D image of the location and in designing drainage lines. No doubt, drones have made a mark in Burgundy, the vine region of France. Bordeaux, France, which has a traditional grape production belt, is also exposed to drones. Farmers are trying to replace traditional methods of crop scouting, application of fertilizers and pesticides with drone-aided procedures. Drones are swift, accurate and could be deployed repeatedly at a lower cost than traditional manual systems (Magrez, 2013). In fact, drones are getting accepted during grape production in several European nations such as France, Germany, Italy, Spain and Portugal (Matese and Di Gennaro, 2014; Colomina and Molina, 2014; Gao et al., 2013). In general, the aerial imagery that shows variations of grape vine vigour is said to be highly helpful in decision-making.

4.8.1.3   MIDDLE EAST

A report from Israel states that drones have a good future in the agricultural crop production. Drones may find immediate acceptance with farmers possessing large units, and with farming companies adopting precision farming procedures. Precision farming needs small drones with high-resolution multispectral sensors. Hence, Israelis have embarked on producing lightweight drones with cameras that operate at various wave bands, from visual (R, G and B) to IR and thermal range. For example, Sensilize, a company dealing with drone technology, has developed a sensor. It is integrated with analytical software for the ortho-mosaic that the drones capture. It is called 'Robin Eye'. It has a series of cameras that operate at eight different bandwidths. They provide detailed images of soil, vegetation and crops. The 'Robin eye system' provides farmers with colour images of field and crop stand within 24 h of drone's flight so that they could adopt them during precision farming (Leichman, 2015). The 'Robin eye' system offers data on website. Hence, data can be accessed at any time by the farmers. Clearly, sensors and appropriate analytical software are critical, particularly if we aim at spreading aerial mapping, variable-rate methods and precision agriculture in any agrarian region.

4.8.1.4   ASIA

Reports suggest that several different models of drones and in large numbers have been exported and sold, to Central Asian nations such as Kazakhstan, Tajikistan and Kirgiz. They are supposed to be tested for use during crop production (Tuttle, 2013). Bendig (2013b) reports that drones with high-resolution sensors have made a mark in the rice fields of Asia, particularly, in Northeast China. Drones are also in use in China's Northeast wheat belt to spray pesticides. Report by Ministry of Agriculture (2013) suggests that low-flying copter drones are utilized to spray pesticides on wheat at short intervals. Drone-based technology has also been adopted to assess crop-N status, in China and other nations of the Fareast. For example, Li et al. (2015) have estimated NBI, using leaf chlorophyll estimates. It could help farmers in channelling fertilizer-N to crops more accurately. The Northeast rice belt of China is a large expanse. Here, farmers adopt intensive farming techniques. It involves high labour and chemical usage. Drones are perhaps the best bets, if labour requirements have to be reduced perceptibly. Farm workers could be kept out of risks of coming into close contact to pesticides, if drones are adopted. Farmers could accomplish tasks rapidly, efficiently and with assured safety to farm labour by using drones.

A few drone companies such as Yintong Aviation Supplies and DJI Inc. have been successful in introducing copters, to spray pesticides and other plant protection chemicals on crops. They have used copters on wheat, rice and soybean grown in China, Korea and Japan (Yintong Aviation Supplies, 2015; Vandermause, 2015; RMAX, 2015). DJI Agras MG-1 is a drone made of dust proof, water resistant and anticorrosive material. It takes a payload of 10-kg pesticide in its tanks. The copter covers about 10 ac h^{-1} spraying pesticides. These drone models are popular and economically efficient for farmers running large farms in the Fareast (Tuttle, 2014). Reports suggest that about 30% of the rice belt in Japan is now adopting drones to conduct various agronomic procedures. Drones, particularly, copters are popular and are used, to spray pesticides and apply liquid fertilizer-N formulation (Bennett, 2013). A report by Cornett (2013) suggests that during past couple of years, over 2500 Yamaha's RMAX drones (a single brand) was sold to farmers in the Japanese rice belt. This step supposedly takes care of pesticide application and fertilizer sprays for over 2 million·ha of rice crop.

There are reports that we can pick spectral reflectance of rice panicles infected with pathogens such as *Ustilaginoidea virens* and pest *Nilaparvata lugens*. We can also distinguish infected panicles from healthy ones (Liu et al., 2010). We should be able to utilize this information and adopt drone technology to study panicle reflectance versus total grain yield harvested. Drone imagery obtained using visual and NIR band width should be able to distinguish panicles with healthy (filled grains) and infected ones. The infected panicles are chaffy and do not contribute to grain harvests. Agronomic procedures to remedy disease/pest affliction of rice panicles could then be based on aerial images from drones.

4.8.1.5 AFRICA

Rice production in West Africa involves series of carefully weighed out agronomic procedures. Precision farming techniques are making inroads to the rice fields. For example, farmers are asked to use computer-based decision support systems such as 'Nutrient Manager for Rice (NMR)'. This software considers fertilizer-N and irrigation needs of low-land rice, cultivated in Senegal. Reports suggest that on an average, rice productivity increases to 1.5–2.25 t·ha^{-1} and net profits increase by 216–640 US$·ha^{-1} (Saito et al., 2015). Now, considering that drones are already tested in several other rice producing regions, it should be a matter of time, before drone-aided imagery of rice is practised regularly, in these West African locations. Drone could collect data about crop growth and nutrient status. Drones have also been applied to assess changes in crop and natural vegetation caused, due to differences in precipitation levels. Drones could cover over 200 ha in a matter 1–2 h (D'Oleire-Oltmanns et al., 2012). Descroix et al. (2011, 2012) have stated that drones have made an entry into African farming regions. They could be used to study changes in the flow of water in rivers, assess total surface flow, and identify regions with floods and drought. Drones could also provide a view about productivity differences, if any, in the natural vegetation and crops. Hetterick and Reese (2013) state that drones have the potential to help farmers in conducting several types of agronomic procedures, during crop production. Simplest among them is the scouting. A manned aircraft costs 3.0 US$·per ha, to scout and process images for nutrient deficiency, diseases and drought effects, whereas a drone does the same for 5–10 cent·per hectare.

4.8.1.6 AUSTRALIA

In the Australian continent, crop production is expansive. Large farming companies dominate the agrarian belts. Major crops such as wheat, maize, legumes and oil seeds are grown, using techniques that are advanced and efficient. Recently, drones have attracted the attention of Australian farmers. Drones are intended mainly to monitor the crop, phenotype them, also decide on agronomic procedures and their timing. Perry et al. (2012a, 2012b) state that drones serve excellently to observe growth, general morphology and yield patterns. These measurements are useful for plant breeders, particularly if they are evaluating genotypes. Drones are apt to survey and detect crop diseases that are common to wheat belt in Australia. Chester (2014) has reported that drones were evaluated successfully for performing tasks, including monitoring farm operations, in general. In particular, drones were tested to monitor ploughing, seeding, irrigation and detection of pests and diseases, in the farms within Queensland. Drones provided images with a resolution of 1.7–3.0 cm. They served the farmers in detecting changes in crop growth and grain yield formation. The program that aims to popularize drone technology in the agricultural farms of Australia is called, 'Eye in the Sky'. Drones have been evaluated in the banana plantations of Queensland. They are utilized to surveillance and identify problems in the plantations, mainly, drought, floods and pests/disease. Satellites have also been used to surveillance banana plantations (Johansen, 2009). However, drone techniques are highly accurate. Even a single banana tree could be marked, monitored and remedial measures could be adopted accordingly.

Drones have also been evaluated for their performance in the apple orchards in Australia. Drones could be used, in future, to obtain data on vigour, disease/pest affliction and nutrient needs of apple trees, using aerial imagery (Packham and Davies, 2013; Wilson, 2014; Townsend, 2013). In mixed farms, drones have also been explored for daily usage, mainly to monitor pastures. Further, drones have been examined to conduct spraying fertilizers and to monitor dairy cattle.

4.8.2   DRONES IN CROP PRODUCTION PROCEDURES: A FEW SPECIFIC EXAMPLES

4.8.2.1   WHEAT (TRITICUM AESTIVUM)

Drones could surely become regular instruments in wheat producing farms. Right now, they are being experimented to scout for several aspects of wheat crop production. They are used to estimate crop stand, seedling growth pattern, biomass formation, LAI, leaf chlorophyll content and grain yield (Haney, 2014). In the Prairie States of Northern United States of America, drones' imagery has been utilized to detect insect pests that attack wheat crop (Basso, 2013). Wheat crop grown in rotation with soybeans in the Southern Plains region of United States of

America has also been assessed, using drone imagery. Drones have been used to assess leaf chlorophyll, canopy-N status and water stress index. It is done to apply fertilizers and irrigation at variable-rates (Cowan, 2015). Crop phenotyping at various stages of growth and mapping diseases/insects are other aspects covered, using drone technology. Drones are definitely useful to detect diseases that afflict wheat crop in the Southern plains regions. For example, Rush (2014) states that there are about 1.0 million·ha of wheat crop in the Texas High plains. They need to be periodically scouted and monitored for occurrence of drought and Russian aphid attack. Russian aphid transmits viral diseases. Actually, aphids need to be monitored and identified at the earliest. Drones fitted with multispectral sensors are best suited to conduct such aerial surveys repeatedly and at low cost. Further, Rush (2014) states that copters are able to provide 3D images of wheat crop and provide data about phenomics, so that fertilizer and irrigation could also be channelled appropriately.

Wheat genotypes grown in European plains have been assessed, using drone technology. Wheat specialists have been able to study phenomics of wheat genotypes and differentiate them, using a set of 4–5 sensors (R, G, B, IR and NIR). Drone-aided data collection has, in fact, shown good correlation with regard to ground level reality data for biophysical traits (Lelong et al., 2008; Perry et al. 2012a, 2012b). Drones have been adopted to quantitatively assess wheat crop, using VIs (Lelong et al., 2008; Abuzar et al., 2009). Wheat genotype selection and scheduling of agronomic procedures based on data from sensors is a clear possibility. Therefore, it should be pursued (Lelong et al., 2008).

Torres-Sanchez et al. (2014) have reported that wheat crop grown in Spain could soon be routinely monitored both spatially and temporally, using drones. Parameters such as VIs, pests, diseases and water status could be estimated, using drones. Drones with multispectral sensors have been tested at experimental stations. The digital data so collected could be used to supply wheat seedlings with fertilizer-N and water accordingly, using precision techniques. Actually, drones offer digital data that could be directly utilized by the VRAs, during wheat production.

As stated earlier, in Australia, wheat crop has been effectively monitored and assessed for its reaction to nitrogen and water stress, using remote sensing (Tilling et al., 2007; Perry et al., 2012a, 2012b). Jensen et al. (2007) have also evaluated the low-flying drones, for their efficiency in obtaining data about wheat crop, at various stages of growth and maturity. It is said that drones could supply researchers/farmers with information about growth rate of seedlings, tillering, photosynthetic efficiency and grain yield. Periodic flights by drones can help farmers to judge the changes in crop growth. Drone imagery was highly correlated to crop growth. Vegetation indices such as NDVI was correlated with grain yield ($R^2 = 0.91$) and grain protein content ($R^2 = 0.66$). Hence, drones could be used regularly, to obtain data from wheat crop, at a swift pace and repeatedly.

4.8.2.2   RICE (ORYZA SATIVA)

Drones have almost found a niche for themselves in the rice belt of Japan. They are used to conduct a range of agronomic procedures. They are utilized to monitor rice crop for growth. They are also used for phenotyping the crop, at various stages, during the crop season. Imagery obtained using R, G, B and NIR band width could be processed and utilized, to supply inputs at variable-rates (RMAX, 2015; Tadasi et al., 2010; Shibata et al., 2002; Ishii et al., 2006). The drone technology supposedly helps production of high-quality rice grains at a better production efficiency. Tadasi et al. (2010) reported that drone and precision techniques helped in reducing fertilizer requirements. Drones were also effective in assessing different cropping systems, particularly those that included rice in the sequence. They say, several combinations of crops could be tested, using drone technology. Therefore, modifications to rice-based cropping systems could be effected, if necessary.

Efforts have been made to assess rice crop's N status, using aerial imagery. It is done by measuring the dark green colour index of canopy/ leaves. Li et al. (2015) suggest that it should be possible to supply fertilizer-N based on NBI values derived, using sensors on drones. The aim is to enhance fertilizer-N efficiency, avoid N accumulation and reduce loss of soil-N.

In the Democratic Republic of Korea, agricultural scientists have evaluated drone imagery integrated with crop models. They have aimed to assess crop-N status. The idea is to reduce on fertilizer-N input and emissions (Ko et al., 2015). The data from simulations and actual growth pattern and grain yield formation coincide. They show a correlation of $R^2 = 0.939$. It is believed that such drone-aided aerial monitoring of rice crop will be helpful in the regulation of rice crop production.

In the wetland regions of Louisiana, United States of America, drones are being examined for their utility during spraying pesticides to rice fields. They seem to be useful while collecting data about phenomics (NDVI, GNDVI, water stress index) and assessing different rice genotypes for their performance. Rice breeding station in Louisiana is trying to adopt drones to accrue data about elite agronomic lines (LSU Ag Center, 2013).

Swain and Zaman (2010) have made a detailed report about various techniques that could be conducted, using drones, during rice production. They have highlighted that in future, drones with their multispectral sensors could be used to assess VIs such as NDVI, GNDVI, leaf chlorophyll and leaf-N status. In future, it should also be possible to apply fertilizer-N to rice crop in India, and in other regions of Asia, using drones. Drones could be used right from soil preparation stage, wherein soil surface and landscape could be imaged. They have also studied rice crop supplied with different levels of fertilizer-N. They have shown that fertilizer-N input has direct impact on NDVI and GNDVI measured, using drones. A step further, NDVI values from sensors on drones is directly related to biomass accumulation by rice crop ($Y = 31.85x - 23.83$; $R^2 = 0.7598$). Further, they have also shown that NDVI values could be used to forecast grain yield and protein content in grains. Indeed, a few years later, when drones and data they collect get utilized routinely, these measurements and their relevance to rice grain yield forecast will receive greater attention.

4.8.2.3 MAIZE (ZEA MAYS)

Regular monitoring and multispectral analysis of maize genotypes is a clear possibility. As stated earlier, maize crop's response to low-N in soil could be assessed, using phenotyping of genotypes. Data about VIs obtained using sensors on drones is of value to farmers. Drones are already in operation in experimental research centres that evaluate large number of maize genotypes. Basically, researchers aim at achieving genetic improvement for low-N tolerance (Zaman-Allah et al., 2015).

Maize production is an important enterprise in several regions of North America. It is most intense in the corn belt covering a few states such as Iowa, Nebraska, Illinois, Minnesota and so forth. (see Krishna, 2012, 2014, 2015). Farmers regularly seek improved methods and genotypes that enhance productivity. Drones are among the most recent methods that seem to be of great use to farmers. Drones are used in scouting, monitoring and obtaining detailed data about crop phenomics. Drone imagery could also help farmers in forecasting grain yield. Most farmers prefer to use drones in conjunction with precision farming methods. They could reap better harvests of corn forage/grain. Demonstrations of drones and their utility during production of maize are becoming popular in the entire corn belt and other regions. A recent report states that drones are useful to monitor and obtain data, so that, agronomic procedures are suitably modified and adopted (Noble, 2014; Krienke and Ward, 2013).

Reports from Illinois State, United States of America suggest that drones are made to fly the same path over corn fields. Drones are flown twice weekly to monitor the progress of seedlings, identify nutrient dearth if any and observe disease/pest incidence. Repeated predetermined flight paths supposedly help in better observation of crop's progress, throughout the season. Drones are actually utilized to scout fields even before they are sown. The field and soil are the first items to be scouted in detail and imaged by the drones (Ruen, 2012a, 2012b; Glen, 2015).

Drones have been effectively tested for their ability to collect data from a range of maize genotypes sown under multilocation trials. Such evaluations are being conducted by experts at the CIMMYT, Mexico. They could also conduct such drone-aided evaluation of maize genotypes in Zimbabwe and Mexico (Hunt et al., 2010; Lumpkin, 2012; Mortimer, 2013). Lumpkin (2012) opines that a sizeable portion of maize belt may get exposed to surveillance by drones. Drones accomplish scouting, imaging for NDVI, leaf chlorophyll and biomass. Drones are also adopted for fertilizer and pesticide application at variable-rates, using drones. Magri et al. (2005) state that potential of drones during maize production begins earnestly from imaging bare soils, monitoring crop throughout the season and up to grain harvest. It is believed that geo-spatial techniques and variable-rate methods could be effectively used to enhance maize grain/forage yield. Drone technology may help to achieve better energy and economic efficiency. We may note that maize is grown in a range of agro-environments. Its cropping systems engulf both low-input subsistence farms and high-input commercial grain producing farms. Commensurate with this reality, Bechman (2014) states that drones are available to farmers at a wide range of cost. It ranges from as low as 1000–10,000 US$ based on purposes that need to be served. Hence, drones could be versatile instruments hovering over maize fields grown with different intensities. One idea that is gaining ground in the North American maize/soybean belt is that we should then be able to conduct crop scouting, monitoring and estimating VIs entirely through drones. Farm labour requirement should be least.

4.8.2.4 BARLEY (HORDEUM VULGARE)

Barley production zones have already been exposed to drone technology. CSMs have been developed, using drone imagery. The CSM and plant height data has the potential to help farmers in routine forecasts about barley grain yield. Measurement made at early seedling stage correlates excellently with final grain yield ($R^2 = 0.74 - 0.83$). Hence, drone imagery could be an important aspect, while deciding fertilizer inputs to barley, also while adopting agronomic procedures related to irrigation, pest control and in fixing yield goals. Farmers may routinely adopt drone imagery prior to conducting crop husbandry procedures (Bendig, 2013a, 2013b; Bendig, 2015; Geipel, 2014; Krienke and Ward, 2013).

4.8.2.5 VEGETABLES

Tomato crops grown in Northern Italy have been assessed using drone imagery. Candiago et al. (2015) have reported that tomato crop could be periodically assessed for leaf area, chlorophyll content and biomass. The VIs such as NDVI, GNDVI and SAVI could be utilized to assess biomass. Interestingly, ortho-images obtained from drones clearly depicted crop zones with low and optimum growth. Based on ground reality assessment, the low growth was attributable to bacterial disease that afflicts tomato crop. They suggest that high-resolution UAV-aided images could be utilized to monitor tomato crop and to take appropriate agronomic measures. The VIs also depicted zones of low growth that was caused due to bacterial spots. Ground reality data showed that bacterial spot disease was also responsible for defoliation. High-resolution imagery clearly showed that low growth rate of tomato crops was due to bacterial disease.

Reports by Agriculture and Agri-Foods Canada (2016) make it clear that drones are gaining ground in the potato fields of Eastern Canada. Farmers with fields of more than 1000 ac are preferring drones. Drones have been used to surveillance and identify nutritional problems of crops, also pest and disease, if any. Actually, extension agencies are trying to demonstrate the utilities of drones to potato farmers. The aim is, of course, to enhance production efficiency.

Regarding potato cultivation in different agrarian belts, Quiroz (2010) states that drones are currently in an early testing stage. Drones could be adopted to get aerial imagery of crops in the South American potato growing regions. Drone-aided aerial imagery could be of great help while evaluating potato genotypes disease/pest resistance and yield. It is said that in future, scientists dealing with potato could exchange drone-derived information, using networks. In Oregon, drones are deemed as highly useful during potato production, particularly to conduct scouting and to obtain data about vegetation indices (Oregon State University, 2014; Hunt and Horneck, 2013). Drones are expected to be used to surveillance large acres of potato crop. In Idaho, a state known for large scale cultivation of potatoes, drones are getting evaluated for their ability to get, accurate multispectral imagery of the crop, at various stages. Drones are actually used to identify regions with water stress effects, in the fields (O'Connell, 2014).

4.8.2.6   CASH CROPS

Drones provide digital imagery and data about sugarcane crop and its growth status. VIs such as NDVI, GNDVI and SPRI are utilized to assess nitrogen status of the crop. These measurements help sugarcane farmers to arrive at decisions to supply fertilizers accurately. The digital data are used while practising precision farming techniques (Lebourgeois et al., 2012). Sugarcane is an important cash crop in several regions of the world. It is a relatively longer duration field crop. It needs high amounts of fertilizer-N and other amendments to reach the elevated yield goals set by farmers. Sugarcane production in Brazil, for example, involves formation of strips and management blocks, depending on soil fertility/yield trends. Sugarcane is supplied with fertilizer-N based on soil tests. Soil chemical assays could be tedious, costly and time consuming. Instead, farmers are being advised to adopt leaf-N meters and ground-based sensors perched at vantage locations in the field (Amaraj et al., 2015). Then, calculate N requirements using appropriate models and software, and yield goals. Sugarcane is provided with fertilizer-N at variable rates. Fertilizer-N supply is done in several split dosages, mainly to enhance N-use efficiency. Drones with a full complement of sensors such as Visual, NIR, IR and thermal sensors are preferred, particularly to estimate crop biomass, N status (deficiency/sufficiency indices), moisture status and maturity of cash crops.

Hoffmann (2008) states that drones could be useful instruments to farmers producing cotton and sorghum, in the Southern Plains of United States of America. Drones, with their ability for multispectral imagery, can easily distinguish healthy from disease/pest attacked patches of cotton. For example, boll weevil affected cotton could be detected, accurately, in a matter of few minutes of drone's flight over the fields. In the general course, it takes several scouts and a few days to map the areas affected by weevils. Cotton and sorghum production is a regular practice in Southern Plains. Procedures such as scouting, identification of drought affected patches, disease/pest attacks and mapping is to be conducted, year-after-year, with good accuracy. Therefore, it is preferable to adopt drones. The idea suggested here is of course applicable to almost all crop belts that support cotton. Drones could turn out to be cost-effective and quick.

4.8.2.7 PASTURES AND RANGE LAND

Pastures and range lands have been exposed to aerial imagery and spectral analysis, using satellites and drones for a long time. Schellberg et al. (2008) have reviewed possible applications of satellite and drone-mediated technology. Natural grass lands, pastures and shrub land used for grazing have all been accurately assessed for growth, biomass accumulation and diseases, using drone technology. Drones are also useful in precision management of turf grass (Stowell and Gelernter, 2013). Drones provide data that could be used during site-specific management of pastures (Bueren and Yule, 2014; Rango et al., 2009). Drones with their ability for high-resolution multispectral imagery can be used to assess botanical diversity of natural grass lands, pastures and man-made mixed pastures. Kutnjak et al. (2015) have shown that drones have a great potential in estimating the botanical composition and surveying grass-legume mixed pastures. Drones could assess pastures for growth, foliage, chlorophyll content, leaf-N and water status. To quote an example, pastures near Zagreb, in Croatia, were assessed for grass and legume species using their spectral signatures. The specific spectral signatures were detected by sensors on drones. The percentage of each grass and legume species found in the mixed-pasture could be detected, mapped and quantified accurately

(Kutnjak et al., 2015). Drones could also be employed to spray liquid fertilizer formulations at variable-rates. Digital data that depict variations in pasture growth is utilized to spray fertilizer formulations. At the bottom line, knowledge about spectral signatures of legumes, grasses and other species that form the pasture is essential.

4.8.2.8 VITICULTURE

Drones have entered the grape vine yards of Europe. For example, in Germany, copter drones such as 'Microdrone' have been utilized to detect pests. The drones equipped with multispectral cameras, determine infestation of vines with fungal diseases like powdery mildew. They can negotiate the steep regions in grape orchards efficiently and bring back excellent images of crops. Such spectral images depict their health, grape bunches and pest attack, if any (Microdrones GMBH, 2014; Mamont, 2014). Reports suggest that German Agricultural Ministry has supported use of drones in vineyards. The intention is to detect pests and spray them with plant protection chemicals. It supposedly helps farmers to reduce loss of fruits to pests to a great extent. In Italy, Primicerio et al. (2012) have shown that drones could be adopted to conduct several of the site-specific agronomic procedures required for vine yard management. A drone known as 'VIPtero', which is a copter fitted with multispectral cameras is being used in Central Italy. Such drone imagery provides farmers with digital maps that show variations in vineyard vigour.

Candiago et al. (2015) have studied the grape vines in Sorrivoli, a village in Forli-Cesna, Italy. They have used imagery from low-flying drones such as hexa-copter ESAFLY A2500. They have estimated the VIs such as NDVI, GNDVI and SAVI. It helps to obtain an idea about grape vine growth. The ortho-images of 5-cm resolution derived at R, G, B and NIR band widths were utilized to assess vegetation and growth. Subsequently, drones were adopted for precision farming. Vegetation indices indicated the regions of low and high growth of vineyards. Ground reality data proved that low vegetation was due to fungal diseases such as *Armillaria* root rot. The grape vine also suffered from trunk diseases. The disease affected regions could be deciphered as low growth zones, using VIs data.

Mathews and Jensen (2013) have utilized high-resolution multispectral analysis to derive values for Vis, also to note the growth rate and vigour of grape vines, periodically. They have used an octocopter with ability for low-altitude flight over grape vines. It has been reported that drones could also be utilized to estimate leaf carotenoid content, in the vine yards (Zarco-Tejada et al., 2013). Matese et al. (2013a, 2013b) have also used low-flying drone to detect crop vigour and VIs. Moreover, they have assessed anthocyanin contents of grapes (fruits). They have tried to compare drone-derived values with ground reality data.

Drones were examined for their utility in vine yard mapping using visible, multispectral and thermal imagery. The data could be utilized to prepare 'digital surface models (DSM)'. The DSMs are of great use to farmers, particularly to those who wish to compare the drone imagery with a few standard models of growth pattern and fruit bearing (Turner et al. 2012). Drones with high-resolution multispectral sensors are offering better aerial imagery about individual plots of grapes. They depict soil surface characteristics, irrigation channels and crop vigour (Magrez, 2013). Drones help in providing data regarding variations in nutrient and water status, so that, digital data could be used during precision techniques (Colomina and Molina, 2014; Matese et al., 2013a, 2013b; Matese and Di Gennario, 2014; Baluja et al., 2012). Sepulcre-Canto et al. (2009) have reported that grape vine's growth status, its LAI, chlorophyll and other VIs tell us about crop vigour. Further, spectral data could be used to supply inputs at variable-rates. Matese et al. (2015) have, in fact, evaluated different modes (drones, aircrafts and satellites) of collecting data, about vine yards, particularly their growth, foliage and fruit bearing pattern. This is with an aim to use the data for making decisions, about fertilizer-N input, pesticide application, irrigation scheduling and timing the harvest of fruits. Among the three systems compared, drones (UAVs) costed less to farmers. The images were of high resolution. Also, the digital data were highly relevant for precision farming methods.

4.8.2.9   CITRICULTURE

Citrus culture is in vogue in several agrarian regions of the world. Citrus production is prominent in Florida and California in United States of America. Citrus is produced in sizeable quantities in the Sao Paolo region of Brazil. Citrus cultivation is also well-distributed in Asian and Euro-pean regions. Citrus production methods have improved vastly during past decades. During recent years, precision farming methods have been evalu-ated. They have used ground-based sensors located at vantage points or handheld leaf meters to map the nutritional status of citrus groves. Soil texture and fertility has been assessed using the chemical analysis. It is tedious and cumbersome. Electrical conductivity that is slightly easier to record, using proximal techniques and GPS connectivity, has been a recent introduction. Management blocks created using such soil/crop traits have been adopted. No doubt, adoption of variable-rate methods depends on soil fertility and citrus productivity maps obtained for yester years. Yield goals set for current year by the farmers are equally important. Variable-rate supply of N, P, K and lime has been done after analysing leaves picked from grids. Sometimes, maps that depict EC, soil-N, soil-P and pH have been overlayered and used for prescribing variable rates. These steps involve computer software and VRAs. During recent years, drones that fly low over the citrus orchards and collect data about leaf chlorophyll, leaf-N, water status and plantation vigour are preferred. Drones can be adopted easily to monitor citrus orchards (Colaco et al., 2014; Colaco and Molin, 2014).

Citrus orchards in Florida are exposed to several diseases and pests that attack different aspects of the canopy and fruits. During recent years, farmers/researchers have shown greater interest in Citrus Greening disease (Huanglongbing disease). This disease affects fruit productivity severely. Drones with multispectral cameras are being adopted, to detect the affected citrus tree (Kumar, et al., 2012; Garcia-Ruiz, 2013; Gmitter, 2015). Ehsani et al. (2012) has shown that it is possible to develop digital data and imagery about a single citrus tree. Drone with high-resolution multispectral sensors is utilized to get such accurate images. Bouffard (2015), in fact, states that soon drones may become part of citrus production systems in Central Florida. They could be effectively utilized to monitor the water status of trees, soil moisture and disease/pest attack on trees and in applying fertilizers. The digital data from drone's sensors could help in variable-rate application of plant protection chemicals and nutrients. Drones could impart greater accuracy to citrus production procedures.

4.8.2.10   APPLE ORCHARDS

Drones are in the forefront of apple orchard management. They are useful in monitoring apple tree growth, its foliage and fruit bearing pattern. Drones are regularly used to surveillance and identify regions/patches that show affliction, with diseases such as 'Apple Scab' caused by fungi. Drones actually offer farmers with images of patches of trees affected by scab. The digital maps showing scab could be utilized during aerial sprays of fungicides. It is said that multiple sprays of fungicides are needed to cure the farm from scab. Obviously, this could be accomplished using drones, repeatedly (Kara, 2013; Homeland Surveillance and Electronics LLC, 2015).

4.8.2.11 OLIVES

Olives are an important orchard crop. Olive fruits offer edible oil. Olive agronomy involves a series of procedures. They are directed at measuring canopy and leaf growth, tree canopy size, fruit bearing intensity and oil content. Olives are exposed to cold injury and droughts. Routinely, olive farmers try to monitor the olive crown growth rates and patterns each season. Monitoring is done as the crop puts forth new flush of foliage and fruits. Diaz-Valera et al. (2015) have reported that olive plantations could be monitored, using drones fitted with commercial grade cameras with facility for visible range imagery. The images obtained could be processed to identify, mark the frame work of the tree canopy, foliage growth and fruits. Generally, the workflow involved imaging the trees, preparing models and pinpointing tree growth pattern, using GPS tags. Drones with sensors provided accurate measurements of tree growth and canopy size. Drones offered greater details about the tree crown. Further, the data collected using drones tallied accurately with ground data. The ground data was obtained using skilled farm workers who measured canopy crowns. Usually, visual and thermal sensors are used and values for CWSI are calculated. Drones may also help farmers in detecting canopy temperature under an olive tree (Berni et al., 2009) and photosynthetic radiation interception (Guillen-Climent et al., 2012a, 2012b). Agricultural drones could therefore be useful in deciding several of the agronomic procedures in the olive orchards, also in carrying out a few of them that involves spraying pesticides, fertilizer formulations and so forth.

4.8.2.12 OIL PALM

Oil palm plantations, it seems, need regular scouting and monitoring of individual trees. Palm trees have to be monitored from bottom to top of the crown for various disease-causing agents, pests and nutrient deficiencies that appear, periodically (MosaicMill, 2015a, 2015b). Efficient airborne and ground scouting is essential. A typical solution would be to use drones that fly low above the tree canopy. The drone could trace and collect data about tree health and nutritional status. Drones are in fact capable of offering excellent, high-resolution 3D images of palm trees for farmers. Farmers may analyse the data, using appropriate computer programs.

Oil palm is a major plantation crop in Malaysia and adjoining tropical countries. Farmers adopt a series of agronomic procedures. Many of them could be accomplished using drones. Drones may after all be more efficient than many of the previous methods. Shafri and Hamadan (2009) have examined drone technology on oil palms in Peninsular Malaysia. They state that oil palms could be regularly scouted for health and diseases. Drone-aided spectral imagery of oil palm and NDVI measurements are done regularly. The aim is to detect plantation growth and to decide fertilizer application. Drones have also been evaluated for their ability to provide maps depicting crop water stress. Ganoderma is a disease on palm. It could be identified, in time, before it causes wide damage.

Currently, there are private agricultural agencies that utilize drones to collect a range of data about oil palms. These agencies offer drone-aided services that include, (a) general evaluation of palm plantation, using aerial imagery via drones, (b) mapping plantation infrastructure and tree spread, (c) mapping tree health of overall orchard and individual trees, using hyperspectral imagery, (d) surveillance of plantation boundaries and answering security concerns, (e) plantation audit using drone images, (f) plantation yield prediction, using spectral images and appropriate computer software, (g) designing, re-designing and planning tree planting programs, using drone-aided aerial surveys, and (h) regular monitoring of trees for drought effects, diseases and pest attack (GeoPrecision Tech, 2015).

4.8.2.13   COFFEE

Drones could be very effective in scouting and noting data from large coffee plantations. They conduct it swiftly and at low cost to farming companies. Drones can fly over the plantations in quick succession. Drones could be adopted to spray pesticides, at variable-rates and at only spots that are infested with pests. In certain regions, coffee plantations are situated in hilly terrain. In such locations, human scouts may find it difficult to navigate and observe the plants in detail. For example, in the hills of southern India, drones could be very effective in informing farmers about the crop status in a matter minutes and at low cost. Coffee plantations in Hawaii, United States of America have been assessed for growth, nutritional status and disease incidence, using drones. Herwitz et al. (2004) have tested drones to derive spectral imagery. Such spectral data is then utilized to prescribe fertilizer supply to coffee plantations. Drones were examined for efficiency, both when their flights were predetermined using computer software and when ground pilots managed their flight pattern. They say, drones could be used to monitor and identify coffee blossoms and detect fruit maturity accurately. The close-up images from sensors on drones offer details on fruits.

4.8.2.14 AVOCADOS

Avocados are grown in several different agrarian belts. Avocados are important tree crops in Southeastern United States of America, particularly in Georgia, Florida, Carolinas and so forth. General reports on commercial avocado production state that farms are affected by several diseases caused by fungi. Recently, 'Laurel wilt' that spreads through a vector called Ambrosia beetle has been devastating the plantations, in Georgia. It is heading into areas with avocado orchards within Florida (Buck, 2015). The Laurel wilt spreads very fast on the tree. The loss could be total in terms of fruit bearing. Hence, drone technology that allows farmers to scout the trees in greater detail seems apt. Drones can pick spectral data as many times during a week or even a day. Drones are helpful in tracing the disease at very early stages of spread. Drones with multispectral cameras flying at 200 ft above the trees can provide data about tree health and disease progress, if any. This allows farmers to thwart the disease (laurel wilt) right at the early stages. Drone-aided spray of chemicals will be effective, if disease inoculum is still small. Also, as sprays could be confined only to trees afflicted by laurel wilt, consumption of plant protection chemicals gets reduced significantly compared with traditional blanket prescriptions.

4.8.2.15 FORESTRY

Forestry is among most often studied item using satellite and drone technology. Drone usage in the forest plantations could start right at the seeding time. Aerial seeding of perennial forest tree species and those meant for agroforestry purposes is in vogue (Wood, 2014). Drones could be used to study the forest canopies, spread of forest plantations, biomass accumulation, diseases/pests, the impact of natural factors such as rain, drought or erosion and so forth. Periodically, drone imagery could be used to assess tree height and quantify its biomass accumulation (Zarco-Tejada et al., 2014; Jakkola et al., 2010). Demonstrations in Australia have shown that forest trees could be monitored effectively, using Lidar placed on UAVs. Periodic flights by such drones could help in assessing tree growth rates, biomass accumulation and canopy closures of Eucalyptus forests (Terraluma, 2014).

4.8.2.16   MIXED FARMING

Mixed farming is a common ploy in many agrarian regions. Such mixed cropping systems usually include cereals/legume rotations or intercrops, strip cropping of food crops with leguminous agroforestry trees and mixtures that include cereal food grains/legumes or oilseed crops. Mono-cropping systems too flourish, but they are often found in rotation with pastures or legume crops. Mixed farms that support a combination of crops, livestock, pastures and agroforestry trees species are easily trace-able in the agrarian belts. So far, we have been discussing drones that scout crops, obtain multispectral images of fields, water sources and so forth. However, we should note that each time a drone flies, it can pick images of most, if not all, components of a mixed farming enterprise. Drones can simultaneously record data about crop growth, livestock population on pastures, even their movements, forest tree conditions and agroforestry tree species that offer forage/wood all in one flight, if needed. Farm instal-lations too could be imaged during the same flight. Drones are indeed versatile as they obtain images through multispectral cameras. Actually, farmers having mixed enterprises can direct a predetermine flight paths of drones to collect data about as many components of a mixed farm (Nowitzki et al., 2014). Further, we may note that a mixed farm with food grain crops, cash crops, fruit trees and livestock generally emits higher amounts of N as N_2O, NO_2, N_2 and C as CO_2, CH_4 from cattle dung and other animal wastes. Farmers can have a record of greenhouse gas emis-sions and loss of N from their farms, by measuring them using atmospheric gas samples. Drone-derived maps of greenhouse gas emission patterns can help farmers to go for mulches, residue reincorporation and reduction in soil disturbance by opting for no-till systems.

As stated earlier, there are drone models that suit mixed farming enterprises. Mixed farms usually produce annual crops, pastures and maintain cattle herds on mixed pastures. Drones that serve a range of purposes within mixed farms are available. They accomplish tasks such as such as scouting crops for growth, collecting data on VIs, monitoring crops for diseases/pests and detect drought stress. These drones are also useful to surveillance cattle herds (Stevenson, 2015).

In summary, drones with ability for autonomous, predetermined flight paths could be guiding the farm community in conducting, a series of agronomic procedures (Wihbey, 2015). The above discussion on agricultural drones makes it clear that they are instruments of great utility to farmers, in future. Drone-collected data pertain to crop fields exactly as they appear and they need not be extrapolated. In addition, drones offer close-up images from vantage locations above the crop canopy. Drone-aided techniques could actually revolutionize the way we sow seeds, monitor, maintain and produce food grains. Worldwide, production agronomy will depend immensely on drone technology as years pass by.

KEYWORDS

- **agronomic procedures**
- **vegetative indices**
- **crop phenotyping**
- **low-flying drones**
- **variable-rate applicators**

REFERENCES

Abuzar, M.; O'Leary, G.; Fitzgerald, G. F. Measuring Water Stress in a Wheat Crop on a Special Scale Using Airborne Thermal and Multispectral Imagery. *Field Crops Res.* **2009,** *112*, 55–65.

Ag Business and Crop Inc. UAS/UAV Drone Information. 2015, pp 1–3. http://agbusiness. ca (accessed May 13, 2015).

Agribotix. *Adventures with Drones for Precision Agriculture*; Agribotix Inc.: Colorado, USA, 2014; pp 1–6. http://agribotix.com/blog/2014/10/misconceptions-about-uav-collected-ndvi-imagery-and-the-Agribotix-experience-in-ground-truthing-these-images-for-agriculture (accessed May 30, 2015).

Agriculture and Agri-Food Canada. Using Research Drones to Increase Potato Production. *Agric. News Lett.* **2016**, 1–7. http://www.agr.gc.ca/eng/about-us/publications/agri-info-news-letter-may-2015/id=1429792312943#g (accessed April 12, 2016).

Aguera, F.; Carvajal, F.; Perez, M. Measuring Sunflower Nitrogen Status from an Unmanned Aerial Vehicle-based System and an on the Ground Device. *International Archives of the Photogrammetry Remote sensing and Spatial Information Sciences* **2011**, *38*, 1–5.

Alchanatis, V.; Cohen, Y.; Cohen, S.; Moller, M.; Meron, M.; Tsipris, J.; Orlov, V.; Charitt, Z. Fusion of IR and Multispectral Images in the Visible Range for Empirical and Model-based Mapping of Crop Water Status. American Society Agricultural Engineers paper No 061171, ASAE, St. Joseph, Michigan, USA, 2006; pp 1–12 (accessed May 10, 2015).

Amaraj, L. R.; Molin, J. P.; Schepers, J. S. Algorithm for Variable Rate Nitrogen Application in Sugarcane, Based on Active Canopy Sensor. *Agron. J.* **2015**, *107*, 1513–1523.

Anderson, W.; You, L.; Anisimova, E. Mapping Crops to Improve Food Security; International Food Policy Research Institute: Washington, D.C., 2014; pp 1–5. http://www.ifpri.org/blog/mapping-crops-improve-food-security?print (accessed May 10, 2016).

Anthony, D.; Elbaum, S.; Lorenz, A.; Detweiler, C. On Crop Height Estimation with UAVs. 2014, pp 1–8. cse.unl.edu/~carrick/papers/AnthonyELD2014 (accessed Sep 18, 2016).

Antic, B.; Culibrk, D.; Crnojevic, V.; Minic, V. An efficient UAV-Based Remote Sensing Solution for Precision Farming. In: *Bioscience—The First International workshop on ICT and Sensing Technologies in Agriculture, forestry and Environment;* Novi Sad, Serbia, 2010; pp 223–228.

Ascending Technologies GMBH. UAVs—Drone-based Precision Agriculture and Smart Farming. 2016, pp 1–4. http://www.asctec.de/en/uav-drone-based-precision-agriculture--smart-farming.htm (accessed Feb 5, 2016).

Azorobotics.com. Explore use of Drones, UAVs and Crop Models at Growing Michigan Agriculture Conference, 2014; pp 1–3. http://www.msue.edu (accessed June 23, 2015).

Babar, S. A. Spectral Reflectance to Estimate Genetic Variation for In-Season Biomass, Leaf Chlorophyll, and Canopy Temperature in Wheat. *Crop Sci.* **2006**, *46*(3), 1046–1049.

Baluja, J.; Diago, M. P.; Balda, P. Assessment of Vineyard Water Status Variability by Thermal and Multispectral Imagery Using an Unmanned Aerial Vehicle (UAV). *Irrigation Science* **2012**, *30*, 511–522.

Baret, F.; Fournier, A.; Verger, A.; Comar, A.; Solan, B.; Burger, P.; Chernon, C. Phenotyping Under Field Conditions: Interest and Limits of UAV Systems for Field High-Throughput Phenotyping. *Proceedings of third International Plant Phenotyping Symposium;* Chennai, India, 2014; p 28.

Barnes, E. M.; Clarke, T. R.; Richards, S. E. Coincident Detection of Crop Water Stress, Nitrogen Status and Canopy Density using Ground Based Multispectral Data. *Proceedings of the Fifth International Conference on Precision Agriculture,* Madison, WI, USA: American Society of Agronomy, Unpaginated CD. 2000; Robert, P. C., Rust, R. H., Larson W. E., Eds.

Basso, B. Montana State Researches Wheat Crop Disease from the Air. DIY Drones, 2013, pp 1–3. http://diydrones.com/profiles/blogs/montana-state-researches-wheat-crop-disease-from-the-air (accessed Sep 26, 2016).

Basso, B.; Cammarano, D.; Carfagna, E. Review of Crop Yield Methods and Early Warning Systems. *Proceedings of the First Meeting of the Scientific Advisory Committee of the Global Strategy to Improve Agricultural and Rural Statistics,* FAO Headquarters, Rome, Italy, 2015, pp 1–58.

Bechman, T. Corn Farmers see UAV Potential in Crop Production. Prairie Farmer, 2014, pp 1–7. http://farmprogress.com/story-corn-farmers-uav-potential-crop-production-15-115093 (accessed June 19, 2015).

Bendig, J.; Bolten, A.; Bareth, G. UAV-Based Imaging for Multi-Temporal, Very High Resolution Crop Surface Models to Monitor Crop Growth Variability. *Photogramm. Fernerkund Geo-Inf.* **2013a,** *13,* 551–562.

Bendig, J.; Willkomm, M.; Tilly, N.; Gnyp, M. L.; Bennertz, S.; Qiang, C.; Miao, Y.; Lenz-Wiedmann, V. L. S.; Bareth, G. Very High Resolution Crop Surface Models (CSMs) from UAB-Based Stereo Images of Rice Growth Monitoring in Northeast China. *International Archives of Photogrammetry and Remote Sensing, Spatial Information Science* **2013b,** *40,* 45–50.

Bendig, J.; Yu, K.; Assen, H.; Bolten, A.; Bennertz, S.; Broscheit, J.; Gnyp, M.; Bareth, G. Combining UAV Based Plant Height from Crop Surface Models, Visible and Infrared Vegetation Indices for Biomass Monitoring in Barley. *Int. J. Appl. Earth Obs. Geo-Inf.* **2015,** *39,* 79–87.

Bennett, C. Drones Begin Decent on US Agriculture. Western Farm Press, 2013, pp 1–3. http://westernfarmpress.com/blog/drones-begin-descent-us-agriculture (accessed Sep 6, 2013).

Berni, J. A. J.; Zarco-Tejada, P. J.; Sepulcre-Canto, G.; Feres, E.; Villalobos, F. Mapping Canopy Conductance and CWSI in Olive Orchards Using High Resolution Thermal Remote Sensing. *Remote Sens. Environ.* **2009,** *113,* 2380–2388.

Bouffard, K. Drones Find Uses on Farms. 2015, pp 1–7. http://www.heraldtribune.com/article/20150326/ARTICLE/303269992.htm (accessed June 20, 2015).

Buck, B. Low Altitude Aerial Images Allow Early Detection of Devastating Avocado Diseases. Growing Florida. 2015, pp 1–3. http://growingfl.com/news/2015/05/low-altitude-aerial-images-allow-early-detection-devastating-avaacado-disease.htm (accessed June 20, 2015).

Bueren, S.; Yule, I. Multipsectral Aerial Imaging of Pasture Quality and Biomass Using Unmanned Aerial Vehicles (UAV). Institute of Agriculture and Environment, Massey University. Internal report, 2014, pp 1–5.

Cammarano, D. Spatial Integration of Remote Sensing and Crop Simulation Modelling for Wheat Nitrogen Management. Ph.D. Thesis, University of Melbourne, Victoria, Australia, January 2010, p 148.

Cameron, L.; Basso, B. *MSU lands first drone.* MSUToday, Michigan State University: Michigan; 2013, pp 1. http://msutoday.msu.edu/news/2013/msu-lands-first-drone/ (accessed May 19, 2015).

Candiago, S.; Remondino, F.; Degiglio, M.; Dubbini, M.; Gatelli, M. Evaluating Multispectral Images and Vegetation Indices for Precision Farming Applications from UAV Images. *Remote Sens.* **2015,** *7,* 4026–4047. DOI: 10.339/rs70404026 (accessed March 10, 2016).

Carlson, G. Precision Agriculture Steering Us into Future. Modern Agriculture—A British Columbia's Agricultural Magazine. 2015, pp 1–4. http://modernagriculture.ca/precision-agriculture-steering-us-future/ (accessed April 9, 2016).

Case, P. Rothamsted Unveils Octocopter Crop-Monitoring Drone. 2013, pp 1–2. http://www.fwi.co.uk.arable/rothamsted-unveils-octocopter-crop-monitoring-drone.htm (accessed June 25, 2013).

Chester, S. UAV: Growing for Australia. 2014, p 102. http://www.spatialsource.com.au (accessed July 23, 2014).

Chevigny, E. Using Aerial Image Analysis as a Tool for Soil and Substrate Mapping in Vineyards (Burgandy, France). 2014, pp 1–5. http://dumas.ccsd.cnrs.fr/GIP-BE/hal-01115522v1.htm (accessed June 26, 2015).

Christensen, S. *Field Phenotyping with the Platforms Phenofield and Hexacopter;* University of Denmark: Copenhagen, 2015; pp 1–3. http://erhverv.ku.ku.dk/english/researchinfra/list/phenomics-infrastructure/ (accessed May 27, 2015).

Ciampitti, I. A.; Wang, H.; Shroyer, K.; Price, K.; Stamm, M.; Prasad, V.; Mangus, D.; Sharda, A. SUAVS Technology for Better Monitoring Crop Status for Winter Canola. *Proceedings of 12th International Conference on Precision Agriculture*, Sacramento, California, USA, 2014 p. 17.

Colaco, A. F.; Molin, J. P. A Five-Year Study of Variable Rate Fertilization in Citrus. *Proceedings of 12th International Conference on Precision Agriculture,* Sacramento, California, USA, 2014; pp 127

Colaco, A. F.; Ruiz, M. A.; Yida, D. Y.; Molin, J. P. Management Zones Delineation in Brazilian Citrus Orchards. *Proceedings of 12th International Conference on Precision Agriculture*, Sacramento, California, USA 2014; p 126.

Colomina, I.; Molina, P. Unmanned Aerial Systems for Photogrammetry and Remote Sensing: A Review. *ISPRS Photogramm. Remote Sens.* **2014,** *92,* 79–97.

Cornett, R. Drones and Pesticide Spraying a Promising Partnership. Western Plant Health Association, Western Farm Press, 2013, pp 1–3. http://westernfarmpress.com/grapes/drones-and-pesticide-spraying-promising-partnership (accessed Aug 30, 2014).

Cowan, D. Remote Sensing Program in Precision Agriculture. AGRIS Co-Operative Ltd. 2015, pp 1–2. http://www.agris.coop/index.cfm?show=10&mid=1156 (accessed May 13, 2015).

Croft, J. Sowing the Seeds for Agricultural Drones. Aviation Week and Space Technology 2013, pp 1–3. http://www.Aviationweek.com/awin/sowing-seeds-agriculture-uavs (accessed June 21, 2015).

Daughtry C. S. T.; Walthall, C. L.; Kim, M. S.; de Coulston, E. B.; McMurtreyll, J. E. Estimating Corn Leaf Chlorophyll Concentration from Leaf and Canopy Reflectance. *Remote Sens. Environ.* **2000,** *74,* 229–239.

Descroix, L.; Bouzou, I.; Genthon, P.; Sighmnou, D.; Mahe, G.; Mamadou, I. Impact of Drought and Land-Use Changes on Surface-Water Quality and Quantity: The Sahel Paradox. INTECH, 2011, pp 243–258. http://dx.doi.org/10.5772/54536 (accessed Sep 9, 2014).

Descroix, L.; Genthon, P.; Amogu, O.; Rajot, J.; Sighomnou, D.; Vauchin, M. Hydrograph: The Case of Recent Red the Niamey Region. *Global Planet. Change* **2012,** *99,* 18–30.

Diaz-Valera, R. A.; De la Rosa, R.; Leon, L.; Zarco-Tejada, P. J. High Resolution Airborne UAV Imagery to Assess Olive Tree Crown Parameters Using 3D Photo-Reconstruction. Application in Breeding. *Remote Sens.* **2015,** *7,* 4213–4232.

Dietz, J. Crop Scouting Drones. 2015, pp 1–2. http://www.agriculture.com/technology/robotics/uas/crop-scouting-drones.587ar478.htm (accessed May 29, 2015).

DMZ Aerial. DMZ Arial; Your Eyes in the sky. DMZ Aerial Autonomous Scouting Robotics. 2013, pp 1–3. http://www.Dmzaerial.com (accessed April 9, 2016).

DMZ Aerial Autonomous Scouting Robotics. Unmanned Aerial Vehicles and Scouting. 2013, p 10. http://www.dmzaerial.com/uavscouting.html (accessed August 15, 2014).

Dobberstein, J. Drones Could Change Face of No-Tilling. No-Till Farmer, 2014, pp 1–7. http://www.no-till-farmer.com/pages/Spre/SPRE-Drones-Could-Change-Face-of-No-Tilling-May-1 (accessed July 26, 2014).

Dobberstein, J.; Kanicki, D.; Zemlicka, J. The Drones are Coming: Where are the Dealers. Farm Equipment. 2014, pp 1–16. http://www.farm-equipment.com/pages/From-theOctober-2013-Issue-The-Drones-are-Coming-Where-are-the-Dealers.php (accessed Sep 6, 2014).

Dobbs, T. Farms of the Future will run on Robots and Drones. 2013, pp 1–14. http://www.pbs.org/wgbh/nova/next/tech/farming-with-robotics-automation-and-sensors/ (accessed July 3, 2014).

D'Oleire-Oltmanns, S.; Marzolff, I.; Peter, K. D.; Ries, J. B. Unmanned Aerial Vehicle (UAV) for Monitoring Soil Erosion in Morocco. *Remote Sens.* **2012,** *4*, 3390–3416.

Dreiling, L. *New Drone Aircraft to Act as Crop Scout;* Kansas State University: Salina, USA, 2012; pp 1–2. http://www.hpj.com/archives/2012/apr1216/apr16/springPlanting-MACOLDsr.cfm (accessed May 30, 2015).

Drone Remote Sensing. Vegetation Indices. 2015, pp 1–18. http://www.droneremote-sensing.com (accessed Jan 10, 2016).

Eble, J. *Penn State Crop Educator Explores Drone-Driven Crop Management;* Pennsylvania State University: USA, 2014; pp 1–7. http://news.it.psu.edu/article/penn-state-crop-educator-expplores-drone-driven-crop-management (accessed May 14, 2015).

Eckelkamp, M. Take a Crop's Temperature. Farm of the Future. 2014, pp 1–2. http://farmofthe future.net/#/article/take-crops-temperature.htm (accessed June 29, 2015).

Ehmke, T. Unmanned Aerial Systems for Field Scouting and Spraying. *CSA News Mag.* 2013, *58*, 4–9.

Ehsani, R.; Sankaran, S.; Maja, S.; Garcia, F. Advanced Stress Detection Techniques for Citrus. Citrus Industry. 2012, pp 1–7. http://www.crec.ifas.ufl.edu/extension/trade_journals/2012/2012_May-tree_stress.pdf (accessed Sep 2, 2014).

Farming Online. Using Drones to Monitor Crops. Rothamsted Experimental Agricultural Station, United Kingdom, 2013, pp 1–2. http://farming.co.news/article/9243 (accessed May 22, 2015).

Fitzgerald, G. J.; Rodriguez, D.; Christensen, L. K.; Belford, R.; Sadras, V. O.; Clarke, T. R. Spectral and Thermal Sensing for Nitrogen and Water Status in Rain Fed and Irrigated Wheat Environments. *Precis. Agric.* **2006,** *7*, 233–248.

French, A. Non-invasive Phenotyping of Crops Using New Imaging Technologies. *Proceedings of a meeting on 'Crop Resource use efficiency and field phenotyping',* Belton, North Grantham, United Kingdom, 2013, pp 12–14.

Gao, J.; Martorell, S.; Tomas, M. High Resolution Aerial Thermal Imagery for Plant Water Status Assessment of Vineyards Using Multi-Copter RPAS. *Proceedings of VII Congreso Iberico de Agroingenieria y Ciencas Horticolas.* Universidad Politecnica de Madrid, Spain, 2013, pp 1–6.

Garcia-Ruiz, F.; Sankaran, S.; Maje, S.; Lee, W.S.; Rassmussen, J.; Ehsani, R. Comparison of Two Imaging Platforms for Identification of Huanglongbing-Infected Citrus Disease. *Comput. Electron. Agric.* **2013,** *91,* 106–115.

Geipel, J.; Link, J.; Claupein, W. Combined Spectral and Spatial Modeling of Corn yield based on Aerial images and Crop Surface Models Acquired with an Unmanned Aircraft System. *Remote Sens.* **2014,** *11,* 10335–10355 DOI: 10.3390/rs61110335 (accessed June 23, 2015).

Genik, W. Case Study: Wild Oat Control Efficiency Using UAV Imagery. AgSky Technologies Inc. 2015, pp 1–5. http;//agsky.ca/case-study-wild-oat-control-efficiency-using-uav-imagery (accessed June 26, 2015).

GeoPrecision Tech. UAV Services for Oil Palm Plantation. GeoPrecision Tech. 2015, pp 1–2. http://www.mygeoprecision.com/services.html (accessed May 14, 2015).

Giles, F. High Flying Crop Protection. 2011, http://www.growingproduce.com/uncategorized/high-flying-crop-protection/ (accessed June 20, 2015).

Gitelson, A. A.; Kauffman, Y. J.; Merzlyak, M. N. Use of Green Channel in Remote Sensing of Global Vegetation from EOS-MODIS. *Remote Sens. Environ.* **1996,** *58,* 289–298.

Glen, B. Rise of the Field Drone. The Western Producer. 2015, pp 1–8. http://www.producer.com/2015/02/rise-of-the-field-drones/ (accessed June 19, 2015).

Gmitter, J. Agritech: Unmanned Aerial Vehicle Update. Central Florida Agnews. 2015, pp 1–4. http://www.centralfloridaagnews.com/agritech-unmanned-aerial-vehicle-update/ (accessed June 20, 2015).

Guillen-Climent, M. L.; Zarco-Tejada, P. J.; Berni, J. A. J.; North, P. R. J.; Villalobos, F. J. Mapping Radiation Interception in Row-Structured Orchards Using 3D Simulation and High Resolution Airborne Imagery Acquired from a UAV. *Precis. Agric.* **2012a,** *13,* 473–500.

Guillen-Climent, M. L.; Zarco-Tejada, M. L.; Villalobos, F. J. Estimating Radiation Interception in an Olive Orchard Using Physical Models and Multi-Spectral Airborne Imagery. *Isr. J. Plant Sci. Spec. Issue on NIR Spectrosc.* **2012b,** *60,* 107–121 DOI: 10.1560/IJPS.60.1-2.107.

Haney, S. Aerial Crop Scouting Winter Wheat with UAV. Real Agriculture. 2014, pp 1–2. http://www.realagriculture.com/2014/04/aerial-crop-scouting-winter-wheat-uav/ (accessed June 21, 2015).

Hansen, J.; Schjoerring, J. K. Reflectance Measurement of Canopy Biomass Status in Wheat Crops Using Normalized Difference Vegetation Index and Partial Least Squares Regression. *Remote Sens. Environ.* **2003,** *86,* 542–553.

Haute, A. R. A Soil Adjusted Vegetation Index (SAVI). *Remote sens. Environ.* **1988,** *25,* 295–309.

Herwitz, S. R.; Johnson, L. F.; Dunagan, S. E.; Higgins, R. G.; Sullivan, D. V.; Zheng, J.; Lobitz, B. M.; Leung, J. G.; Gallmeyer, B. A.; Aoyagi, M.; Slye, R. E.; Brass, J. A. Imaging from an Unmanned Aerial Vehicle: Agricultural Surveillance and Decision Support. *Comput. Electron. Agric.* **2004,** *44,* 49–61.

Hetterick, H.; Reese, M. Drones can be Positive and Negative for the Agricultural Industry. *Ohio Country J.* **2013,** *78,* pp 7–9.

Hoffmann, W. C. *Aerial Application Research for Efficient Crop Production;* USDA Agricultural Research Service: USA College Station, Texas, USA, 2008; pp 1–14.

Homeland Surveillance and Electronics LLC. Apple Orchards and UAV technology. 2015, pp 1–4. http://www.hse-uav.com/apple_orchrds_uav_technolgoy (accessed April 2, 2016).

Huang, Y.; Hoffmann, W. C.; Lan, Y.; Thomson, S. J.; Fritz, B. K. Development of Unmanned Aerial Vehicles for Site-Specific Crop Production Management. *Proceedings of 10th International Conference Precision Agriculture, Denver, Colorado, CDROM.* 2010a, p 1. http://www.ars.usda.gov/research/publictions/publictions.htm?SEQ_NO_115=253718 (accessed March 20, 2014).

Huang Y.; Thomson, S. J.; Lan, Y.; Maas, S. J. Multispectral Imaging Systems for Airborne Remote Sensing to Support Agricultural Production Management. *Int. J. Agric. Biol. Eng.* **2010b,** *3,* 50–62.

Hungry. S.; Thomson, S. J.; lan, Y.; Maas, S. J. Multispectral Imaging Systems for Airborne Remote Sensing to Support Agricultural Production Management. *Int. J. Agric. Biol. Eng.* **2010,** *3,* 50–62.

Hunt, R.; Horneck, D. The 2013 UAS/Precision Agriculture Experiment, at Hermiston, Oregon. 2013, pp 1–24. http://www.uasag.com (accessed Feb 5, 2016).

Hunt, E. R.; Hively, W. D.; Daughtry, C. G. T.; McCarty, G. W.; Fujikawa, S. J.; Ng, T. L.; Tranchitella, M.; Linden, D. S.; Yoel, D. W. Remote Sensing of Crop Leaf Area Index Using Unmanned Airborne Vehicles. *Proceedings of a Conference Pecora-The future of Land Imaging,* Denver, Colorado, USA, 2008, pp 1–12. http://www.asprs.org/a/publication/proceeding/pecora17/18.pdf (accessed Sep 12, 2016).

Hunt E. R.; Hively, W. D.; Fujikawa, S. J.; Linden, D. S.; Daughtry, T.; Craig, S.; McCarty, G. W. Acquisition of NIR-Green-Blue Digital Photographs from Unmanned Aircraft for Crop Monitoring. *Remote Sens.* **2010,** *2,* 290–305. DOI: 10.3390/rs2010290.

Hunt, E. R.; Horneck, D. A.; Hamm, P. B.; Gadler, D. J.; Bruce, A. E.; Turner, R. W.; Spinelli, C. B.; Brungardt, J. J. Detection of Nitrogen deficiency of potatoes using small Unmanned Aircraft Systems. *Proceedings of 12th International Conference on Precision Agriculture,* Sacramento, California, USA, 2014, p 16.

Ishii, K.; Suguira, R.; Fukagawa, T.; Noguchi, N.; Shibata, Y. Crop Status Sensing System by Multispectral Imaging Sensor (part 1)-Image Processing and Paddy Field Sensing. *J. Jpn. Soc. Agric. Mach.* **2006,** *68,* 33–41.

Jakkola, A.; Hyyppa, J.; Yu, X.; Kaartinen, H.; Lehtomaki, M.; Lin, Y. A Low Cost Multi-Sensoral Mapping System and its Feasibility for Tree Measurements. *ISPRS J. Photogramm. Remote Sens.* **2010,** *65,* 514–522.

Jensen, T.; Apan, A.; Young, L.; Zeller, L. Detecting the Attributes of a Wheat Crop Using Digital Imagery Acquired from Low-Altitude Platform. *Comput. Electron. Agric.* **2007,** *59,* 66–77.

Johansen, K.; Phinn, S.; Witte, C.; Seasonal, P.; Newton, L. Mapping Banana Plantations from Object-Oriented Classification of SPOT-5 Imagery. *Photogramm. Eng. Remote Sens.* **2009,** *75,* 1069–1081.

Jones, C. L.; Maness, N. O.; Stone, M. L.; Jayasekara, R. Chlorophyll Estimation Using Multispectral Reflectance and Height Sensing. *Proceedings of American Society of Agricultural Engineers and Canadian Society for Engineering in Agricultural, Food and Biological Systems.* Paper No. 043081 2004, pp 1–14.

Kara, A. Low Cost UAV Fights Disease Devastating Apple Crops. 2013, p 109. https://www.linkedin.com/groups/Low-Cost-UAV-Fights-Disease-53140.S.5809026007457361924 (accessed Jan 15, 2015).

Knoth, C.; Prinz, T. UAV Based High Resolution Remote Sensing as an Innovative Monitoring Tool for Effective Crop Management. *Proceedings of a Meeting on 'Crop Resource*

Use Efficiency and Field Phenotyping. Belton, North Grantham, United Kingdom, 2013, pp 1–4.

Ko, J.; Jeong, S.; Yeom, J.; Kim, H.; Ban, J. O.; Kim, H. Simulation and Mapping of Rice Growth and Yield Based on Remote Sensing. *J. Appl. Remote Sens.* **2015**, *9*, 1–3 (accessed June 30, 2015).

Krienke, B.; Ward N. Unmanned Aerial Vehicles (UAV's) for Crop Sensing. West Central Crops and Water Field Day. University of Nebraska Extension Service, 2013, p 1–2.

Krishna, K. R. *Maize Agroecosystem: Nutrient Dynamics and Productivity;* Apple academic Press Inc.: Waretown, New Jersey, USA, 2012; p 346.

Krishna K. R. *Agroecosystems: Soils, Climate, Crops, Nutrient Dynamics and Productivity;* Apple Academic Press Inc.: Waretown, New Jersey, USA, 2014; p 546.

Krishna, K. R. *Agricultural Prairies: Natural Resources and Crop Productivity;* Apple Academic Press Inc.: Waretown, New Jersey, USA, 2015; p 499.

Krishna, K. R. *Push Button Agriculture: Robotics, Drones, Soil and Crop Management;* Apple Academic Press Inc.: Waretown, New Jersey, USA, 2016; pp 131–260.

Kumar, A.; Lee, W. S.; Ehsani, R.; Albrigo, C.; Yang, C.; Mangan, R. L. Citrus Greening Disease Detection Using Aerial Hyperspectral and Multispectral Imaging Techniques. *J. Appl. Remote Sens.* **2012**, *6*, 1–7.

Kutnjak, H.; Leto, J.; Vranic, M.; Bosnjak, K.; Perculija, G. Potential of Aerial Robotics in Crop Production: High Resolution NIR/VIS Imagery Obtained by Automated Unmanned Aerial Vehicle (UAV) in Estimation of Botanical Composition of Alfalfa-Grass Moisture. *Proceedings of 10th International Symposium on Agriculture.* Opatija, Croatia 2015, pp 349–353.

Lan, Y.; Thomson, S. J.; Huang, Y.; Hoffmann W. C.; Zhang, H. Current Status and Future Directions of Precision Aerial Application for Site-Specific Crop Management in the USA. *Computers and Electronics in Agriculture* **2010**, *24*, 34–38.

Lebourgeois, V.; Begue, A.; Labbe, S.; Houles, M.; Martine, J. F. A Light Weight Multispectral Aerial Imaging System for Nitrogen Crop Monitoring. *Precis. Agric.* **2012**, *13*, 525–541.

Leichman, A. K. Drones -the Future of Precision Agriculture. 2015, pp 1–4. http://www.israel21c.org/headlines/drones-the-future-of-precision-agriculture (accessed June 21, 2015).

Lelong, C. D. D.; Burger, P.; Jubeline, G.; Roux, B.; Labbe, S.; Baret, F. Assessment of Unmanned Aerial Vehicles Imagery for Quantitative Monitoring of Wheat Crop in Small Plots. *Sensors* **2008**, *8*, 3557–3585.

Liu, Z.; Shi, J.; Zhang, L.; Huang, J. Discrimination of Rice Panicles by Hyperspectral Reflectance Data Based on Principal Component Analysis and Support Vector Classification. *J. Zhejiang Univ. Sci. B* **2010**, *11*, 71–76.

Lopez-Granados, F. Weed Detection for Site-specific Weed Management: Mapping and Real Time Approaches. *Weed Res.* **2011**, *51*, 1–11.

Li, J.; Zhang, F.; Qian, X.; Zhu, Y.; Shen, G. Quantification of Rice Canopy Nitrogen Balance Index with Digital Imagery from Unmanned Aerial Vehicle. *Remote Sens. Lett.* **2015**, *6*, 183–189.

LSU Ag Centre. *AgCenter Researchers Study Use of Drones in Crop Monitoring;* Louisiana State University Ag Centre: Baton Rouge, Louisiana, USA 2013; pp 1–2. http://text.lsuagcenter.com/news_archives/2013/december/headline_news/AgCentre-researchers-study-use-of-drones-monitoring.htm (accessed March 20, 2014).

Lumpkin, T. CGIAR Research Programs on Wheat and Maize: Addressing Global Hunger. International Centre for Maize and Wheat (CIMMYT), Mexico. DG's Report, 2012, pp 1–8.

Magrez, B. Magrez Embraces Drone Technology in Bordeaux. 2013, pp 1–6. http://www.wine-searcher.com/m/2013/12/drone-technology-takes-off-in-bordeaux. (accessed Sept 18, 2016).

Magri, A.; Van Es., Glos, M. A.; Cox, W. Soil Test, Aerial Image and Yield Data as Inputs for Site-Specific Fertility and Hybrid Management Under Maize. *Precis. Agric.* **2005,** *6,* 87–110.

Mamont, A. Drones Could Become Familiar Sight Over Wine Country Vineyards. 2014, p 1. http://thetruthaboutdrones.com (accessed July 6, 2015).

Matese, A.; Di Gennario S. F. Technology in Precision Viticulture: A State of the Art Review. Dovepress. 2014, pp 1–11. http://www.dovepress.com/technology-in-precision-viticulture-a-state-oftheart-review-peer-reviewed-fulltext-article-IJWR (accessed July 7, 2015).

Matese, A.; Capraro, F.; Primecerio, J.; Gualato, G.; Di Gennaro, S. F.; Agati, G. Mapping of Vine Vigour by UAV and Anthocyanin Content by a Non-Destructive Fluorescence Technique. In *Precision Agriculture*; Wageningen Academic Publishers: Lledia, Spain, 2013a; pp 201–208.

Matese, A.; Primecerio, J.; Di Gennaro, S. F.; Fiorillo, E.; Vaccari, F. P.; Genesio, L. Development and Application of an Autonomous and Flexible Unmanned Aerial Vehicle for Precision Viticulture. *Acta Hortic.* **2013b,** *978,* 63–69.

Matese, A.; Toscano, P.; Di Gennaro, S.; Genesio, L.; Vaccari, F. P.; Primecerio, J.; Belli, C.; Zaldei, A.; Bianconi, R.; Gioli, B. Inter-Comparison of UAV, Aircraft and Satellite Remote sensing platforms for Precision Viticulture. *Remote Sens.* **2015,** *7,* 2971–2990.

Mathews, A. J.; Jensen, J. L. R. Visualizing and Quantifying Canopy LAI Using an Unmanned Aerial Vehicle (UAV) Collected High Density Structure from Motion Point Cloud. *Remote Sens.* **2013,** *5,* 2164–2183.

Melchiori, R. J. M.; Kermer, A. C.; Alberenque, S. M. Unmanned Aerial System to Determine Nitrogen Status in Maize. *Proceedings of 12th International Conference on Precision Agriculture,* Sacramento, California, USA, 2014, p 18.

Microdrones GMBH. Flying Farmers-Future Agriculture: Precision Farming with Microdrones. 2014, pp 1–5. http://www.microdrones.com/en/applications/growth-market/microdrones-in-agriculture/ (accessed June 25, 2015).

Ministry of Agriculture. Beijing Applies "Helicopter" in Wheat Pest Control. Ministry of Agriculture of the Peoples of china—A Report. 2013, pp 1–3. http://english.agri.gov.cn/news/dqn/201306/t20130605_19767.htm (accessed Aug 10, 2015).

Misopolinos, L.; Zalidis, C.; Liakopoulus, V.; Stavridou, D.; Katsigiannis, P.; Alexandridiis, T. K.; Zalidis, G. Development of a UAV System for VNIR-TIR Acquisitions in Precision Agriculture. Proceedings SPIE 9535 Third international conference on Remote Sensing and Geo-information of the Environment. 2015, pp 1–7. Doi: 10.1117/12.2192660 (accessed June 26, 2015).

Modern Agriculture. Precision Agriculture Steering us in the Future. British Columbia's Agricultural magazine. 2015, pp 1–15. http://modernagriculture.ca/precision-agriculture-steering-uus-future (accessed May 15, 2015).

Moorehead, S. Precision Agriculture: Farmers See Drones in Their Future. 2015, pp 1–8. http://www.thedroneinfo.com/2015/03/22/precision-agriculture-farmers-see-drones-future/ (accessed June 20, 2015).

Mortenson, E. Forum to Examine Use Drone of Drone Technology in Agriculture. Capital Press: The Wests Ag. 2013, p 108. http://www.capitalpress.com/article/20131122/ARTICLE131129962/1318 (accessed May 16, 2015).

Mortimer, G. 'Skywalker': Aeronautical Technology to Improve Maize Yields in Zimbabwe. International Maize and Wheat Centre, Mexico, DIY drones. 2013, pp. 1–6. http;//www. ubedu/web/ub/en/menu_eines/notices/2013/04/006.html (accessed Feb, 10, 2016).

MosaicMill. EnsoMosaic Quadcopter-Complete UAV system. 2015a, pp 1–2. http://mosaicmill.com/products/complete_uav.html?gclid=CjwKEAjw96aqBRDNh (accessed July 20, 215).

MosaicMill EnsoMosaic for Plantations. MosaicMill. 2015b, pp 1–2. http://mosaicmill.com/applications/appli_plantation.html (accessed May 19, 2015).

Noble, D. Drone Scouting Demonstration is Used at Field Day. AmericanFarm.com. 2014, pp 1–2. (accessed March 10, 2016).

Nowitzki, J.; Bajwa, S. G.; Schatz, B. G.; Anderson, V.; Harnisch, W. L. Verify the Effectiveness of UAS-Mounted Sensors in Field Crop and Livestock Production Management Issues. *Proceedings of 12th International Conference on Precision Agriculture*, Sacramento, California, USA, 2014, p 41.

O'Connell, J. *Idaho Drone Project Studies Potato Stress*; Capital AG Press. 2014, pp 1–4.

Oregon State University. Drones to Check Out Acres of Potato. 2014, pp 1–2. http://www.cropandsoil.oregonstate.edu/context/drones-check-out-acres-potato (accessed July 22 2014).

Packham, C.; Davies, E. Robots to Drones, Australia Eyes High-Tech Farm Help to Grow Food. 2013, pp 1–6. http://www.euters.com/article/us-australia-farm-frobots-idUS-BRE94POE120130526.htm (accessed March 28, 2016).

Paskulin, A. Drone Plus Wine: How UAVs can Help Farmers Harvest Grapes. 3D Robotics, San Diego, California, USA. 2013, pp 1–4. http://www.3drobotics.com/2013/10/drones-wine-how-uavs-can-help-farmers-harvest-grapes/ (accessed Aug 2, 2014).

Patas, M. Drones for Farms a Challenge, but Popular Topic at Precision Ag meet. Agweek. 2014, pp 1–2. http://www.agweek.com/event/article/article/id/22532 (accessed Dec 3, 2014).

Paul, R. 2014 In: Drone could change face of No-till. Dobberstein, J. (Ed.). No-Till Farmer. 2013, pp. 1. http://www.no-tillfarmer.com/pages/Spre/SPRE-Drones-Could-Change-Face-of-No-Tilling-May-1,–2013.php, (accessed July 26, 2014).

Pauly, K. Applying Conventional Vegetation Vigor Indices to UAS-Derived Ortho-Mosaics: Issues and Considerations. *Proceedings of 12th International Conference on Precision Agriculture*, Sacramento, California, USA, 2014, p 44.

Pefia, J. S.; Torres-Sanchez, J.; Isabel de Castro, A.; Kelly, M.; Lopez-Granados, F. Weed Mapping in Early-Season Maize Fields Using Object-based Analysis of Unmanned Aerial Vehicles (UAV) Images. Institute for Sustainable Agriculture. CSIC Spain. 2013, pp 1–16. DOI: 101371/journal.pone.0077151 (accessed Aug 23, 2014).

Perry, E. M.; Brand, J.; Kant, S.; Fritzgerald, G. J. A Field Based Rapid Phenotyping with Unmanned Aerial Vehicles (UAV). 2012a, pp 1–5. http://www.regional.org.au/au/asa/2012/precision-agriculture/7933_perry.htm (accessed Aug 23, 2014).

Perry, E. M.; Fitzgerald, G. J.; Nutall, J. G.; O'Leary, M.; Schulthess, U.; Whitlock, A. Rapid Estimation of Canopy Nitrogen of Cereal Crops at Paddock Scale Using a Canopy Chlorophyll Content Index. *Field Crops Res.* **2012b**, *118*, 567–578.

Primicerio, J.; Fillipo Di Gannaro, S.; Fiolrillo, E.; Lorenzo, G.; Lugata, E.; Matese, A.; Vaccari, F. P. A Flexible Unmanned Aerial Vehicle for Precision Agriculture. 2012, DOI: 10.1007/s1119-012-9257-6 (accessed Feb 20, 2015).

Pinto R. F. Heat and Drought Adaptive QTL in a Wheat Population Designed to Minimize Confounding Agronomic Effects. *Theor. Appl. Genet.* **2010**, *121*(6), 1001–1021.

Plaven, G. Drones to Quinoa, Field Day Showcase Research Center. 2016, pp 1–3. http://www.opb.org/news/articles/dsrones-to-quinoa-field-day-show-cases-research-center (accessed April 12, 2016).

Prasad M. M. Genetic Analysis of Indirect Selection for Winter Wheat Grain Yield Using Spectral Reflectance Indices. *Crop Sci.* **2007**, *47*(4), 1416.

Pudelko, R.; Stuzynski, T.; Borzecka-Walker, M. The Suitability of Unmanned Aerial Vehicle (UAV) for the Evaluation of Experimental Fields and Crops. *Agriculture* **2012**, *99*, 431–436.

Quiroz, R. *Creating a Community in Sub-Sahara Africa to Utilize Unmanned Aerial Vehicle-based Tools for Agricultural Development.* Internal Report; International Potato Centre, Lima, Peru. 2010, pp 1–7.

Rango, A.; Laliberte, A.; Herrick, J. E.; Winters, C.; Havstad, K.; Steele, C.; Browning, D. Unmanned Aerial Vehicle-based Remote Sensing for Rangeland Assessment, Monitoring and Management. *Proceedings of Society of Photo-Optical Instrumentation Engineers.* 2009, pp 1–15. DOI: 10.1117/1.32168221.

Reynolds M. Phenotyping Approached for Physiological Breeding and Gene Discovery in Wheat. *Ann. Appl. Biol.* **2009**, *155*(3), 309–320.

RMAX. RMAX Specifications. Yamaha Motor Company, Japan. 2015, pp 1–4. http://www.max.yamaha-motor.Drone.au/specifications (accessed Sept 8, 2015).

Rovira-Mas, F.; Zhang, Q.; Reid, J. F. Creation of Three-Dimensional Crop Maps on Arial Stereo-Images. *Biosyst. Eng.* **2005**, *90*, 251–259.

Rush, C. Researchers Using UAV to Track Disease Progression in Wheat. AERC News Network. 2014, pp 1–3. http://agrilifeextension.tamu.edu/ (accessed March 10, 2016).

Ruen, J. Tiny Planes Coming to Scout Crops. Corn and Soybean Digest. 2012a, pp 1–4. http://wcorn and soybeandigest.com/corn/tiny-planes-coming-scout-crops (accessed May 29 2015).

Ruen, J. Put Crop Scouting on Auto-Pilot. Corn and Soybean Digest 2012b, pp 1–4. http://www.suasnews.com/2012/put/put-cropp-crop-scouting-on-auto-pilot (accessed April, 2016).

Saito, K.; Diack, S.; Dieng, I.; N'Diaye, M. K. On-Farm Testing of Nutrient Management Decision Support Tool for Rice in the Senegal River Valley. *Comput. Electron. Agric.* **2015**, *116*, 36–44.

Sepulcre-Canto, G.; Diago, M. P.; Balda, P.; De Toda, M.; Morales, F.; Tardaguila, J. Monitoring Vineyard Spatial Variability of Vegetative Growth and Physiological Status Using an Unmanned Aerial Vehicle (UAV). *Proceedings of 16th International Symposium of GIESCO,* Davis, California, USA, 2009, pp 1–7.

Shannon, K. Making Better Management Decisions with Precision Ag Data: Today and into the Future. University of Missouri Extension Service. 2014, pp 1–14 http://

wwwextension.missouri.edu/boone/documents/PrecisionAg/MakingBetterMgtDecisionsUsing PrecAgData_2014.pdf pp 1–58 (accessed May 23, 2015).

Schellberg, J.; Hill, M. J.; Gerhards, R.; Rothmund, M.; Braun, M. Precision Agriculture on Grassland: Applications, Perspectives and Constraints. *Eur. J. Agron.* **2008,** *29,* 59–71.

Schultz, B. *Ag Centre Researchers Study Use of Drones in Crop Monitoring.* Internal Report; Louisiana State University, Ag Centre. 2013; pp. 1–2.

SenseFly. Drones for Agriculture. SenseFly—A Parrot Inc. Chessaeux-Lousanne, Switzerland. 2015, pp 1–5. http://www.sensfly.com/applications/agricuture.html (accessed May 23, 2015).

Shafri, H. Z. M.; Hamdan, N. Hyperspectral Imagery for Mapping Disease Infection in Oil Palm Plantation Using Vegetation Indices and Red Edge Techniques. *Am. J. Appl. Sci.* **2009,** *6,* 1031–1035.

Shibata, Y.; Sasaki, R.; Toriyama, K.; Araki, K.; Asano, A.; Hirokawa, M. Development of Image Mapping Technology for Site-Specific Paddy Rice Management. *J. Jpn. Soc. Machinery.* **2002,** *64,* 127–135.

Starek, M. TAMUCC Agrilife Research Approved for Drone Based Plant Health Study. Texas A and M Agrilife, Corpus Christi, Texas, USA. 2015, pp 1–3.

Stevenson, A. Drones and the Potential for Precision Agriculture. Altech Inc. 2015, pp 1–2. http://www.altech.com/blogposts/drones-and-potential-precision-agriculture.htm (accessed June 26, 2015).

Stowell, L. J.; Gelernter, W. D. Unmanned Aerial Vehicles (Drones) for Remote Sensing in Precision Turf Grass Management International Annual Meetings of ASA-SSSA-CSSA, Tampa, Florida, USA. 2013, p 1.

Swain, K. C.; Zaman, Q. U. Rice Crop Monitoring with Unmanned Helicopter Remote Sensing Images. Remote Sensing of Biomass-Principles and Application. 2010, pp 1-253–272. http://www.intechopen.com (accessed May 10, 2014).

Tadasi, C.; Kiyoshi, M.; Shigeto, T.; Kengo, Y.; Shinichi, L.; Masami, F.; Kota, M. Monitoring Rice Growth Over a Production Region Using an Unmanned Aerial Vehicle: Preliminary Trial for Establishing a Regional Rice Strain. *Agricontrol.* **2010,** *3,* 178–183.

Tattaris, M.; Reynolds, M. Applications of an Aerial Remote Sensing Platform. Proceedings of the International TRIGO (Wheat) yield workshop. Reynolds, M., Mollero, G., Mollins, J., Braun, H. Eds.; International Maize and Wheat Centre (CIMMYT), Mexico, 2015, pp 1–5.

Taylor, D. L. Salinas-Based Drone Company Wows Tech Summit Visitors. 2015, pp. 1–3. http://www.thecaliforniaan.com/story/news/2015/03/26/salinas-based-done-company-wows-tech-summit-visitors.htm (accessed May 29, 2015).

Terraluma, Applications Selected Case Studies. 2014, pp. 1–8. http://www.terraluma.net/showcases.html (accessed Aug. 28, 2014).

Thamm, H. P. SUSI A Robust and Safe Parachute UAV with Long Flight Time and Good Payload. International Archives of Photogrammetry. *Remote Sens. Spat. Inf. Serv.* **2011,** *38,* 1–6.

Thamm, H. P; Judex, M. The Low Cost Drone—An Interesting Tool for Process Monitoring in a High Spatial and Temporal Resolution. In: International Archives of the Photogrammetry, Remote Sensing and Spatial Information Science, ISPs commission 7th Midterm Symposium. Remote Sensing: From Pixels to processes. Enschede, The Netherlands, 2006, *36,* 140–144.

Thenkabail, P. S.; Smith, R. B.; Pauw, E. D. Hyperspectral Vegetation Indices and Their Relationship with Agricultural Crop Characteristics. *Remote Sens. Environ.* **2000,** *71,* 152–182.

Thomson, S. J.; Sullivan, D. G. Crop Status Monitoring Using Multispectral and Thermal Systems for Accessible Aerial Platforms. American Society of Agricultural and Biological Engineers Paper No. 061179: 2006, pp. 1–14.

Tilling, A. K.; O'Leary, G. J.; Ferwarda, J. G.; Jones, S. D.; Fitzgerald, G. J.; Rodrigues, D.; Belford, R. Remote Sensing of Nitrogen and Water Stress in Wheat. *Field Crops Res.* **2007,** *104,* 77–85.

Torres-Sanchez, J.; Penia, J. M.; De Castro, A.I.; Lopez-Granados, F. Multispectral Mapping of the Vegetation Fraction in Early Season Wheat Fields, Using in Images from UAV. *Comput. Electron. Agric.* **2014,** *103,* 104–113.

Torres-Sanchez, J.; Lopez-Granados, F.; Pena, J. M. An Automatic Object-Based Method for Optimal Thresholding in UAV Images: Application for Vegetation Detection in Herbaceous Crops. *Comput. Electron. Agric.* **2015,** *114,* 43–52.

Townsend, S. Drones Dull the Drudgery of Aussie Farming. The Daily Telegraph. 2013, pp 1–3. http://www.thetelegrah.com.au (accessed May 10, 2015).

Trimble Inc. Soil Information System. Trimble Inc. Sunnyvale, California, USA. 2015a, pp 1–2. http://www.trimble.com/Agriclture/sis.aspx (accessed May 20, 2015).

Trimble Inc. Agronomic Services. Trimble Inc. Sunnyvale, California. 2015b, pp 1–2. http://www.trimble.com/Agriculture/agronomic-services.aspx (accessed May 20, 2015).

Trimble Inc. Nutrient Management VRA. Trimble Inc., Sunnyvale, California. 2015c, pp 1–3 (accessed May 20, 2015).

Turner, D.; Lucier, A.; Watson, C. Development of an Unmanned Aerial Vehicle (UAV) for Hyper Resolution Vineyard Mapping Based on Visible, Multi-Spectral and Thermal Imagery. 2012, pp. 1–12. http://citeseerx.ist.psu.edu/viewdoc/summary? DOI: 10.1.1.368.2491, (accessed Sept 4, 2014).

Tuttle, R. Flight Plan for UAVs. 2014, pp 1–3. http://www.precisionag.com/equipment/light-plan-foruavs (accessed Nov 3, 2014).

United Soybean Board. Agriculture Gives UAVs a New Purpose. 2015, pp. 1–2. http://www.agroprofessional.com/news/agriculture-gives-unmanned-aerial-vehicles-a-newpurpose-255380931.html (accessed June 21, 2015).

UNSCAM. Background on Appropriate Precision Farming for Enhancing the Sustainability of Rice Production. United Nations Centre for Sustainable Agricultural Mechanization (UNSCAM) and Malaysian Agricultural Research and Development Institute (MARDI), Selangor, Malaysia, 2015, pp 1–9. http://unscam.org/publication/PreRice-Farm.pdf (accessed Feb 5, 2016).

USDA. Hyperspectral and Multispectral Image Analysis of Potatoes Under Different Nutrient Inputs with Centre Pivot Irrigation. United States Department of Agriculture-Agricultural Research Service, Beltsville, USA, 2010, p 1.

Vandermause, C. DJI Launches Smart, Crop-Spraying UAV. Precision Farming Dealer. 2015, pp. 1–3. http://www.precisionfarmingdealer.com/articles/1838-dji-launches-smart-crop-spraying-UAV.htm (accessed Dec 6 2015).

Weir, J.; Herring, D. Measuring Vegetation (NDVI and RVI). 1999, Online: http://Earthobservatory.nasa.gov (accessed Nov 17, 2014).

Wihbey, J. Agricultural Drones may Change the Way we Farm: Major Innovation in Farming are at Hand, Thanks to a Rapidly Evolving Industry. Kennedy School of Government, Policy, Schoenstein Centre, Harvard, MA, USA, 2015; pp 1–3. http://www.bostonglobe.com/ideas/2015/08/2.2/agricultural-drones-change-way-farm/WTpOWMV9j4C7k-chrbmm/P4J/storey.htm (accessed April 12, 2016).

Weckler, P. R.; Maness, N. O.; Jones, R.; Jayasekara, R.; Stone, M. L.; Cruz, D.; Kersten, T.; Remote Sensing to Estimate Chlorophyll Content Using Multispectral Plant Reflectance. *Proceedings of American Society of Agricultural Engineers*, Annual Meeting, Las Vegas, Nevada, USA, 2003; pp. 1–12, Paper No. 031115.

Wilson, K. Farmers Gear Towards Robots. 2014, pp 1–12. http://www.farmweekly.com.au/news/agriculture/machinery/general-news/farmers-gear-towards-robots/2683809.aspx (accessed Nov 2, 2015).

Wood, E. GPS and Drones can Improve Tree Planting. Cross Trade. 2014, p 1 http://www.fordaq.com/fordaq/news/Elmia_drone_GPS_treeplanting_32856.hhtm (accessed May 25, 2015).

Wulfsohn, D.; Lagos, I. Z. The Use of a Multirotor and High Resolution Imaging for Precision Horticulture in Chile: An Industry. *Proceedings of 12th International Conference on Precision Agriculture,* Sacramento, California, USA, 2014, p 42.

Yintong Aviation Supplies. Agriculture Crop Protection UAV. 2014, pp 1–3. http://china-yintong.com/en/producectsow_asp?sortid=7&id=57 (accessed Sept 22, 2016).

Zaman-Allah, M.; Vergara, O.; Araua, J. L.; Trakegne, A.; Magarokosha, C.; Zarco-Tejada, P. J.; Hornero, A.; Hernandez Alba, A.; Das, B.; Crauford, P.; Olsen, M.; Prasanna, B. M.; Cairns, J. Unmanned Aerial Platform-Based Spectral Imaging for Field Phenotyping of Maize. *Plant Methods.* **2015**, *11*, 35–39. http://www.plantmetods.com/content/11/1/35 (accessed June 30, 2015).

Zarco-Tejada, P. J.; Guillen-Climent, M. L.; Hernandez-Clementi, R.; Catalina, A.; Gonzalez.; Martin, P. Estimating Leaf Carotenoid Content in Vineyards Using High Resolution Hyperspectral Imagery Acquired from an Unmanned Aerial Vehicle (UAV). *Agric. For. Meteorol.* **2013**, *172*, 281–294.

Zarco-Tejada, P. J.; Diaz-Verala, R.; Angileri, V.; Loudjani, P. Tree Height Quantification Using Very High Resolution Imagery Acquired from an Unmanned Aerial Vehicles (UAV) and Automatic 3D Photo-Reconstruction Methods. *Eur. J. Agron.* **2014**, *55*, 89–99.

Zhang X.; Shi, L.; Jia, X.; Seielstad, G.; Helgason, C. Zone Mapping Application for Precision-Farming: A Decision Support Tool for Variable Rate Application. *Int. J. Adv. Precis. Agric.* **2009**, *11*, 103–114. DOI: 10.1007/s11119-009-9130-4.

Zhang, C.; Walters, D.; Kovacs, J. M. Applications of Low Altitude Remote Sensing in Agriculture upon Farmer's Requests: A Case Study in Northeastern Ontario, Canada. *PLoS One.* **2014**, *9*, 1–10. DOI: 10.1371/journal.pone.0112894 (accessed June 28, 2015).

CHAPTER 5

DRONES IN IRRIGATION AND WATER MANAGEMENT DURING CROP PRODUCTION

CONTENTS

5.1 INTRODUCTION

Agricultural crop production depends immensely on water resources. Water received through various sources such as through precipitation, from rivers, lakes, ponds, dams and their canals contributes to crop production. It seems agricultural crops garner major share of water available on land surface. No doubt, satellite imagery is utilized regularly to assess irrigation potential of different agrarian regions. Global irrigation maps showing spatial and quantitative variations are available. For example, Aquastat, FAOSTAT and Achtinich all provide information on irrigation trends. Since past few years, India (501,020 km^2), China (460,030 km^2) and United States of America (234,934 km^2) are the top three nations in terms of cropping area under irrigation (Wood et al., 2000; Faures et al., 2010; Krishna, 2014; Siebert et al., 2006). Together, they constitute 47% of total irrigated agricultural land. To a certain extent, food grain production seems directly related to quantity of precipitation received and irrigation water at farmer's disposal. Irrigated land area is expanding based on the availability of water resources, the crop species grown and the intensity of cropping systems adopted. At this juncture, we should note that, irrigation also has an impact on extent of fertilizer usage and its efficiency. Water resources are precious and need for grains/forage is ever increasing; therefore, water-use efficiency, as a concept, is of utmost value to farmers. The soil moisture status, its fluctuations during the season, precipitation-use efficiency, availability of stored water and irrigation methodology adopted are important parameters which affect the cropping systems preferred by farmers, crop species that dominate an agrarian region and yield goals envisaged by the farmers.

Monitoring the in-season irrigation needs greater attention. Satellite imagery can provide us insights only at a larger scale, say a district, state or large agrarian belts as it lacks in resolution. Instead, drones could be considered and they could play a vital role to assess water resource and its usage in a large farm, as well as in individual farm unit with small acreage. Drones, with their ability for accurate imagery using visual, NIR and IR sensors, seem apt to effectively assess and map water bodies, mark irrigation channels, depict crop water status and analyse drought or flood effects on an entire large farm or a village. Such data is acquired by drones in a matter few minutes of flight above the fields. Drones could aid in matching crop species and its genotype with water resources. Knowledge about distribution of water resources within a large farm is essential. Based on it, 'management blocks' could be formed. Such blocking and adoption of precision techniques (variable-rate irrigation) helps in improving water-use and maximizing crop productivity. During practical crop production, farmers and companies with large cropping areas adopt several different types of irrigation methods. All of them need around-the-clock observation, particularly, when water is distributed on to fields. In this regard, drones could be effective in reducing costs on farm scouting. Periodic flights by drones with high-resolution sensors are required. The imagery derived using predetermined flight paths can help farmers and irrigation technicians. Drones actually offer excellent information about progress of irrigation. The water flow in irrigation channels and distribution of water via centre-pivot sprinklers can also be monitored. In fact, drones could be of immense utility, in monitoring and coordinating movements of sprinklers located in large farms. If the farm size is small, drones could even be used to aerially sprinkle water and nutrients, such as urea in dissolved state (foliar application). Large farms need swarms of drones. Using IR imagery, drones could be of value in monitoring water resources, surveillance of dams, lakes, irrigation canals, and mapping soil moisture variation. Drones could also help in assessing crop water stress index (CWSI), periodically. They could also be used to detect loss of irrigation water and soil nutrients via surface flow, erosion, seepage and so forth. Drones may be of use to farmers, irrespective of irrigation method adopted, such as sheet irrigation, furrow irrigation, drip irrigation, aerial sprays or centre-pivot sprinkler systems. These could also offer help to agricultural experts with data about dynamics of water in a field at various points of time during a crop season/year. The water cycle, its storage and use during crop production is generally complex. However, foremost, we have to be able to estimate moisture in the surface soil and rooting zone, then, map its variations. This step allows us to match the inherent soil moisture status, irrigation water applied and demand for water by a crop species. Crop productivity depends on the accuracy with which we match these factors. Drone's imagery may add to accuracy of judgement about irrigation requirement. Drones with ability to offer digital data about soil moisture and CWSI may actually play crucial role particularly, during precision farming that adopts variable-rate application of water.

Globally, crop production is dependent on natural precipitation pattern and supplemental irrigation. There are also areas that are predominantly irrigated. Whatever be the crop species, its demand for water and method of irrigation, bottom-line requirement is to improve water-use efficiency. We have to preserve water received as precipitation and allocate it judiciously to crops. Demand for water is directly proportional to the demand for food grains increase in future. McKee and Torres-Rua (2015) state that, we have to develop methods that can detect water requirements of crops accurately and also identify critical stages of crops that need irrigation. Then accordingly, adopt efficient methods of irrigation (e.g. precision irrigation; sub-surface drip irrigation etc.). They further state that, we have to apply irrigation water in such a way that it is not lost, to atmosphere and subsurface soil. Precision land levelling, accurate identification of soil moisture variation, use of variable-rate applicators, sprinklers, avoiding overflow of furrows and few other measures are commonly prescribed. Now, since drones are capable of aerial imagery, we should be able to accurately scout for water deficits on crops and collect digital data about water deficit by employing thermal sensors. As a consequence, water supply should become more efficient in future years. Drones, with their small tanks, could also be used to supply water at crucial stages via foliar mode. Swarms of drones could be effective in overcoming severe drought effects particularly, if drought is suspected to affect crop stand, biomass and grain formation severely. Drones could also be very useful in rapid aerial irrigation. McKee and Torres-Rua (2015) opine that over all, drones with their ability for rapid and timely detection of water needs (via thermal IR sensors) are useful. They could serve farmers and irrigation agencies remarkably well in future.

5.2   DRONES TO DETECT SOIL MOISTURE AND CROP WATER STATUS

Knowledge about moisture content in the surface soil and in the rooting zone is important. We should have an accurate idea of soil moisture and its variation prior to fixing the quantity of irrigation. Surface soil moisture (SSM) could play a vital role during crop growth and yield formation. It is a key component of soil water balance and needs to be estimated. Actually, there are several previous reports stating that, we can estimate soil moisture content, using thermal remote sensing. SSM estimates using NIR and IR cameras has often correlated with those estimated using ground instruments (Jackson, 1986; Quattrochi and Luval, 1999; Kaleita et al. 2005). In the past couple of years, Esfahani et al. (2014, 2015) have attempted to apply drone technology to estimate SSM (0.76 cm depth) using NIR and IR sensors fitted on drones. They found that spatial data from visual and IR sensors correlated with ground data ($R^2 = 7.7$–8.8). Furthermore, they have suggested that data pertaining to SSM could be utilized to adjust appropriately particularly, when fields are large and centre-pivot sprinklers are used. The data could also be utilized to feed variable-rate applicators. Use of drones to obtain SSM data is relatively less costly and offers high-resolution spectral images of crop fields (Esfahani et al., 2014, 2015; Yang et al., 2015).

Goli (2014) has conducted some novel experiments using drone technology. Drones may have relevance to soil moisture estimation in agricultural crop fields. He has tried to detect heat signatures of soils using a commercial brand UAV (drone) fitted with IR camera. The assumption here is that soil moisture affects spectral signatures of soil heat. The amount of heat reflected is proportional to changes in soil moisture. Using IR spectral data, it was possible to obtain a map depicting soil heat and moisture variations. The drone system was also attached with multispectral cameras for visual imagery. This helps in superimposing and corroborating IR imagery, as well as soil heat/soil moisture data with crop growth pattern. Goli (2014) argues that a healthy seedling is a sign of optimum soil moisture in the below ground. We can agree to the fact that crop growth is directly related to soil moisture content (i.e. soil heat signatures). The technique perhaps needs calibrations to fit a crop species and soil type. Actually, we should know the limits of soil moisture that drone imagery can detect and map. Such digitized soil moisture maps could be used to calculate soil moisture needs accurately. The soil heat/moisture maps could be digitized and used in variable-rate applicators. This way, wastage of irrigation water could eventually be minimized. Further, Goli (2014) found that there is a possibility to arrive at estimates about moisture in the rooting zones of crops. Prior knowledge about rooting depth of a crop species or genotype should be available.

At present, we have several different methods to estimate soil and crop water status. Such data will eventually help us in deciding about the amount of irrigation water to be applied. There are many techniques based on proximity of sensors to soil or crops. Plantation crops, in particular, have been monitored for soil moisture and tree water stress index, using proximal methods. Usually, such data is utilized along with others such as photosynthetically active radiation (PAR), canopy temperature, ambient temperature, humidity and so forth. Sensors mounted on different field vehicles are also used to obtain soil moisture data. They adopt proximal techniques most of which are still tedious and costly particularly, if grids are too many and sampling is intense. They are used during variable-rate application of water to crops such as pecan, grapes, almonds, walnuts, hazelnuts and so forth (Upadhyaya, 2015). Drones, with their ability for thermal imagery, could be useful and efficient, in the identification of soil moisture heterogeneity and its influence on tree crop water status. Drones may also have a role to play in monitoring irrigation sources, spray equipment and assessing tree crop water status all through the year. In general, measuring variations in tree/orchard water status is not easy (Gonzalez-Dugo et al., 2013). However, we have to realize that knowledge about spatial and temporal variations in soil moisture becomes critical when irrigation is limited. Actually, water supply has to be highly focussed and efficient. The water status of individual tree crowns may have to be clearly estimated and mapped before applying irrigation at variable rates. Drones seem apt for such situations (Sepulcre-Canto et al., 2009; Berni et al., 2009b).

In the general course, aspects such as soil sampling, estimation of moisture and mapping are time and cost consuming. It involves tedious task of obtaining large number of soil cores. We may have to measure soil moisture tension at each spot and then arrive at appropriate soil moisture maps. Many of the fruit tree crops are high value ones; therefore, water status of orchards should be measured frequently. But, it could turn out to be costly, cumbersome and time consuming to farmers. However, Gonzalez-Dugo et al. (2013) believe that, remote sensed indicators (parameters) derived using drones could be useful, particularly, in mapping water status of fruit tree plantations, such as apple, citrus, avocado and so forth. There are several indicators that could be measured to decipher soil and tree crop's moisture status. For example, canopy temperature is known to be a good indicator. It serves well in monitoring crop water status. It is a parameter that has been adopted by plantations world over, since 1970s. The CWSI is actually the difference between air temperature (T_a) and tree canopy temperature (T_c). Keeping above points in view, Gonzalez-Dugo et al. (2013) have experimented with five different tree crops using drones. They have obtained the CWSI values. They maintained tree crop units (farms) with and without irrigation to test the accuracy of drone-derived data. Tree plots were imaged using a drone, fitted with visual and thermal IR cameras. A drone was flown three times in a day to record canopy and air temperature. The difference in canopy and air temperature was well correlated with CWSI. It was also linked to stomatal activity in response to ambient air temperature and hygroscopic conditions. They have reported that ample variability was detected in the orchards. Plots with and without irrigation were clearly identified regarding water status using drone imagery. The data allows us to fix thresholds of water stress index values and associated risks. The data could be used to channel irrigation. It should be noted that skilled farm workers need several days to map the tree plantation in entirety for canopy and air temperature. Then they prepare CWSI maps. Whereas a drone flight throughout the tree plantation takes just an hour. The drone imagery is so much accurate that each individual tree could be assessed for canopy and air temperature. These changes could be marked using GPS co-ordinates.

Satellite techniques have been in vogue to assess soil heat and moisture, but, at a large scale of say several 100 ha resolution. They lack in accuracy since small farm units may be smaller than the threshold of resolution. On the other hand, ground-based methods are highly tedious. They need any number of farm scouts and regular drudgery in fields to collect soil heat signature. Drones fly closer to crops and soil, and can picturize soil surface as well as collect spectral reflectance data of crops. It can be done rather frequently, quickly and at much lower cost. Spectral data from drone's cameras are high in resolution. They are available to farmers immediately and as needed. The basic idea here is, of course, to utilize drones to obtain data about soil heat variations and moisture content. Crop producers could benefit, particularly, when they have to schedule irrigation to crops.

Stomatal regulation has direct impact on photosynthetic efficiency and water relations of crops. Actually, stomatal conductance to water vapour (g_c) and transpiration (E) are highly relevant to study plant water relations, and also to decide on quantity of irrigation. However, sampling leaves and assessing leaf temperature, stomatal conductance and gas exchange could be really tedious. Costa et al. (2013) have reviewed the relevance of remote sensing and thermal imagery in studying the plant water relationship. They have stated that thermal imagery has been successfully utilized to record canopy and air temperature. It is suitable in wide range of environmental conditions. Remote sensing assesses CWSI rapidly using thermal cameras (NIR, IR). It gives an idea about plant water status. Such data in digital form can be used by variable-rate irrigation water suppliers.

Walker (2014) suggests that some of the drone systems produced recently and those meant for agricultural purposes are excellent in identifying crop water stress. The spatial variation in crop stress measured via CWSI ($T_a - T_c$) is highly relevant to farmers, particularly, when they try to supply water using precision techniques. Drones can cover an area of 1000 ac in a matter of few hours. This saves high costs incurred due to employing, a large number of skilled farm scouts.

Reports suggest that sensors placed on ground vehicles are being evaluated, standardized and in some cases effectively utilized (Upadhyaya, 2014). On the other hand, the major lacunae with satellite-guided remote sensing are its low resolution and uncongenial turn-around time. However, results obtained with low flying drone are applicable everywhere. Hence, drones are preferred to monitor crop stress, study the symptoms and accrue pertinent data (Berni et al. 2009a). Measuring CWSI is an important aspect that affects irrigation schedules and the final water use efficiency. Usually, crop canopy and ambient temperatures are computed and CWSI is calculated. Farmers then react with appropriate irrigation levels. Bellvert et al. (2013) calculated CWSI as follows:

$$\mathrm{CWSI} = (T_c - T_a) - (T_c - T_a)_{\mathrm{LL}} / (T_c - T_a)_{\mathrm{UL}} - (T_c - T_a)$$

Where, T_c is canopy temperature; T_a is air temperature; LL is lower limit of $(T_c - T_a)$ values and UL is upper limit of $(T_c - T_a)$ values.

Innova (2009) states that Andalusian scientists, from the Institute for Sustainable Agriculture and University of Cordoba, Spain, are highlighting the benefits of drone imagery. They are campaigning for adoption of thermal IR spectral data derived using drones (remote sensing). It is in order to estimate crop water status. In other words, they are asking farmers to estimate CWSI, using canopy and air temperatures and calculate crop's water needs accordingly. Let us consider an example dealing with grape vines. Identification of spatial and temporal variability of crop water status in grape orchards is an essential step, if irrigation has to be applied judiciously and efficiently. It ensures that yield goals envisaged are attained. Knowledge about CWSI throughout the season is almost mandatory in any plantation. Bellvert et al. (2013) have explored the possibility of measuring CWSI within grape vine yards located at Raimat, Lleida, Spain. A thermal sensor (Miricle 307 by Thermoteknix Systems ltd, Cambridge, UK) was installed into a drone. It was used to conduct remote sensing of canopy and soil temperatures. They have actually aimed at applying precision techniques (variable applicators) to grape orchards, using multispectral and thermal data. They measured grape canopy temperatures and ambient temperatures in the farm. Later, they tried to compare it with ground-based estimates of leaf water potential (Ψ). Incidentally, there are reports indicating that CWSI and leaf water potential do correlate with drone-derived data (see Acevedo-Opazo et al. 2008a, 2008b; Bellvert et al. 2013; Moller et al. 2007; Berni et al. 2009a, 2009b). In the above case, Bellvert et al. (2013) noticed that $T_c - T_a$ was related to leaf water potential (Ψ) through an equation $y = -6.266 \times -9.156$ at $R^2 = 0.46$, if imagery was obtained at 9.30 a.m. in the morning. However, data on Tc−Ta obtained at 12.30 p.m. correlated highly ($R^2 = 0.71$) and was related through an equation $y = -7.425 \times -5.815$. They have suggested that correlation of drone-derived data of CWSI with ground data was affected, by the diurnal time when the thermal imaging was conducted. At early morning hours, up to 07.00 h, thermal images were not able to distinguish soil and crop canopy temperatures with much accuracy. However, when canopy temperatures were measured at 12.30–14.00 h, soil factors that affect thermal data could be easily detected and eliminated, while computing canopy temperature. It helps in developing accurate digital maps showing variations in canopy temperature. Therefore, data derived from thermal cameras on drones in the afternoon correlated best with leaf water potential. Thermal imagery obtained at 0.3 megapixel using drone-based cameras showed significant correlation ($R^2 = 0.71$) with leaf water potential compared to those measured at 0.6–2.0 megapixel ($R^2 = 0.29$). The crop water needs could then be calculated and applied. Since, digital data showed variations in CWSI for the entire field, variable-rate applicators on sprinklers could utilize such data.

Ortega-Farias (2012) has evaluated drones for their ability to obtain high-resolution multispectral imagery including at thermal IR bandwidth. It was done in order to identify variability of water status. The correlation of CWSI with stomatal conductance (g_s) and stem water potential (Ψ) was studied. The thermal indices obtained from aerial imagery correlated with vineyard water status. Further, they tried to corroborate thermal indices data with visual imagery, the NDVI and leaf chlorophyll index. In summary, thermal and visual imagery helps in assessing and mapping spatial variation of water status.

5.2.1 DRONES TO MONITOR IRRIGATION IN FIELDS

A wide range of drone models belonging to both flat-winged and rotor (copters) types have been tested, tried and adopted, to obtain digital imagery using appropriate sensors. In particular, let us list a few examples, where in, drone models have been utilized to assess soil moisture, CWSI and a few other parameters relevant to crop productivity. Valencia et al. (2008) employed flat-winged drone with a wingspan of 2.5 m. This drone has a flight endurance of 45 min. During the flight, this drone could get digital data and images of soil moisture variation, soil salinity affected locations and crop growth. The data retrieval and processing was possible instantaneously.

Now, let us consider roto-copters used to assess soil moisture/crop water status. Turner et al. (2011) reported adoption of electric battery powered multi-copter drone, to map vineyards. They used multispectral and IR thermal sensors to map CWSI in grape orchards. Archer et al. (2004) experimented with 'Auto-copter-xl', which is a small drone capable of rapid imagery, using visual and thermal sensors. The drone system, including ground station computers and ortho-mosaic processing facility, was utilized to estimate soil moisture of crop fields. Imaging was done prior to sowing and then during in-season to assess fluctuations in CWSI. Tsouvaltsidis et al. (2015) have utilized low-cost copter with vertical take-off and landing (VTOL), a landing gear for smooth touchdown and a power source made of electric batteries. The copter has 10–15 min endurance. It had a payload of sensors to assess soil moisture and CWSI. MacArthur et al. (2014) suggest that we can adopt rotary copter drones fitted with VNIR, IR sensors (e.g. QE, Ocean Optics) that operate, at 400–1000 and 700–1200 nm bandwidth. Using such a system, it is possible to obtain thermal imagery and assess soil and crop water status. Practically, it should be possible to obtain thermal imagery of soil and crops using a simple drone fitted with a range of sensors such as visual, NIR and IR thermal bandwidth. Visual images are required to superimpose and mark the regions that need irrigation. In fact, Gago et al. (2015) have clearly mentioned that data derived from IR and NIR sensors on drones, show positive correlations with water potential (Ψ) and stomatal conductance (g_s). Therefore, thermal imagery is becoming a common remote sensing technique that helps to assess CWSI. However, thermal indices difference between canopy of crops and air need to be estimated, accurately. Often, drone imagery developed using visual bandwidths are superimposed to understand the effects of drought on growth and grain productivity. Generally, irrigation scheduling could be more accurate, if thermal images are consulted periodically.

Reports and reviews by National Centre for Engineering in Agriculture, Australia suggest that precision irrigation is still in the early stages of usage, in Australia. However, promotion and propaganda to adopt precision irrigation is increasing. Precision instruments such as variable applicators, centre-pivot sprinklers and computer-based decision-support systems are being popularized (NCEA, 2015). At the same time, drones with visual and IR thermal sensors are being touted to assess soil moisture variation and CWSI. Both, flat-winged and copter drones are being evaluated. Drones are supposed to help farmers and irrigation agencies with real-time data about crop water stress. Consequently, need for irrigation and its timing could be decided.

Zhang et al. (2014) have explained the advantages of using drones to monitor irrigation equipment and tile drainage system. They state that in Ontario (Canada), fields are clayey and flat leading to stagnation of water and loss of crop yield. Crop loss from excess water has to be avoided. Hence, monitoring the fields regularly for stagnation of water is needed. The tile drainage system has to be routinely monitored and corrected if soil is to be kept well drained. Drones supposedly help in identifying patches of a field prone to stagnation of irrigation water and they also help in monitoring tile drainage system and its functioning.

5.3   DRONES IN EXPERIMENTAL EVALUATION OF CROPS FOR THEIR RESPONSE TO IRRIGATION

As stated earlier in Chapter 4 on 'Production Agronomy', Tataris and Reynolds (2015) too have opined that drones with their ability for crop phenotyping, regular estimation of canopy temperature and NDVI could be effective instruments, in experimental farms. Drones could be used to assess and evaluate crop genotypes in experimental plots. The genotypes of cereals such as wheat and maize could be evaluated using canopy temperature and NDVI values. A preliminary evaluation in the experimental plots of International Maize and Wheat Centre (CIMMYT), at Obregon in Mexico has shown that canopy temperature measured using drones and that from ground measurements correlate ($R^2 = 0.71$). Similarly, NDVI derived from drone and ground measurements are also well correlated ($R^2 = 91$). Hence, drones could detect differences in genotype performance (Tattaris and Reynolds, 2015). Thermal imagery obtained using IR cameras placed on drones could be used effectively, to estimate water content of natural vegetation and crops. Drones that fly very close to crop canopy can relay data pertaining to a field crop or tree using IR cameras with GPS tags on each image. In a field, IR images depict variation in plant water status. The data about CWSI is useful as it helps in judging water needs of a crop. Such data can be utilized by variable-rate applicators during precision farming procedures. Labbe et al. (2012) state that, thermal IR imagery obtained using drones helps in monitoring irrigation. The thermal imagery has to be coupled with images from visual band cameras and superimposed. It then provides a correct idea about variations in water status. The data has to be calibrated and corrected for interference from atmosphere. The atmosphere between crop canopy and sensors may alter reflectance measurements. Hence, a correction factor has to be introduced (Labbe et al., 2007; Lebourgeois et al., 2008).

5.3.1   IDENTIFYING DROUGHT TOLERANT GENOTYPES

Drones are gaining acceptability in the various experimental techniques that agricultural researchers adopt. In particular, drones are finding a role in evaluation of large number of genotypes for characteristics such as NDVI, water stress index and drought tolerance (Berni et al. 2009a, 2009b). Drones are opted because they are relatively low-cost instruments. They could be used repeatedly in experimental farms. They are used to obtain data about crop canopy temperature and water stress index. A large number of genotypes could be assessed for drought and heat tolerance, in one go. Berni et al. (2009a) have reported that a thermal camera such as 'Thermovision A40M' equipped with field of view lens or 'ThermoMAP' along with multiSPEC 4C (Canon cameras) sufficed, to provide thermal imagery of crop genotypes. Simultaneously, multispectral imagery of the same crop species (or genotypes) too could be obtained, superimposed and corroborated. Multispectral imagery aids in analysing and confirming the crop genotype's reaction to heat and drought stress. Simple fact is that drought/heat stress results in relatively sparse growth and small canopy; whereas, crop canopy and growth are optimum, if soil water availability is optimum. Irrigation could be channelled accordingly, using thermal imagery as a guide. Drones cost less to conduct experimental evaluation of crop genotypes for drought tolerance.

According to Berni et al. (2009a), one of the major applications of drones in plant breeding is their ability to obtain data about canopy temperature. Drones are rapid and data could be obtained repeatedly. In future, plant breeders intending to screen crop genotypes for drought stress could make use of drones. They may find it easy to identify genotypes that tolerate drought and heat stress. Cultivars that avoid drought effects through drought escape methods could also be identified.

As mentioned earlier, Costa et al. (2013) have stated that stomatal conductance and evapo-transpiration are key factors affecting plant water relationships, in different environments. Thermal imagery is a technique that helps to study and quantify CWSI. Data on CWSI could be used to classify genotypes. In addition, Costa et al. (2013) opine that in future, thermal imagery can be of use to identify heat tolerant and drought tolerant crop genotypes. We can study the vulnerability of crop species and their genotypes to extended periods of water scarcity. Thermal imagery can also be an excellent tool to study crop water status and forecast irrigation requirements. In addition, as a matter of routine, we can record thermal images of crop genotypes in an experimental plot. It helps to identify/ classify drought tolerant and susceptible genotypes. Of course, we can also conduct well directed multilocational experimental trials, to evaluate crop genotypes using thermal imagery. Data derived using drones can be quickly retrieved, transmitted and exchanged electronically.

McCabe (2014) reported that data about genetically advanced corn lines could be obtained regularly and evaluated. In the experimental plots, corn lines that tolerate drought/heat stress could be judged accurately using data obtained by sensors. Crop breeders in Nebraska, USA have utilized an octocopter, to obtain data such as NDVI, chlorophyll content, leaf area, leaf and canopy temperature and ambient air temperature. Drones are flown 4–5 times in a season and corn genotypes are ranked based on their ability to tolerate drought and heat stress. Such data from drones also helps breeders to identify critical stages for each genotype when irrigation has to be mandatorily applied. Genotypes that avoid or overcome intermittent droughts better could also be selected.

Drones fly close to crop canopy; therefore, they can be utilized effectively to compare as many crop species and their genotypes grown in a field. The water status of each individual genotype can also be measured and compared. Consequently, 'drought tolerant genotypes' could be identified with better accuracy. Further, the performance of crop genotypes exposed to different quantities of irrigation can be also studied using drones. Gago et al. (2015) have clearly stated that rapid phenotyping and estimation of thermal indices of crop canopy of different genotypes is useful. It can help us in identifying genotypes that show better performance under drought stress.

5.4   DRONES TO MONITOR WATER RESOURCES AND IRRIGATION EQUIPMENT

Agricultural drones are used to monitor the source and storage facilities of irrigation water, meant for crop production (Achtilik, 2015; Plate 5.1). Reservoirs meant for irrigation need 24-h surveillance. Water storage levels

PLATE 5.1   An Agricultural drone inspecting a water reservoir in Germany.
Note: Regular surveillance and maintenance of water reservoirs are essential aspects of irrigation in an agricultural region.
Source: Dr. Micheal Achtilik, Ascending Technologies, (http://www.astec.de/en/ January 24, 2016); and Prof. Guido Morgenthal, Technologien im Bauwesen, Germany.

are to be monitored. Similarly, depletion of water level in the reservoirs and canals need regular monitoring. It is done with greater accuracy and at any time of the year, using drones. Drones can fly over the water bodies to check for floating weeds, dam condition and so forth. They can keep a watch on irrigation channels that act as conduit to crop fields. Clogging of channels, undue seepage loss of water or spillage into barren regions could be identified instantaneously, if drones are used to surveillance the dams (Plate 5.1). In fact, we can regulate water flow from dams using drone imagery. Agricultural drones have been adopted to sample the dam/lake water. McCabe (2014) reports that drones could regularly surveil the water bodies used for irrigation of crops. They can also pick water samples for analysis of nutrients and contaminants, if any (Detweiler and Elbaum, 2013).

We know that drones are useful in performing tasks such as aerial imaging, mapping, monitoring and keeping a vigil on crop fields. They also help in assessment of soil fertility and crop's nutritional status. They can also spray fertilizers. In addition, drones have a clear role to play in irrigation of crops, particularly, as precision farming methods get more common across different agrarian regions. Drones are efficient in aerial

PLATE 5.2   A cereal field being irrigated using centre-pivot sprinklers.
Note: Such fields are required to be monitored periodically for accurate distribution of water through variable-rate sprinklers. Clogging or excessive leakage of water from the system could be identified, using close-up pictures obtained by drones. Drones can also monitor movement of sprinkler set-up in the field.

surveys and mapping field topography. They offer three-dimensional (3D) pictures of entire farm in one go to the farmer. This is something impossible using just a few farm workers or skilled technicians. Drones, therefore, help farmers in accurate designing and layout of irrigation channels (Microdrones, 2015).

Drones are excellent flying machines in farms. When fitted with sensors, they can monitor crops and their water status. Mapping of the areas that need irrigation and those that need to be drained could also be done. Drones are apt to monitor and watch the irrigation equipment and their functioning during irrigation (Grassi, 2014). Drone images reveal the state of functioning of centre-pivot irrigation equipment and their movement, in the large fields. They show up nozzles that are in working condition and those that need repair (Plate 5.2). We have to note that monitoring irrigation equipment and their functioning in a large farm is tedious and laborious. It involves walking and scouting for many hours. Maps drawn later could be less accurate. Skilled farm workers who scout the crop for irrigation requirement cost much higher than a drone; whereas, drones accomplish these tasks in a matter of 15 min to 1 h for 200 ac of land. Just the fact that drones are quick and reduce human drudgery seems to suffice for their deployment in large farms.

Reports suggest that large farms usually have 5–20 different sections, each section being irrigated by one or two centre-pivot sprinkler equipment.

Sprinklers apply irrigation water either adopting precision techniques or using blanket prescription. In either case, it is a vast area to surveil using ground vehicles. Watching each centre-pivot system for clogged nozzles and accurate movement in the field is a tedious task. They say it is difficult even to watch the tall centre-pivot sprinkler system located in a maize field. On the contrary, a drone flying at a vantage height above the maize canopy provides detailed picture of all the nozzles and helps in locating clogged nozzles. Therefore, we attend to only areas that specifically need correction (Heck, 2016; Plate 5.2).

In Australia, drones have been adopted to design irrigation systems. Drones' imagery showing topography and 3D relief of farm land allows farmers, to design and lay irrigation pipes/channels accordingly. In addition, once the crop is established, drone imagery is also utilized to monitor irrigation. In particular, zones prone to drought or flooding could be detected and corrected (Rennie and Chelard, 2016). In fact, geo-referenced 3D data obtained using drones helps in identifying drought affected and flooded locations prominently. Drone images could be obtained quickly since turn-around time is short (Peek Drone Ltd, 2016). Such 3D maps of fields could also be procured from private agricultural consultancy agency that operates drones. Such imagery is provided prior to sowing a crop, particularly, to lay irrigation lines and to arrange for proper drainage. This facility is needed, if flooding or stagnation is suspected immediately after a rainfall event.

KEYWORDS

- **crop water status**
- **soil moisture**
- **irrigation**
- **drought-tolerant genotypes**
- **stomatal conductance**

REFERENCES

Acevedo-Opazo, C.; Tisseyere, B.; Guillame, S.; Ojeda, H. The Potential of High Resolution Information to Define Within Vine Yard Zones Related to Vine Water Status. *Precis. Agric.* **2008a,** *9,* 285–302.

Acevedo-Opera, C.; Tessier, Ojeda, H.; Ortega-Farias, S.; Guillemets. It is Possible to Estimate the Spatial Variability of Vine Water Status. *Journal International des Sciences de la Vigne et du Vin* **2008b**, *42*, 203–291.

Archer, F.; Shutko, A. M.; Coleman, T. L.; Haldin, A.; Novichikhin, E.; Sidorov, I. Introduction, Overview, and Status of the Microwave Autonomous Copter System (MACS). *Geoscience and Remote Sensing Symposium, 2004. IGARSS'04, Proceedings. of 2004 IEEE International*. 2004, 5, 3574–3576.

Achtilik, M. UAV Inspection and Survey of Germany's Highest Dam. Ascending Technologies. 2015, pp 1–3. http://www.asctec.de/en/uav-inspection-survey-of-germanys-highest-dam. (accessed June 26, 2016).

Bellvert, J.; Zarco-Tejada, P. J.; Girona, J.; Fereres, E. Mapping Crop Water Stress Index in a 'Pinot-noir Vineyard: Comparing Ground Measurements with Thermal Remote Sensing Imagery from an Unmanned Aerial Vehicles. Precision Agriculture 2013, pp 1–6. DOI 10.1007/s11119-013-9334-5 (accessed April 10, 2016).

Berni, J. A.; Zarco-Tejada, P. J.; Suarez, L.; Fereres, E. Thermal and Narrowband Multispectral Remote Sensing for Vegetation Monitoring from an Unmanned Aerial Vehicle. IEEE Xplore 2009a, pp 722–738. DOI: 10.1109/tgrs.20082010457 (accessed Jan 20, 2016).

Berni, J. A. J.; Zarco-Tejada, P. J.; Sepulcre-Canto, G.; Fereres, E.; Villalobos, F. Mapping Canopy Conductance and CWSI in Olive Orchards using High Resolution Thermal Remote Sensing. *Remote Sens. Environ.* **2009b**, *113*, 2380–2388.

Costa, J. M.; Grant, O.; Chaves, M. Thermography to Explore Plant-Environment Interactions. *J. Exp. Bot. 2013*. DOI: 10.1093/jxb/ert029 (accessed August 29, 2016).

Detweiler, C.; Elbaum, S. UNL Developing Water-Collecting Drones for Tests, Remote Locales-UNL Crop Watch September 4, 2013-Archives. University of Nebraska-Lincoln Crop Watch. 2013, pp 1–4. http://cropwatch.unl.edu/archive/-asset_publisher/VHeSpt-v0Agju/content/unl-developmeng...html (accessed Oct 9, 2015).

Esfahani, H.; Torres-Rua, L; Jensen, A.; McKee, M. Opsoil Moisture Estimation for Precision Agriculture using Unmanned Arial Vehicle multispectral imagery. *Proceedings of the IEEE International Geoscience and Remote Sensing Symposium (IGARSS)*. 2014, pp 1–7 http://aggieair.usu.edu/node/292 (accessed Jan 28, 2016).

Esfahani, H.; Torres-Rua, A.; Jensen, A.; MacKee, M. Assessment of Surface Soil Moisture using High Resolution Multispectral Imagery and Artificial Neural Networks. *Remote Sens.* **2015**, *7*, 2627–2646.

Faures, J. M.; Hoogeveen, J.; Bruinsma, J. The FAO Irrigated Area Forecast for 2030, Land and Water 2010, pp 1–4 ftp://ftp.fao.org/agl/aglw/docs/fauresetalagadir.pdf (accessed April 3, 2014).

Gago, J.; Douthe, R. E.; Coopmen, R. E.; Gallego, M.; Ribas-Carbo, J.; Flexas, J.; Madrano, H. UAVs Challenge to Assess Water Stress for Sustainable Agriculture. *Agric. Water Manage.* **2015,** *153*, 9–19.

Goli, N. Researcher uses Drones to Measure Soil Moisture. University of Alabama at Huntsville: USA, 2014; pp 1. http://www.chargetimes.com/2092/science-and-technology/researcher-uses-drones-to-mesure-soil-moisture.html (accessed January 26, 2016).

Gonzalez-Dugo, V.; Zarco-Tejada, P.; Nicola's, E.; Nortes, P. A.; Alarco, J. J.; Intrigliolo, S. S.; Fereres, E. Using High Resolution Thermal Imagery to Assess the Variability in the Water Status of Five Fruit Tree Species within a Commercial Orchard. *Precis. Agric.* DOI; 10.1007/s11119-013-9322-9.

Grassi, M. 5 Actual uses for Drones in Precision Agriculture Today. *Drone Life.* 2014, pp 1–4 http://dronelife.com/2014/12/30/5-actual-uses-drones-precision-agriculture-today/ (accessed May 13, 2015).

Heck, K. Is that a Hummingbird Outside My Window. Heck Land Company. 2016, pp 1–2 http://hecklandco.com/agricultural-commodity/helicopter-drone/ (accessed April 28, 2016).

Innova, A. Estimating Crop Water Needs Using Unmanned Aerial Vehicles. Science Daily 2009 https://www.sciencedaily.com/releases/2009/07/090707094702.htm*? (accessed April 25, 2016).*

Jackson, T. Soil Water Modeling and Remote Sensing. *IEEE Trans. Geosci. Remote Sens.* **1986,** *24,* 37–46.

Kaleita, A.; Tian, L.; Hirschi, M. Relationship Between Soil Moisture Content and Soil Surface Reflectance. *Trans. Am. Soc. Agri. Eng.* **2005,** *48,* 1979–1986.

Krishna, K. R. *Agroecosystems: Soils, Climate, Crops, Nutrient Dynamics and Productivity.* Apple Academic Press Inc.: Waretown, New Jersey, USA, 2014; pp 459–478.

Labbe, S.; Roux, B.; Lebourgeois, V; Begue, A. An Operational Solution to Acquire Multi-Spectral Images with Standard Light Cameras: Spectral Characterization and Acquisition Guidelines. In: *ISPRS Workshop on Airborne Digital Photo-grammetric Sensor Systems.* New Castle: England, 2007; pp 23–28.

Labbe, M.; Lebourgeois, V.; Jolivot, A.; Marti, R. Thermal Infra-Red Remote Sensing for Water Stress Estimation in Agriculture. In *The Use of Remote Sensing and Geographic Information System for Irrigation Management in Southwest Europe;* Erena, M., Lopez-Francos, Montesinos, S., Berthoumieu, J. P., Eds.; Zaragoza: Europe. CHEAM. Options Mediterranean's. Series B 2012, pp 175–184.

Lebourgeois, V.; Labbe, S.; Begue, A; Jacob, F. Atmospheric Corrections of Low Altitude Thermal Airborne Images Acquired Over a Tropical Cropped Area. *IEEE International Geoscience and Remote Sensing Symposium,* Boston, Massachusetts, USA 2008, pp 1–18.

MacArthur, A.; Robinson, I.; Rossini, M.; Davis, N.; MacDonald, K. A Dual-Field-of-view Spectrometer System for Reflectance and Fluorescence Measurements (Piccolo Doppio) and Correction of Etaloning. 5th International Workshop on Remote Sensing of Vegetation Fluorescence. Quoted in: Tsouvaltsidis, C., Zaid Salem, N., Bonari, G. Vrekalic, D., Quine, B. 2015 Remote spectral imaging using low cost UAV system. *International Archives of the Photogrammetry, Remote Sensing and Spatial Information Sciences* 2014, XL, 25–34.

McCabe, D. Sky is the Limit for UAVs in Agriculture. *Nebraska Farmer* **2014,** *75,* 34.

McKee, M; Torres-Rua, A. Technologies will Tackle Irrigation Inefficiencies in Agriculture's Drier Future. The conversation. 2015, pp 1–4 http://theconvrsation.com/technologies-will-tackle-irrigation-inefficiencies-in-agriculture-drier-future-40601 (accessed June 27, 2015).

Microdrones. Flying Farmers-Future Agriculture: Precision Farming with Microdrones. 2015, pp 1–7. http://www.microdrones.com/en/applications/growth-markets/microdrones-in-agriculture. (accessed May 30, 2015).

Moller, M.; Alchanatis, V.; Cohen, Y.; Meron, M.; Tsipiris, J.; Naor, A. Use of Thermal and Visible Imagery for Estimating Crop Water Status of Irrigated Grape Vine. *J. Exp. Bot.* **2007,** *58,* 827–838.

NCEA. Review of Precision Irrigation Technology and Their Application. National Centre for Engineering in Agriculture, Australia. 2015, pp 1–5. http://www.ncea.org.au/index. php?option=com_content&task=view&id=298&Itemd.htm (accessed June 3, 2015).

Ortega-Farias, S. Assessment of Vineyard Water Status Variability by Thermal and Multi-spectral Imagery Using an Unmanned Aerial Vehicle (UAV). *Irrig. Sci.* **2012**, *30*, 511–522.

Peek Drone Ltd. Drone UAV Services by Peek. 2016, pp 1–4. http://peekdrone.com/ndvi. php (accessed April 28, 2016).

Quattrochi, D.; Luvall, J. Thermal Infrared Remote Sensing for Analysis of Landscape Ecological Processes: Methods and Applications. *Landscape Ecol.* **1999**, *14*, 577–598.

Rennie, J.; Chelard, H. The Use of Unmanned Aerial Vehicles (UAVs) in Agriculture in Regulation, Challenges and Opportunities. Grain Research and Development Corpora-tion, Australian Government. 2016, pp 1–3. http://grdc.com.ac/researchandDeelopment/ GRDC-update-paper/2015/02/thhe-use-of-unmanned-aerial-vehicles (accessed April 28, 2016).

Siebert, S.; Hoogenveen, S.; Frenken. K. Irrigation in Africa, Europe, Latin America. Update of Global Digital Map. Food and Agricultural Organization of the United Nations. Rome, Italy Frankfurt Hydrology Paper No. 05: pp 137.

Sepulcre-Canto, G.; Diago, M. P.; Balda, P.; De Toda, M.; Morales, F.; Tardaguila, J. Moni-toring Vineyard Spatial Variability of Vegetative Growth and Physiological Status, using an Unmanned Aerial Vehicle (UAV). *Proceedings of 16th International Symposium of GIESCO,* Davis, California, USA, 2009, pp 1–7.

Tattaris, M.; Reynolds, M. Applications of an Aerial Remote Sensing Platform. *Proceed-ings of the International TRIGO (Wheat) Yield Workshop.* Reynolds, M., Mollero, G., Mollins, J., Braun, H., Eds.; International Maize and Wheat Centre (CIMMYT), Mexico, 2015; pp 1–5.

Turner, D.; Lucieer, A.; Watson, C. Development of an Unmanned Aerial Vehicle (UAV) for Hyper Resolution Vineyard Mapping Based On Visible, Multispectral, and Thermal Imagery. *Proceedings of 34th International Symposium on Remote Sensing of Environ-ment.* 2011; pp 342–347.

Tsouvaltsidis, C.; Zaid Salem, N.; Bonari, G.; Vrekalic, D.; Quine, B. Remote spectral imaging using low cost UAV system. *International Archives of the Photogrammetry, Remote Sensing and Spatial Information Sciences* 2015, XL, 25–34.

Upadhyaya, S. K. Precision Canopy and Water Management of Speciality Crops Through Sensor Based Decisions. University of California, Davis, California, USA. 2015, pp 1–8   http://portal.nifa.usda.gov/web/crisprojectpages/0222356-precision-canopy-and-water../ (accessed June 3, 2015).

Valencia, E.; Acevo, R.; Bosch-Lluis, X.; Aguasca, A.; Rodriiguez-Alvarez, N.; Ramos-Perez, I.; Camps, A. Initial Results of an Airborne Light-Weight L-Band Radiometer. *Geoscience and Remote Sensing Symposium, 2008. IGARSS 2008. IEEE International* 2008, 2, 1170–1176.

Walker, M. Oregon Company Nabs Funding for Water Saving Drones. 2014, pp 1–4. http:// www.bizjournals.com/portals/blog/sbo/2014/03/oregon-company-nabs-funding-for. html (accessed April 21, 2016).

Wood, S.; Sebastian, K.; Scherr, S. J. A Pilot Analysis of Global Ecosystems. Internationals Food Policy Research Institute: Washington D.C. USA, 2000; pp 185.

Yang, Y.; Guan H.; Long D.; Liu, B.; Qin, G.; Qin, J.; Batelaan, O. Estimation of Surface Soil Moisture from Thermal Infrared Remote Sensing Using an Improved Trapezoid. *Remote Sens.* **2015,** *7*, 8250–8270; DOI: 10.3390/rs70708250; www.mdpi.com/journal/ remotesensing (accessed April 23, 2016).

Zhang, C.; Walters, D.; Kovacs, J. M. Application of Low Altitude Remote Sensing in Agriculture Upon Farmer's Request—A Case Study in Northeastern Ontario, Canada. *PLoS One* **2014,** 1–9. http://journals.plos.org/plosone/article?id=10.137/jurnal.pone.0112894 (accessed June 25, 2015).

CHAPTER 6

DRONES TO MANAGE WEEDS IN CROP FIELDS

CONTENTS

6.1 INTRODUCTION

Weeds are endemic to agrarian regions worldwide. In nature, including within crop fields, weeds express great adaptability and diversity. They affect cropping belts in many ways. The type and extent of detriment caused to the main crop and the farming enterprise may vary. It depends on several factors related to geography, soil, weather, weed species, cropping systems adopted, agronomic procedures, farm equipment employed and so forth. Weeds occur at different intensities. Hence, their influence on crop production strategies and production efficiency attained also varies proportionately.

In the agrarian belts, both natural and man-made factors affect weed flora, its intensity and the crop productivity. In the present context, our focus is on methods utilized by farmers, to eradicate or regulate weed population in the crop fields. Ploughing is the basic method that reduces weeds in the crop fields. Through the ages, both light and deep ploughing have helped farmers in digging, disturbing and turning the soil, so that weed seeds are destroyed and their sprouting is suppressed. Nowadays farmers adopted application of pre- and post-emergent chemicals known as 'herbicides', to suppress weed growth and proliferation. Otherwise, hand weeding periodically in short intervals was the main method. It is still in use in many areas. Weeds compete with crop plants for resources such as anchorage, soil moisture, nutrients, photosynthetic light, canopy space and so forth. Weeds may become collateral hosts for disease-causing agents or pests that affect crop productivity. Weed control measures often aim at their eradication at the earliest stages, beginning with the suppression of germination of weed seeds, uprooting at early seedling stage and curbing growth. Hence, rapid and accurate weed detection is an essential process. Farmers often employ skilled scouts to inspect the fields and record the types of weeds, their distribution and intensity all across the large fields. This is a time-consuming process. It needs good skills in collecting correct data and mapping weed distribution. Often, weed scouting and preparing maps for variable-rate herbicide applicators could be costly, particularly if human scouts are employed. Research personnel at agricultural research stations and farm companies have consistently tested several different pre- and post-emergence herbicides to control the weeds. Innumerable books, journals, research papers, farm manuals and information brochures are available that tell us about herbicides and their influence on weeds. A long list of farm machinery that is utilized to eradicate weeds can be traced in any agrarian location. Most, if not all, farm stores and industries right now show up ground vehicles, sprayer types and herbicide chemicals with a range of active ingredients. They prominently show up efficacy, target weeds and productivity gains.

Adoption of herbicides to curb weeds became a prominent agronomic procedure during mid-1900s. There was a rapid increase in herbicide application to cereal crop fields in the United States of America between 1950s and 1960s. This phenomenon actually reduced cost on human farm workers, hired mainly to pull weeds and burn them. The cost of production of crops reduced significantly. At present, herbicide sprays are common in agrarian regions that practise intensive crop production tactics. A really wide range of herbicides are used by farmers worldwide (Syngenta, 2015). In the United States of America, for example, about 87 million·ha of crop land is sprayed with nonselective or selective herbicides. According to reports by USDA-ERS (2010), scouting fields for weeds and adoption of control measures using post-mergence herbicides has increased, since 1996. Such increases in herbicide usage were marked on corn, maize, cotton, soybeans and winter wheat crops. In particular, soybean and cotton fields were sprayed with higher dosages. It is attributable to the advent of herbicide-tolerant crop genotypes. Winter wheat fields received higher quantities of pre-emergent herbicides. Application of broad-spectral herbicides such as glyphosate and atrazine increased to 85 million kg in the United States of America in 2007. In the European cropping belts also, weeds are among major grain yield retardants. Reports suggest that each year, about 87 billion·€ are lost due to weed infestation of cereal crop production zones in the European plains (Bayer Crop Science, 2015). Weeds are equally, if not more, detrimental to crop yield in the intensive cropping zones of South Asia and Far East.

Farmers have accepted genetically modified (GM) crops resistant to herbicides (e.g. [RR]Soybean, [RR]Cotton, [RR]Maize). Currently, 72% of GM crop acreage in the United States of America has 'herbicide tolerant (HT) gene' embedded into them. Actually, rapid adoption of HT crop genotypes has increased herbicide usage markedly, in the past two decades (Institute for Science in Society, 2010; GMO Compass, 2015). For instance, herbicide usage in crop belts of the United States of America has increased significantly by 37 million·kg/year. It is a serious environmental hazard in the making. Moreover, weeds that tolerate higher levels of herbicide might appear in fields. Reports suggest that glyphosate-tolerance in crops/ weeds has increased the use of this herbicide perceptibly. It has also led to appearance of super weed species such as pigweeds and horseweeds. Worldwide, there are at least 125 resistant biotypes of 68 species of weeds that infest crop land. Hence, HT technique seems to be prone to perpetually increased use of herbicides. In the United States of America, annual application of glyphosate (round-up) has tripled since 2007 (Institute for Science in Society, 2015). Yet another cause for enhanced use of herbicides during crop production is the recent trend to adopt, no-tillage systems. No-tillage induces rapid rise of weeds in the fields, particularly, after immediate sowing of seeds. Relatively higher amounts of herbicides are, therefore, applied to restrict weed build up in no-till plots.

Precision farming uses weed distribution maps. It prescribes application of herbicides only at the infested spots, thus avoiding excessive use of herbicides. Herbicide application rates match the intensity of weed infestation. Hence, precision technique is gaining in popularity across many agrarian belts. Larger farms prefer precision techniques to reduce on herbicide usage. Therefore, precision methods avoid contamination of ground water. Precision techniques involve detailed scouting of crop fields for weed infestation, preparation of maps and use of digitized data to feed the variable-rate herbicide applicators. A more recent method is to adopt 'agricultural drones'. These scout the field swiftly and accrue accurate data about weed infestation. Farmers are provided with weed infestation maps. We can say precision techniques thwart environmental deterioration.

In the past 5 years, a wide range of agricultural drone models have been produced and advertised. Information, along with relevant data offered, tells us that these autonomous aerial robots are efficient in detecting weeds. Drones cover large fields of hundreds of hectares in one go and collect digital imagery through multispectral sensors. Using the imagery, we can prepare weed distribution maps, and depict weed species and their intensities. Such information could be relayed to farmers (see Meier, 2014b; Krishna, 2016). High-resolution spectral imagery and versatile computer decision-support systems that accurately identify spectral signatures of weed species and crops are essential. The accuracy and speed with which agricultural drones operate and offer data is noteworthy and such spectral data from drones is usually accurate. It is devoid of human errors and fatigue related problems. Such data could be used to guide ground robots or aerial sprayers (drones).

Agricultural engineers and field agronomists alike believe that autonomous robots which perform the hard task of weeding may hold the sway in near future. They could effectively replace drudgery by human farm labour. Aerial robots (drones) could add accuracy to herbicide application by reaching and applying herbicide only at locations infested with weeds. Drones leave out the rest (unifested) of the area. This is different from the present blanket application of herbicides done by ground vehicles. Such a practice of variable-rate techniques is collectively called as 'precision farming'. However, during precision farming, accurate detection, identification of weeds, discriminating different types of weeds, their growth stage and intensity, and calculating herbicide requirement remains a challenge (Slaughter et al., 2008). Equally important is the collection of digital data (spectral reflectance) and preparation of maps that could be used on the agricultural drones, tractors or with other variable-rate herbicide applicators. Agricultural drones are indeed very handy and low-cost robots. They help farmers to obtain digitized data about weeds. To quote an example, in Japan, farmers are known to use Yamaha's RMAX copter drone fitted with variable-rate nozzles, to spray pre- and post-emergent herbicide on to rice fields.

According to Ball et al. (2012), farmers in the agrarian belts of Australia are moving towards using larger machinery. Their traffic in field is usually controlled electronically. Farm vehicles are guided by the digital data and maps during precision farming. Agricultural crop production is being exposed to intelligent technology. They adopt robotics to accomplish a variety of tasks, including weed control. Precision guidance using Global Positioning System (GPS), machine vision and computer-based decision-support system is critical. Incidentally, both ground and aerial robotics are becoming part of precision farming procedures practised in Australia. Agricultural drones are useful in gaining digital data and preparing maps rapidly for use of ground vehicles with variable-rate applicators.

Zero-tillage systems often allow weed growth and proliferation because disturbance to soil is minimal. Therefore, weeds have to be identified and eradicated rapidly after seeding. There is need for alertness and swift action. A project in Queensland (Australia) aims exclusively on devising techniques to control weeds that are resistant to herbicides. The project involves use of robots. In the present context, we may note that using aerial robots (drones) to detect weeds in cereal fields kept under zero-tillage is an excellent proposition. Within zero-tillage fields, drones that take flight over the seedlings, say, immediately within a week after seeding will be able to collect spectral images. Such digital images are useful in deciphering weed infestation, their distribution and intensity at a very rapid pace. Human scouts may not be able to match the drone imagery, particularly, with regards to speed and accuracy. Drones cover hundreds of acres in a matter of 45-min flight over the crop. Same task requires over a week or even more for human crop inspectors.

According to Young et al. (2014), in future, crop production and weed control procedures may involve 'unmanned aerial systems' regularly. According to a 'Drone Production Industries' forecast, precision farming (including weed control) may become prominent, by the year 2025 (Jenkins and Vasigh, 2013). The efficacy of such unmanned aerial vehicles (UAVs) or agricultural drones will largely depend on their ability, to hover over crops. Drones could focus at desired spots in the fields and collect imagery from large areas rapidly. For this to happen, they will need sophisticated hyperspectral sensors that will distinguish weeds and crop. Matching computers and 'digital data banks' that will help in arriving at accurate judgments about weeds and crop canopy is essential. Current trends show that flat-winged drones are well suited for rapid scouting of crop fields mainly to identify weeds (Young et al., 2014). Flat-winged drones are swift and possess sufficient endurance. Copter drones too could be used, particularly, when the intention is to spray herbicides based on blanket recommendation or at variable rates. Moreover, it is believed that 'drone swarm technology' may be beneficial, particularly if the intention is to accrue digital data about weed infestation as rapidly as possible in a large area. Matching computer programs that read the images (spectral signatures) and offer a cohesive digital map seems equally essential and important.

Weed control methods vary depending on several factors related to geographic location, cropping system followed, current crop species, weed species traced and their intensity, and economic considerations. Weed control procedures could be grouped broadly into mechanical, chemical and biological methods. Weed management methods have evolved a great deal. Earliest procedures involved hand weeding, ploughing, inter-culture and earthing-up. Commercial farms often adopt chemical control using selective and nonselective herbicides based on weed flora. However, it is common to suggest weed control by adopting integrated procedures that may involve a few different methods. In the present context, drone technology essentially involves mechanical, optical and chemical procedures. It integrates all these to eradicate weeds.

6.2   DRONES TO DETECT OCCURRENCE AND MAP THE SPREAD OF WEED SPECIES IN CROP FIELDS

Reviews published a decade ago about 'Remote sensing of weeds in crop fields' had indicated that, images of weed infestations that occur in large patches could be obtained. Specific spectral responses of weeds have to be analysed. We need to be careful in reading the spectral signatures of weeds/ crops. Spectral reflectance of even a single crop/weed species or genotypes could change with growth stages, canopy structure and a variable soil background. However, Thorp and Tian (2011) had clearly stated that, spectral differentiation of weeds and crops should consider spectral responses of crops at various stages of growth, plant canopy and biomass. Over all, they had remarked that remote detection of weeds in agrarian regions is possible by using ground, satellite and drone platforms. However, large scale use of aerial imagery during variable-rate application of herbicides is still rudimentary. Precise location of weeds is possible with the recently popularized 'drone technology'. Therefore, it is believed that aerial imagery from drones could be used. They could be used during weed management in crop fields. Basically, remote sensing techniques allow us to detect weed growth and their spread in crop fields. Satellite imagery with high-resolution spectral sensors has been already used to detect and map weed-infested patches in crop fields (Lopez-Granados, 2011, De Castro, 2013; Pena-Barragan et al., 2006; Pena et al., 2013; Thorp and Tian, 2011). Satellite imagery could be hazy as it lacks in resolution. Whereas, drones that fly close to the crop canopy offer us with accurate images of weed infestation. Drones have been able to detect weed population that infests wheat fields right at early stages of the crop. Drones allow close-up shots of crop fields and weed species that infest the crops. In fact, during no-till precision farming procedures, detection of early-season weed growth is an important aspect. Small sized weeds have to be detected early and eradicated. It then reduces cost on weeding during mid-season of the crop. In fact, in most cases, weedicide recommendations are meant as pre-emergent soil injection that avoids weed seed germination. Next, herbicide recommendations are aimed at eradicating weeds right at early seedling stage. Detection at early stages and spraying herbicide using drones is said to reduce on herbicide requirements, by 50% at later stage (Torres-Sanchez et al., 2013). Early-season weed detection and site-specific weed management actually avoids, several other difficulties that occur during late stages of the crop. In cereals, for example, late-season crop canopy masks weeds occurring both in intra and inter-row space. Weed detection using drones fitted with spectral sensors may become difficult. Often, detecting and mapping occurrence of grass weeds in a well-grown cereal crop field such as wheat or barley, is not easy. Similarly, detecting leguminous weeds in broad-leaved pulse crop fields may become difficult. Spectral imagery has to be sharp and computer decision-support has to distinguish the spectral signatures of weeds and crops accurately. Much depends on data banks of spectral signatures that computers consult to identify and discriminate weeds versus crops. To quote another example, it has been observed that spectral data derived from drones and computer decision-support may find it difficult to detect and separate out broad-leaved *Chenopodium* species and well-grown Sunflower crop mainly because spectral signatures of crop and weed species overlap, particularly, at mid-stages of growth (Torres-Sanchez et al., 2013; Thorp and Tian, 2011). Therefore, to apply precision farming methods on to a mid- or late-stage cereal or legume crop, we need drones fitted with high-resolution visual and near infrared (NIR) cameras. Such cameras should provide images at very small pixels of 2–5 cm (Lopez-Granados, 2011). Torres-Sanchez et al. (2013) have reported that currently sensors fitted to

UAVs (drones) are capable of providing such high-resolution images. Such digital images discriminate weeds and crops. Lopez-Granados et al. (2006) have reported that multispectral imagery using visual and NIR bandwidths is able to identify weeds such as wild oats, canary grass and rye grass found in wheat fields. Spectral data allows farmers to discriminate late-season weeds with an accuracy of 80–90%. Genik (2015) has shown that drones could be efficient in detecting wild oats that are rampant in parts of North American wheat belt. Drone imagery could be helpful in reducing cost on weed control. There is actually a strong need to develop a library or data bank of spectral signatures of various weed species, as observed using drones. It will allow drone technologists to be efficient in detecting and mapping a wide range of weeds in a crop field (Hung et al., 2014). Thomson et al. (2014) have reported that cameras with facility for color-infrared (CIR), thermal infrared (TIR) and visible NIR (VNIR) bandwidth imagery could be adopted to judge the effect of weedicide glyphosate. We can also prepare a map of weeds tolerant to glyphosate and those susceptible to it.

Torres-Sanchez et al. (2013) state that, weed control during precision agriculture involves detection of weeds early in the season, using drone imagery. Without doubt, conducting weed control measures at early stages leads us to greater success. Torres-Sanchez et.al. (2013) further state that, during a drone-aided weed detection program, we can envisage at least the following three stages: (a) mission planning, (b) drone flight over crop to acquire imagery and (c) spectral image processing. These steps will help skilled technicians in streamlining the flight route, area covered, cameras and imaging intervals, if any. Flight altitude and sets of multispectral cameras are crucial, for proper detection of weeds in a crop field. Drone altitude above the crop has to be kept low if high resolution close-up shots are required. Detection and accurate discrimination of, say, a grass weed in cereal field may need such high-resolution close-up shots. Reports suggest that detection of weeds in the inter-row space is also easier, using visual and NIR cameras. Spectral imagery provides good distinction among soil surface, weeds in the inter-row and crop in the rows (Torres-Sanchez et al., 2013).

'TOAS Project' is a Spanish project aimed at generating geo-referenced weed infestation maps of a few cereal crops and woody plantations. This project employs drones as platforms to place the visual and NIR sensors. Researchers have evaluated a series of cameras (sensors) with different specifications, particularly, regarding resolution of images obtained. Drone imagery is actually aimed at studying the phenology and rate of spread of weeds in cereal fields. Weed maps produced offer a robust basis for farmers to prepare herbicide spray schedules and a generalized weed control program (TOAS Project, 2015). Further, this project involved identification and mapping of weeds using object-based image analysis (OBIA). Crop-row identification using drone imagery was done with 100% accuracy. Detection of weed distribution in the inter-row space was achieved at more than 80% accuracy. The OBIA technique involves procedure known as segmentation. In addition, it uses weed traits such as location, proximity and hierarchical relationships with the spectra of the object (weed/crop). Such a system offers more accurate identification and mapping of weeds in crop fields.

There is no doubt that in any cropping belt, detection and accurate mapping of weeds are essential steps, during weed management. Weed scouting by field workers is the most common method. It has helped in locating weeds, noting the intensity of infestation and then mapping them. However, the present trend is to design and adopt drones with hyperspectral sensors. Drones provide digital data and maps showing weeds in the fields. Therefore, development of drones, their sensors, appropriate computer software and decision-support systems are crucial. Often, it takes several trial-and-error steps before drones could be effective. Let us consider an example. Jones (2007) evaluated drones fitted with visual cameras, for their ability to detect specific weeds, such as musk thistle (*Carduus nutans*) and Dalmatian toadflax (*Linaria dalmatica*). Initial efforts at Camp Williams in Utah, USA have shown that certain weeds are rapidly and accurately identified in a cereal field. In the above study, musk thistle was accurately located and mapped by drones. However, detection of Dalmatian toadflax by visual and IR cameras was not clear. This weed species was confused with sweet clover (*Melilotus officinalis*). Jones (2007) states that computer software that analyses the postflight imagery is crucial. In the past few years, several advancements have occurred with regard to weed/crop discrimination, using spectral signatures. Further, accrual of digital signatures of weed species in the 'big data bank' and usage of high-resolution spectral sensors has also improved. Therefore, drone-aided detection and mapping of a wider range of weed species should be possible.

Reports from Nevada, USA, suggest that a hexa-copter such as 'Aeroscout' with a weight of 2.2 kg and endurance of 12 min can be efficient. The drone can detect weed growth and flora in crop fields. This particular drone covers an area of 40 ac in 10 min of flight. It flies at 20–30 m height above the crop. The digital imagery is then utilized by an aerial drone (copter) or ground vehicle (sprayer) fitted with herbicide tanks (Meier, 2015b).

Gray et al. (2008) report that several different weed species such as hemp sesbania, prickly sida, sicklepod, small flower morning glory, *Ipomea* sp and a few others commonly infest soybean fields in the Mississippi, USA. These species could be identified, differentiated and mapped using multispectral cameras. High-resolution multispectral imagery actually allows farmers to distinguish soil (unplanted area), soybean crop and weed patches of different species (Gray et al., 2008). Farmers could actually classify weeds and group them with 60% accuracy using remote sensing. Hence, it has been suggested that by using a flat-winged drone, farmers could obtain digital data and imagery of soybean fields showing weed distribution. Herbicide sprays could be based on such weed maps. Such an effort reduces on herbicide requirements since, only spots with weeds are sprayed. It also reduces contamination of irrigation channels and ground water with herbicide.

In a study aimed at crop/weed discrimination, Jones et al. (2008) have reported that a good accuracy could be attained even by using ground robots equipped with spectral cameras. It has to be followed by image processing. Software to accurately discriminate the weed species is needed. Weed detection accuracy may reach 80% for inter-row weeds. However, intra-row weeds are difficult to identify because of crop canopy that masks them during imagery.

Currently, there are several agricultural drone related companies that offer processed aerial images showing weed infestation, crop growth status and yield forecasts. In fact, high-resolution imagery of crop/weed in fields is sought periodically by farmers, in order to make decisions about weedicide sprays (SoyL, 2013).

McGowen et al. (2014) opine that weed management on a large scale requires thorough knowledge of weed abundance. Further, periodic changes in weed flora and distribution that occur in a crop field should also be known. In the New South Wales region of Australia, conventional techniques of weed mapping are costly. Frequent scouting of large farms is usually not feasible. Remote sensing offers a low-cost alternative. Remote sensing can be useful in preparing weed maps. Target weed species with specific spectral reflectance pattern could be picked accurately by the multispectral imagery. For example, Scotch thistle could be mapped with 80–86% accuracy. Similarly, serrated tussock was detected and mapped with an accuracy of 72–82%. They have suggested that reliability of remote sensing could be improved using high-resolution cameras. Agricultural drones with high-resolution cameras are perhaps apt for detecting and mapping the weed flora repeatedly and to note changes, if any.

Drones could also be directed to detect and map a specific weed. It could be an invasive species that affects crops and natural vegetation. There are several weed species that spread rapidly, season after season, within in a large agrarian belt. For example, in the United States of America, cogon grass (*Imperata cylindrica*) invades pastures, natural and planted forests, riverine regions and croplands. It affects agrarian regions in Arkansas, Alabama, Mississippi, Georgia and Texas (University of Arkansas, 2015). It spreads through rhizomes and fluffy seed. Spectral signatures of this grass and its variants need to be accrued and stored in computer decision systems. Then, we have to quickly identify the occurrence of cogon grass patches. Monitoring this grass periodically using drones is a clear possibility. We can pinpoint the locations (hot spots) that harbour cogon grass. Then, eradicate the weed in time to stop its spread further into cropland.

6.2.1   DRONES USED TO SPRAY HERBICIDES

Application of different plant protection chemicals is an established agronomic procedure to control weeds, insects or diseases. Several different types of farm vehicles are used to apply herbicides. Herbicides are applied to field either as pre-emergent spray to soil or incorporated at a depth (soil injection). Farmers practising precision techniques utilize weed infestation maps of previous seasons. Then, they apply herbicides only at locations prone to weed infestation. Post-emergent spray is done after obtaining aerial imagery of weed spread and their intensity. Helicopters could be used to spray liquid formulations or to dust crops with herbicides. Dusting could be done in powder or granular form. Generally, drones are easy to operate for crop dusting, if the fields are located in plains region. Farmers need to be very well skilled in handling drone's flight pathway and remote control methods. Drones could also be adopted to dust crops growing in undulated or hilly terrains. For example, dusting rice crops grown on mountainous region or terraces, say in Philippines or Japan, is a clear possibility. It will need skilled farm technicians to handle drone's path and control dusting sprees (University of California, Davis, 2014;

Diaz-Varela et al., 2014). Reports suggest that in Japan, currently 40% of rice crop is dusted, using helicopter drones. These drones move swiftly over submerged rice fields. They apply herbicide as liquid, powder or granules. Drones have also been utilized to spray herbicides and other plant protection chemicals into fields supporting crops such as wheat, oats, radish, grape vines and citrus (University of California, Davis, 2014; CREC, 2015; Gmitter, 2015). Drones have also been used to spray herbicides on an intense weed growth in wasteland. A drone such as Yamaha's RMAX is efficient in distributing herbicides rapidly and accurately. Such autonomous copters could cover 1.0 ha cropped land for every 8 min of flight in the air (Staff-writer, 2013; Meier, 2014b; Table 6.1).

Herbicide spray using drones could be made efficient and accurate. We have to carefully consider the targeted weeds, extent of control intended and ultimate purpose. Drone-aided herbicide spray may also have to be modified to suit the natural vegetation or crop species that gets sprayed. Williams (2013) reported that weed killing drones are on the rise in Denmark. The weed eradication system actually aims at reducing herbicide usage. It avoids rampant application of larger dosages of herbicides all over the crop fields. Drones release appropriate quantities of herbicides only at spots infested with weeds. Drones actually spray herbicides in 'targeted bursts' using nozzles. The activity of nozzles is regulated through computer decision-support systems. In fact, a steady improvement in spraying techniques has occurred. It has been possible through computer-controlled precision application of herbicides and other chemicals. For example, Zhu et al. (2014) reported that, pulse width modulation controller helps in higher precision during chemical spray using drones.

6.3   DRONES IN PRACTICAL WEED CONTROL: FEW EXAMPLES

Agricultural drones have been utilized to detect weeds that infest wheat fields in the Canadian Prairie region. Reports by Alberta's Wheat Commission suggest that small, lightweight drones are able to photograph the wheat fields. Drones provide multispectral imagery of weeds that infest the cereal and lentil fields (Glen, 2014). They say, '*weeds cannot duck or hide when drones fly very closely, just above them, in the crop fields*'. Their aim is to compare the drone image-aided weed control with traditional techniques that involve human scouts. During practical farming, it is

TABLE 6.1 Drones Utilized to Spray Herbicides on Crop fields: Few Examples. (*Sources:* Cornett et al. 2013; RMAX, 2015; Yintong Aviation Supplies, 2014)

Yamaha's RMAX (Yamaha Motor Co., Iwata, Shizuoka, Japan)	Drone weighs about 94 kg at take off; maximum payload is 31 kg and chemical tanks hold 21–24 kg herbicide granules or dust or liquid; RMAX drone sprays for 10–15 min at a stretch and covers 12 ac with liquid or granule herbicides (Cornett, 2013; RMAX, 2015)
Yintong UAV YT, P5 and P10 (Yintong Aviation Supplies, Zhubai City, China)	Drone weighs 15 kg at take off; maximum payload is 5 kg or 10 kg herbicides depending on model; sprayers cover about 45–60 ac·h^{-1}; spraying perimeter is 2.8 m, spraying area is 2.25 ac·min^{-1}; herbicide tank holds 5 kg chemical (Yintong Aviation Supplies, 2014)
Agricultural Helicopter Crop Duster-AG-RHCD-80-15 (Homeland Surveillance Electronics LLC, United States of America)[a]	Drone's payload is 15 kg; sprayer capacity is 8 ha·h^{-1}; spray width is 306 m depending on altitude of flight; pesticide box has a volume of $545 \times 358 \times 85$ mm and spray speed is 0–6 m·sec^{-1};
Rotomotion's SR200 (Rotomotion LLC, Charleston, SC, USA)	The total payload is 2.7 kg; spray tank weighs 1 kg and holds 5 kg or 1.5 gallon chemicals (herbicide); sprayer on the drone covers 14 ha per load at low-volume spray rate of 0.3 L·ha^{-1} (Huang, 2014; Huang et al., 2009)
AG-V8A Octo-Copter UAV (Duster Sprayers) (Homeland Surveillance Electronics LLC, United States of America)[b]	These are drones with a payload of 23 kg. Flight endurance is 40 min and it covers about 500 ac·hr^{-1}. Drone tanks hold about 8–10 kg chemicals and the spray rate is 33.5 ac·hr^{-1} or 1000 ac for every 30 h of flight.
UAV Helicopter (Belize)[c]	This copter weighs 35 kg at take off; herbicide tank holds about 10 kg granules/liquid; drones spray for 8–10 min per flight at a stretch and cover about 10,667 m². They can spray at a rate of 1332 m² per minute with fine spray.

[a]http://www.uavcropdustersprayers.com/agriculture_uav_crop_duster_8015.htm
[b]http://www.uavcropdusterssprayers.com
[c]http://www.b-oilbelize.com/uavhelicopters.html

always possible to club a few related agronomic procedures. In the present case, it has been suggested that in addition to weed detection, drone images also depict the occurrence of late blight of potatoes in the fields. Over all, ability of farmers to scout large areas of wheat crop for weeds, note their intensity and to spray herbicides will be enhanced. Farmers have to utilize spectral data offered by the sensors on drones. It is said that adoption of drone to image crops/weeds at early stages of crop is important. In fact, no-till fields are prone to infestation by weeds rather quickly usually immediately after sowing seeds. Therefore, earliest detection of weeds is a necessity.

Shaner and Beckie (2014) have forecasted a few aspects about future for weed control programs in Canada. They have focused on herbicide application to weed populations that have developed a certain degree of resistance to weedicides. They suggest that an integrated approach is required. It should involve drones among several other control measures. Detection of herbicide-resistant weed species using drone imagery is a clear possibility. It helps in reducing herbicide quantities that otherwise has to be sprayed, to quell such herbicide-tolerant weed populations. Farmers currently use GM cultivars (e.g. RRSoybean) that tolerate herbicide sprays at higher concentrations so they have to spray herbicides frequently and in higher concentrations. It does not affect crops but only kills weeds. Such a procedure induces build-up of herbicides in soil, contaminates irrigation channels and ground water. Excessive use of herbicides is after all detrimental to soil and even above-ground environment. Adopting drones to spray crops when weeds are still succulent and susceptible to lower levels of herbicides is appropriate. A study of corn fields infested with Canadian thistle clearly proves the net advantage of aerial imagery. About 122 ac of corn grown in a location in Northeast Kansas was surveyed, using drones fitted with sensors. Aerial imagery showed that only 0.6 ac were affected with weeds (thistle). A routine traditional weed control procedure would have required 1212 gallons of diluted herbicide. The herbicide prescription is 10 gallon herbicide per acre of land. Therefore, only 6.6 gallons are required to be sprayed to control thistle. Indeed, about 99% of herbicide could be saved. Adopting precision techniques and drones could reduce application of herbicides significantly (HSE, 2015).

German agricultural engineers at Leibnitz Universtitat have tried to provide a unique alternative, to reduce the rampant usage of herbicides in crop fields. They propose to use laser-armed ground robots and drones. This aspect is not yet feasible in crop field. It has not been fully standardized and repeatedly evaluated. Yet, it seems a good concept to mention and discuss here. Christenson (2015) states that, drones fly over the crop fields to capture high-resolution spectral data. Such spectral data show up crop status and weed growth. Laser guns possess the ability to destroy weeds. A few stages of development of 'laser gun attached drone' are being tested to eradicate weeds. A few other steps are in the process of development. But, some are still in the drawing board stage. Currently, researchers are adopting CO_2 lasers that emit beams at mid-IR bandwidth. Drone's imagery has to be used shrewdly to assess the age of weed, thickness of the stem and other parts of the plant that needs to be destroyed. This is to ensure that weed dies. The height of the weed and crop species should be carefully considered. In certain cases, stem needs to be destroyed. Actually, laser beam has to hit the area most vulnerable in the weed plant. Laser techniques avoid herbicide usage. However, it requires a period of development, testing and economic evaluation. Drone-based methods with laser gun can easily become a part of the organic farming, as it will reduce on herbicide sprays. Clearly, they are environmentally friendly methods.

Reports from Spain suggest that remote imagery using UAVs has immense potential to control weeds. Drones are apt, particularly, while performing various procedures related to site-specific management of crops (i.e. precision farming). Pena et al. (2013) believe that UAVs may have a strong role in post-emergent weed control. Drones offer better resolution and accurate data at a cost much lower than manual, airplane or satellite-aided methods. Even then, drone usage to control weeds in agricultural expanses is still rudimentary. Pena et al. (2015) further state that accurate map of weed infestation, early during the crop season is important, if variable-rate herbicide application has to be effective. Early detection of weed species is itself dependent on few factors such as foliage formation by weed species, temporal variations in weed eruption, resolution of multispectral sensors, drone's flight altitude and resolution of spectral data and so forth. Usually, it is easy to detect weeds that are aged enough and where foliage is large. Drones flown at an altitude of 40–50 m above the crop canopy provide high-resolution images of weeds. It is easy to discriminate aged crop and weed. However, early detection of weeds is essential for effective weed eradication. Sensors operated at visual bandwidth should provide data about weeds, right at early stage of the weed growth.

There are a few reports that depict the performance of drones on crops grown in agricultural experimental stations. However, in certain regions, such as in Japan and other Far-Eastern nations, agricultural drones are popular. They are used regularly by farmers to spray plant protection chemicals, including herbicides. Actually, drone technology has been utilized regularly to control weeds in farmer's rice fields. Copter drones such as RMAX, Yintong or AutoCopter are utilized to dispense quantities of herbicides suggested as blanket recommendation. These models of drones are also preferred to obtain digital maps first and then, to dispense herbicide through variable-rate nozzles connected to herbicide tanks.

Xiang and Tian (2015) have reported development of a UAV. It is based on a small transportable helicopter platform. It weighs about 14 kg and is equipped with multispectral cameras and autonomous system. This UAV is capable of obtaining multispectral images of crops and turf grass, at any desired location. It flies based on instructions from ground control station. However, its navigation route above turf grass could also be predetermined. Further, they state that the above drone is good to monitor weed growth and to apply herbicide such as glyphosate. Basically, sensors provide highly useful spatial and temporal data about weed infestation on turfs.

Now, let us consider an example that describes post-emergent weed control in maize fields. These fields were located at Arganda del Rey near Madrid, Spain. Pena et al. (2013) have utilized a method known as OBIA. They used an 'md-4-1000 Quadcopter' fitted with Tetracam Mini-MCA-6 camera, which has a lightweight sensor. They have used both visual and NIR sensors that operate at 530–570 and 700–800 nm bandwidths. The drone imagery was conducted at an altitude of 30 m above the maize crop's canopy. A barium chloride Spectralon panel was placed in the field. It acted as standard check during imaging by the cameras on drone. The drone imagery was later calibrated and corrected. Further, Pena et al. (2013) state that often spectral characters of crop plant and weed species prevalent in the fields could be similar. Seedlings of maize and weeds may have similar reflectance traits. Therefore, it was necessary to enhance the effectiveness of crop/weed discrimination. It could be achieved by classifying crop/weed rows using OBIA methodology. This system involves a composite of spectral data. Spectral signatures of weeds and crop species, and the location of crop/weed in the field are utilized. Actually, contextual traits and morphological information of weed and crop plants are traced, using sensors on drones. There are commercial software programs that help farmers to distinguish the crop/weed, using OBIA. For example, 'eCognition Developer-8' is a software used, to analyse UAV imagery and develop images based on OBIA procedure. The OBIA essentially involves following steps: (a) classification of crop rows and their identification, (b) discrimination of weeds and crop species using their location in the row or in intra-row space and (c) generation of a weed infestation map using grid and rows as reference. Crop row identification done within the imagery derived by drones is a crucial aspect. It helps in judging occurrence of weeds accurately using the location as a trait. However, intra-row weeds may get overlooked, if they are not imaged properly. Usually, after discriminating all the crop rows and weed growth, a weed distribution map for the entire field is prepared. Such a map is utilized during herbicide application (Pena et al. 2013; Instituto de Agricultura Sostenible, 2015). Weed mapping could also be done using grids of different sizes. Then, the maps could be adopted during variable-rate application of herbicides. Field trials near Madrid in Spain has shown that weed coverage observed on the ground by human scouts and that obtained using drone imagery correlate highly ($R^2 = 0.89$). Pena et al. (2013) conclude that high resolution spectral imagery, verification of spectral signature of crop species and weeds and use of OBIA methods helps in preparation of weed maps. Such weed maps are highly useful during dusting of crops with herbicides. Incidentally, weed distribution within digital maps is categorized based on intensity of infestation. They are grouped into low-, medium- and high-infestation regions. Such categorization may be helpful while marking 'management blocks' and during spraying herbicides.

Wheat culture in the European plains has been exposed to site-specific techniques. Precision techniques actually help to reduce on herbicide application to nontarget regions. Otherwise, herbicide spray to nontarget areas occurs, if a single dosage or blanket recommendation is practised. Lopez-Granados et al. (2006) suggest that in Spain, it is common to see patchy distribution of late-season weeds in the wheat fields of Cordoba region. There are actually several reports from different agrarian regions suggesting that, site-specific weed management (precision technique) reduces the herbicide quantities required to control weeds.

A few experiments conducted in the North American plains indicated that cost savings on herbicide usage is 60–90%, if precision techniques are used. It is dependent on weed flora encountered in the wheat fields (Timmerman et al., 2003). The crux of the precision technique is in the rapid and accurate detection of weeds, their species and the intensity. This is accomplished usually by employing sensors that operate at visual (blue, green and red) and NIR bandwidths. Flat-winged drones carrying a set of multispectral sensors can easily offer images of weed patches, in a matter of minutes. Such drones save cost on detailed and time-consuming scouting by human skilled farm workers. Lopez-Granados et al. (2006) state that weed flora made of canary grass and rye grass in the wheat fields could be deciphered accurately using multispectral imagery. The weed detection accuracy ranged from 85–90%. In some regions, the accuracy ranged from 65–71% for weeds such as wild oats, canary grass and rye grass. In their experiments, spectral data were obtained using satellites and aerial photography. Drones with short flight and fitted with high-resolution spectral cameras are perhaps the best bets. They could reduce cost on obtaining aerial photography to much lower levels. Drones offer details about weed distribution in wheat fields. They allow farmers to obtain real-time data on weed distribution. This is unlike satellite imagery for which farmers may have to wait. We may note that, farmers could employ drones, for example, a copter such as RMAX or Yintong. Such drones have facility to hold herbicide (i.e. a tank) and spray the wheat fields. Drones utilize digital data and signals from computer-based decision-support systems.

Drones have also been tested for use in the vegetable fields of Nordic region. In Denmark, weed infestation could be identified, analysed and mapped using drones (New Scientist Tech, 2013). Drones flying at low altitude above the crop pick images through visual and NIR sensors. Appropriate decision-support systems then detect and discriminate crops and weeds. The computers actually utilize differential spectral signatures of crops and weeds. Data bank that stores spectral data of large number of weed species or at least those dominant in that region is essential. It allows efficient detection and mapping of weed-infested zones, at any time. Drone imagery, generally, allows farm technicians to easily identify weed species and their growth stages. In case of farmers in the Nordic region, they were provided with digital data that helps the drone to spray pesticides only at regions infested with weeds. Drones restrict or totally abstain from regions not showing-up weeds. This is unlike blanket recommendations of herbicides on vegetable crops. As stated earlier, herbicide quantity required under site-specific systems is much less compared to traditional blanket recommendations. Agricultural Scientists at Aarlog University, Denmark have expressed that it is easier to fly a drone, to prepare digital map of weed spread and then apply herbicides using variable-rate applicators (New Scientist Tech, 2013). However, digital data from drones could also be embedded into ground robots and semi-autonomous herbicide sprayers. Weedicides are distributed as accurately as possible using such systems. Ground vehicles apply herbicides only at spots required, and not on the entire field. Researchers in Denmark have also attempted to reduce use of herbicide by concentrating on weed-infested zones. According to La Cour-Harbo (2013), the ASETA project aims at identifying weed-infested zones on the basis of their spectral signatures. Drones then spray appropriate herbicides. Reports suggest that such an effort has been effective in managing thistle (weed) population. Thistle absorbs yellow colour preferentially in a field of sugar beet. This trait has been used effectively. Over all, drones that fly repeatedly above the fields and keep a watch on weed infestation seem a reality. Drones are destined to be used in greater number in Denmark's farm lands (Rychia, 2015).

Rice is a dominant cereal crop in the tropical Asia. Rice production strategies are generally intensive. Farmers try to reap higher yield, by applying larger quantities of fertilizers and other amendments. Weed infestation of such low-land intensive rice fields is highly detrimental to optimum grain and forage yield. Weeds could actually divert a large fraction of fertilizer-based nutrients and water that is meant otherwise for the main crop, that is, rice (Krishna, 2012, 2013, 2015, 2016). Farmers indeed adopt several different types of weed control measures. Hand weeding is costlier. Procuring farm workers, in time, to restrict weed growth is crucial. Generally, herbicide usage in rice fields is rampant. It is done in order to restrict weeds right at an early stage of the crop. Without doubt, herbicides sprayed in excess under blanket recommendations contaminate irrigation channels and ground water. Hence, precision techniques that adopt weed maps and application of weedicides at variable-rates are preferred. Ground robots that spray weedicides need to be provided with aerial imagery. They need the digital data to regulate the spray quantity. Okamoto et al. (2007) have reported that low-flying drones fitted with high-resolution spectral cameras and ground-based tractors fitted with Specim ImSpector V9 could provide excellent data about weed infestation in the rice fields. Reports from Japanese rice belt state that nearly 40% of crop dusting that includes herbicide spray is currently conducted using copter drones.

In Australia, rotary copter drones are employed to obtain crop/weed imagery, detect weeds and to spray weedicides (Cornett, 2013). Weed control in crop fields is achieved first by identifying areas with weed infestation. Then herbicides are sprayed using low-flying pilotless copters.

Meier (2014a) reports that, such drones could serve the farmer as both rapid and efficient weed searchers and accurate sprayers. Recent reports from New South Wales (Australia) indicate that, crop fields and pastures could be monitored for weeds using drones. Consequently, accurate weed maps could be offered to farmers. Basic requirement is high-resolution multi-spectral sensors placed on small copter drones (McGowen et al. 2014). Agricultural drones are particularly useful as practical weed mapping using human scouts is costly in New South Wales. Also, accuracy is affected by fatigue and other human errors. Most researchers believe that monitoring annual weed species and periodic surveys are economically less efficient, if conducted manually. Instead, drones that detect weeds and identify the patches accurately with GPS coordinates are really efficient.

A report from Queensland in Australia suggests that, Natural Resource Department intends to use drones (Yamaha's RMAX) to spray weedicides all across weed-infested zones with an intent to clear up the unwanted vegetation (Staff-writer, 2013). Drone-aided herbicide sprays are opted in cropping zones where conventional weeding programs are difficult due to detection of high-intensity weed growth. Spectral signatures of weeds are consulted to decide the areas that need herbicide sprays. It is said that drones are highly efficient since they cover about 1.0 ha of crop land every 8 min of spray time. Drone-aided weed clearance programs are expected to cover 250,000 ha each year. Swain (2014) states that drones could be controlled from 200 m and guided using digital imagery. It helps farmers to spray herbicides exactly on weed-infested zones. Drones are particularly apt for use in undulated regions, hilly terrains and inaccessible areas. In Queensland, drones help rangers to spray blackberry bushes and convert waste land into crop fields. Incidentally, blackberry-infested zones are spread over 9 million·ha land in Australia. If left unattended, they could smother natural vegetation and also affect crop production zones. They could even induce forest (bush) fires. A few other reports suggest that in Canberra, Australia, rangers have used drones with ability to spray herbicides on crop fields and natural vegetation. Yamaha's RMAX fitted with herbicide tanks of 10 L volume and spray nozzles are being used, to apply herbicides. It seems blackberry infestation in the cropland requires accurate spray of herbicides.

Over all, we may note that drone-aided crop scouting, aerial imagery and accrual of digital data about weed infestation reduce costs on skilled farm labour. It improves accuracy and makes it easier for farmers to conduct weed control related procedures. Precision techniques such as variable-rate application of herbicides further reduce cost on herbicides. Moreover, it restricts chemical contamination of soil and water resources in a farm. Therefore, drone techniques together with precision farming methods offer a good chance for us to be efficient, reduce chemical application, thwart environmental deterioration and at the same time reap better grain harvests.

KEYWORDS

- **genetically modified crops**
- **weeds**
- **digital data banks**
- **weed control**
- **hyperspectral sensors**
- **herbicide-tolerant crops**

REFERENCES

Ball, D.; Ross, P.; English, A.; Patten, T.; Upcroft, B.; Fitch, R.; Sukkarieh, S.; Wyeth, G.; Corke, P. *Robotics for Sustainable-Acre Agriculture;* Project Report on Robotics for Zero-Tillage Agriculture; Queensland University of Technology: Australia, 2012; pp 1–32.

Bayer Crop Science. Why Herbicides Matter Us. 2015, pp 1–3. http://www.cropscience. bayer.com/en/Products-and-Innovtion/Brands/Herbicides.aspx (accessed Oct 3, 2015).

Christensen, B. Laser-Armed Robots to do Weed Control. Technovelgy.com 2015, pp 1–3. http://www.technovelgy.com/ct/Science-Fiction-News.asp?NewsNum=3889 (accessed Sept 29, 2015).

Cornett, R. Drones and Pesticide Spraying: A Promising Partnership. Western Plant Health Association. Western Farm Press. 2013, pp 1–3. http://westernfarmpress.com//grapes/ drones-and-pesticides-spraying-partnership (accessed Aug 30, 2014).

CREC. UAV Application in Agriculture. Citrus Research and Education Centre, Lake Alfred, Florida. 2015, pp 1–3. http://www.crec.ufl.ifas.edu/publications/news/PDF/ UAVwsflyer3-pdf.pdf (accessed Sept 30, 2015).

De Castro, A. L.; Lopez-Granados, F.; Juardo-Exposito, M. Broad Scale Cruciferous Weed Patch Classification in Winter Wheat Using QuickBird Imagery for In-Season Site-Specific Control. *Precis. Agric.* **2013**, 12. DOI: 10.1007/s11119-013-9304-y (accessed Sept 25, 2015).

Diaz-Varela, R. A.; Zarco-Tejada, P. J.; Angileri, V.; Loudjani, P. Automatic Identification of Agricultural Terraces Through Object Oriented Analysis of Very High Resolution

DSMs and Multi-spectral Imagery from an Unmanned Aerial Vehicle. *J. Environ. Manage.* **2014**, *134*, 117–126.

Genik, W. Case Study: Wild Oat Control Efficiency Using UAV Imagery. AgSky Technologies Inc. 2015, pp 1–5. http://agsky.ca/case-study-wild-oat-control-efficiency-using-uav-imagery (accessed June 26, 2015).

Glen, B. Drones Put to Work Hunting Weeds. 2014, pp 1–3. http://www.producer.com/2014/drones-put-to-work-hunting-weeds/ (accessed Sept 28, 2015).

Gmitter, J. Unmanned Aerial Vehicle Update! 2015, pp 1–4. http://www.centralfloridaagnews.com/gritech-unmanned-aerial-vehicles-update (accessed Oct 2, 2015).

GMO Compass. Herbicide Resistant Crops. 2015, pp 1–2. http://www.gmo-compass.org/eng/agri_biotechnology/breeding_aims/146.herbicide_resisitant-genes (accessed Oct 3, 2015).

Gray, C. J.; Shaw, D. R.; Gerard, P. D.; Bruce, L. M. Utility of Multi-spectral Imagery for Soybean and Weed Species Differentiation. *Weed Technol.* **2008**, *22*, 713–718.

Instituto de Agricultura Sostenible. Remote Sensing for Precision Agriculture and Weed Science. CSIC, Spain 2015, pp 1–5. http://www.ias.csic.es/en/crop-production/remote-sensing-precision-agriculllture-weed-science (accessed July 7, 2015).

HSE. Invasive Species Detection: Canadian Thistle. 2015, pp 1–5. http://www.uavcrop-dusterssprayers.com/agriculture_delta-fw70_fixed_wing_uav.htm (accessed May 19, 2015).

Huang, Y. Development of Unmanned Aerial Vehicles for Crop Protection and Production. Symposium on Agricultural Remote Sensing with UAVs: Challenges and Opportunities. 2014, pp 1. https://dl.sciencesocieties.org/publications/meetings/2014am?13358/88127 (accessed July 4, 2015).

Huang, Y.; Hoffman, W. C.; Lan, Y.; Wu, W.; Fritz, B. K. Development of a Spray System for an Unmanned Aerial Vehicle Platform. *Am. Soc. Agric. Biol. Eng.* **2009**, *25*, 803–809.

Hung, C.; Xu, Z.; Sukkarieh, S. Feature Learning Based Approach for Weed Classification, Using High Resolution Aerial Images from a Digital Camera Mounted on a UAV. *Remote Sens.* **2014**, *6*, 12037–12054.

Institute of Science in Society. GM Crops Increase Herbicide use in the United States. 2015, pp 1–4. http://www.i-sis.org.uk/GMcropsIncreasedHerbicide.php (accessed Oct 3, 2015).

Jenkins, D.; Vasigh, B. The Economic Impact of Unmanned Aircraft Systems Integration in the Air Space. United States Association for Unmanned Vehicle Systems International, Arlington, Virginia, USA Internal Report, 2013, pp 1–38.

Jones, B. T. An Evaluation of a Low-Cost UAV Approach to Noxious Weed Mapping. Brigham Young University BYU Scholars Archive, 2007; Paper No. 1220, p 61.

Jones, G.; Gee, C.; Truchetet, F. Crop/Weed Discrimination in Simulated Images. 2008, *pp 1–7.* spie.org/newsroom/1226-detecting-crops-and-weeds-in-precision-agriculture *(accessed Sept 5, 2016).*

Krishna K. R. *Precision Farming: Soil Fertility and Productivity Aspects*; Apple Academic Press Inc.: New Jersey, USA, 2012; p 189.

Krishna K. R. *Agroecosystems: Soils, Climate, Crops, Nutrient Dynamics and Productivity*; Apple Academic Press Inc.: Waretown, New Jersey, USA, 2013; pp 54–64.

Krishna, K. R. *Agricultural Prairies: Natural Resources and Crop Productivity*; Apple Academic Press Inc.: Waretown, New Jersey, USA, 2015; pp 313–399.

Krishna, K. R. *Push Button Agriculture: Robotics, Drones and Satellite-Guided Soil and Crop Management*; Apple Academic Press Inc.: Waretown, New Jersey, USA, 2016; p 412.

La Cour-Harbo, A. The ASETA Project: Precision Herbicide Drones Launch Strikes on Weeds. Technology News. 2013, pp 1–4. http://www.newscientist.com/article/dn23783-precision-herbicide-drones-launch-strikes-on-weeds (accessed Sep 29, 2015).

Lopez-Granados, R. Weed Detection for Site-Specific Weed Management Mapping and Real Time Approaches. *Weed Res.* **2011,** *51,* 1–11.

Lopez-Granados, F.; Juardo-Exposito, M.; Pena-Barragan, J. M.; Lopez-Granados, F. Using Remote Sensing for Identification of Late-Season Grass Weed Patches in Wheat. *Weed Sci.* **2006,** *54,* 346–353.

McGowen, I.; Frazier, P.; Richard, P. Remote Sensing for Broad Scale Weed Mapping: Is it Possible? *Geo-spatial Inf. Agric.* **2014,** 1–19. http://www.regional.org.au/gia/14/283mcgovern.htm (accessed Sep 28, 2015).

Meier, W. The Future of Drones in Agriculture. Airborne. 2014a, pp 1–4. http://www.airbornedrones.com/blogs/news/15025505-the-future-of-dronesin-agriculture (accessed Sep 21, 2014).

Meier, W. Ag Drones: Future and Present. Change is in the Air. ASPRS UAS Technical Demonstration and Symposium, Reno, Nevada, USA, 2014b, pp 1–14. http://www.suas-news.com/2014/08/30393/ag-drones-future-and-precent (accessed Sep 29, 2015).

New Scientist Tech. Precision Herbicide Drones Launch Attack on Weeds. 2013, pp 1–3. http://www.newscientist.com/article/dn23783-precision-herbicides-drones-launch-strikes-on-weedshtml#VACJrvmSxOw (accessed Aug 30, 2014).

Okamoto, H.; Murata, T.; Kataoka, S.; Hata, S. Plant Classification for Weed Detection Using Hyperspectral Imaging with Wavelength Analysis. *Weed Biol. Manage.* **2007,** *7,* 31–37.

Pena, J. M.; Torres-Sanchez, J.; Isabel de Castro, A.; Kelly, M.; Lopez-Granados, F. Weed Mapping in Early-Season Maize Fields Using Object-based Analysis of Unmanned Aerial Vehicle (UAV) Images. *PLoS One* **2013,** *8*(e77151), 1–11. http://www.plosone.org (accessed Sep 25, 2015).

Pena, J. M.; Torres-Sanchez, J.; Serrano-Perez, A.; de Cstro, A. I.; Lopez-Granados, F. Quantifying Efficacy and Limits of Unmanned Aerial Vehicle (UAV) Technology for Weed Seedling Detection as Affected by Sensor Resolution. *Sensors* **2015,** *15,* 5609–5026.

Pena-Barragan, J. M.; Lopez-Granados, F.; Juardo-Exposito, M.; Garcia Torres, L. Mapping *Ridolfia segetam* Patches in Sunflower Crop Using Remote Sensing. *Weed Res.* **2006,** *47,* 164–172.

RMAX. *RMAX Specifications*; Yamaha Motor Company: Japan, 2015, 1–4. http://www.max.yamaha-motor.Drone.au/specifications (accessed Sep 8, 2015).

Rychia, L. More Drones to Combat Weeds on Danish Fields. 2015, pp 1–4. http://cphhpost.dk/news//moe-drones-to-combat-weeds-on-danishfields.html (accessed Oct 18, 2015).

Shaner, D. L.; Beckie, H. J. The Future for Weed Control and Technology. *Pest Manage. Sci.* **2014,** *70,* 1329–1339.

Slaughter, D. C.; Giles, D. K.; Downey, D. Autonomous Weed Control Systems: A Review. *Comput. Electron. Agric.* **2008,** *61,* 63–78.

SoyL. SoyLsight Drone Technology at Lamma 2014. Frontier Agriculture Ltd. 2013, pp 1–2. http://www.soyl.com/index.php/8-latest-news/56-soylsight-drone-technology-at-lamma-2014 (accessed Oct 3, 2015).

Staff-Writer. Queensland Deploys Drones to Kill Weeds. 2013, pp 1–7. http://www.itnews. com.au/news/queensland-deploys-drones-to-kill-weeds-363484 (accessed Sep 28, 2015).

Swain, S. Canberra Rangers to Use Helicopter Drones to Fight Weeds. 2014, pp 1–2. http:// ww.smh.com.au/it-pro/government-it/canberra-rangers-to-fight-weeds (accessed Sep 29, 2015).

Syngenta. Herbicides. 2015, pp 1–3. http://www.syngentta.com/global/corporate/en/products-and-innovation/product-brand (accessed Oct 3, 2015).

Thomson, S.; Huang, Y.; Reddy, K. N.; Fisher, D. K. Advanced Remote Sensing from Agricultural Aircraft and New Unmanned Aerial Platforms in Plant Health Assessment. *Proceedings of 12th International Conference on Precision Agriculture,* Sacramento, California, USA; 2014, pp 175.

Thorp, K. R.; Tian, L. F. A Review on Remote Sensing of Weeds in Agriculture. *Precis. Agric.* **2011,** *5,* 477–508.

Timmerman, C.; Gerhards, R.; Kuhbauch, W. The Economic Impact of Site-Specific Weed Control. *Precis. Agric.* **2003,** *4,* 349–360.

TOAS Project. *New Remote Sensing Technologies for Optimizing Herbicide Applications in Weed-Crop Systems;* CSIC: Madrid, Spain. 2015, pp 1–2. http://toasproject.wordpress.com/ (accessed Sep 29, 2015).

Torres-Sanchez, J.; Lopez-Granados, F.; Isabel de Castro, A.; Pena-Barragan, J. M. Configuration and Specification of an Unmanned Aerial Vehicle (UAV) for Early Season Site-Specific Weed Management. *PloS One* **2013,** *8,* 1–3 e58210. http://www.ncbi.nlm.nih. gov/pmc/articles/PMC3590160 (accessed June 27, 2015).

Williams, S. Weed killing Drones: The Answer to Toxic Herbicide Use. 2013, pp 1–6. http://www.care2.com/causes/weedkilling-drones-the-answer-to-toxic-herbicide-use/ (accessed Sep 29, 2015).

Xiang, H.; Tan, L. Development of a Low-Cost Agricultural Remote Sensing System Based on Autonomous Unmanned Aerial Vehicle (UAV). *Biosyst. Eng.* **2015,** *108,* 174–190.

Yintong Aviation Supplies. Agriculture Crop Protection UAV. 2014, http://china-yintong. com/en/producectsow_asp?sortid=7&id=57 (accessed June 10, 2016).

Young, S. L.; George, M.; Wayne, W. Future Directions for Automated Weed Management in Precision Agriculture. West Central Research and Extension Centre, North Platte; paper 79. University of Nebraska, Lincoln, Nebraska, USA, 2014, pp 1–19. http://digitalcommons.unl.edu/westcentrexst/79 (accessed Aug 14, 2016).

USDA-ERS. Pest Management. United States of Agriculture Department, Beltsville, USA. 2010, pp 1–3. http://www.ers.usda.gov/topics/farm-practies-managment/crop-livestock-practices (accessed Oct 18, 2015).

University of Arkansas. Arkansas Invasiveness: Cogon Grass. Division of Agriculture; University of Arkansas: Fayetteville, Arkansas, USA, Internal report 2015, pp 1–22.

University of California, Davis. UC Davis Investigates Using Helicopter Drones for Crop Dusting. *Agric.-UAV Drones.* 2014, pp 1–7. http://www.agricultureuavs.com/uc_davis_ uav_crop_dusting.htm (accessed June 21, 2015).

Zhu, H.; Lan, Y.; Wu, W.; Hoffman, C.; Huang, Y.; Xue, X.; Liang, J.; Fritz, B. Development of a Precision spraying controller for Unmanned Aerial Vehicles. DOI: 10.1016/ S1672-66529(10)60251-X. 2014, pp 1–8 (accessed May 25, 2015).

CHAPTER 7

DRONES IN CROP DISEASE AND PEST CONTROL

CONTENTS

7.1   INTRODUCTION

Crop production tactics in the agrarian belts have evolved constantly since several millennia. Agronomic procedures devised and practised have actually been dynamic. They have basically kept pace with crop's requirements, soil and environmental conditions, yield goals and economic gains possible in a given geographic location. In addition, crops in any agrarian region are exposed to detrimental factors such as disease-causing microbes, parasites and insect pests. Such biotic detriments could be endemic or may occur as epidemics and hinder the grain productivity. To control such onslaughts by diseases and pests, farmers adopt a series of integrated procedures. First, they opt for locations free of disease/pest inoculum. Then, they opt for crop genotypes that are genetically resistant to disease-causing agents and insect pests, prevalent in the region. Agronomic procedures such as earthing-up, thinning and culling of affected seedlings are invariably conducted at the right time. Application of plant protection chemicals is the most common procedure. Often, chemical sprays are adopted on priority by farmers. Indeed, plant protection chemicals constitute a major portion of total crop production cost for farmers. Periodic scouting of crop fields for diseases and pests, and preparation of maps showing their devastation is an important procedure. It involves farmer's investment. Often, crop scouting for diseases and pests requires highly skilled farm workers. Also, such crop scouting has to be repeated frequently and maps have to be updated. Progress of crop disease/pest attack could be either slow or rapid, depending on several natural factors. Therefore, it is mandatory to keep track and map the progression of disease/pest attack, if any, in a field. There are indeed several procedures and their variations adopted by farmers, to control diseases/pests. Most recent and the one gaining in popularity is the use of 'agricultural drones' to scout the crop aerially for disease and pest attack. Then, map their progression and provide digital data or maps that help farm technicians. Digital maps are used during spraying of chemicals. Precision techniques that involve variable-rate application of plant protection chemicals are also gaining in acceptance. Let us now consider details and discuss aspects related to drones and their usage in crop protection. The focus of first section of this chapter is about crop disease scouting, mapping its spread and intensity and application of chemicals using drones. Second section deals with use of drones to scout for insect pests, obtain digital imagery and data about pests. Then, spray pesticides to crop fields, based on blanket prescriptions or at variable rates.

7.2   DRONES TO CONTROL DISEASES ON AGRICULTURAL CROPS

7.2.1   DRONES TO AID RAPID DETECTION AND MAPPING OF CROP DISEASE INFESTATION

It seems, there are about 50,000 parasitic and non-plant parasitic diseases noticed and recorded in the United States of America alone. We have no idea how many of the diseases/pests show up easily recognizable symptoms on the crops. The sensors on drones should image the symptoms effectively and with good resolution. A data bank for all diseases that are possible to be detected and effectively mapped using drones is a necessity. It allows us to derive full potential of drones. Generally, early detection of any disease helps in efficient control. Drones perform detection and mapping with a greater accuracy and rapidity. At the same time, we ought to realize that several of the diseases that take to epidemic proportions are usually non-native imported strains of causal agents. For example, in the United States of America, annual loss of food crop productivity due to pathogens is 33 billion US$. About 65% loss equivalent to 21 billion US$ is attributable to introduced pathogens (see Sankaran et al., 2010). Recent examples are the Huanglongbing (HLB) Greening disease on citrus, a few scabs, blights and rusts. We need to match the situation by accruing data about diseases, particularly symptoms induced by the recently introduced diseases. It requires efficient and accurate use of drone imagery and computer-based digital support systems. A project to obtain accurate spectral data about diseases and their symptoms is useful. Spectral signatures of disease symptoms that can be detected by drone's sensors have to be stored in data banks. Such an exercise could be repeated for each agro-zone. Intercontinental transfer of spectral data about various crop diseases is a boon to local farmers. Farmers who adopt drone technology to scout, detect and map various diseases and pests could benefit from such spectral data transfers. They say, in any location, crop canopy and leaf colour could be among the best early warning signs about crop health (Lyseng, 2006; West et al., 2003).

Application of remote sensing to detect occurrence of disease on crops is not a new procedure. According to Huang and Thomson (2015), aerial photography was used to detect root rot disease on cotton way back in 1920s. However, in the succeeding period, development of remote sensing techniques to monitor and manage crop diseases has been indeed very gradual. Research on crop disease management aimed more on understanding the causal agent, its physiology, life cycle, mode of infection, its spread and finally methods to control it. Disease surveillance was conducted manually using skilled field workers. During recent years, several rapid techniques based on biotechnology have been devised. Many of these techniques depend on regular farm drudgery. They involve scouting the large tracts of crop fields first, and then picking samples and analysing them in laboratory. Molecular techniques may offer greater accuracy. However, most recent and rapid method to trace a disease on crops is based on multispectral imagery. It involves collecting digital images of crop canopy using close-up photography from low-flying drones.

At present, drone technology using small-sized copter/flat-winged systems seems apt. It is good enough to offer farmers with requisite crop imagery. Such imagery depicts crop's growth stage, its health and disease affliction, if any. Drones could offer digital data about disease spread and intensity. This procedure is effective enough during application of plant-protection chemicals (Ehsani, 2015).

Huang (2014) opines that development of drone, that is, unmanned aerial vehicle (UAV)-based technology to surveil diseases on crops received a definite impetus in the 1980s. Prior to it, research progress and adoption of drones in practical agriculture was literally slow. However, during this past decade, since 2010, there has been greater emphasis on studying crop disease, using remote sensing. In particular, studying occurrence and spread of crop damage using drones has gained, in popularity. Drones have gained acceptance as a means to obtain aerial imagery of crops and diseases. Drones fitted with multispectral sensors are used to gather digital data and prepare maps. Further, such data depicting disease spread and its intensity could be used in the decision-support computers. The computers could be located on the drone itself or in a ground station. Drones with variable-rate applicators could then be employed to spray plant protection chemical, accurately. Huang (2014) cautions that UAVs employed in crop disease scouting are small. They are fitted with multispectral sensors that have limited ability regarding resolution. Many of the rotary drones are slow. They have a low ceiling, regarding altitude during their flight above the crops. They possess a short endurance. Scouting for crop disease using drones could be improved, by fitting the copter/flat-wing aerial vehicles with high-resolution multispectral sensors. Flight endurance too has to be enhanced. High-performance cameras are almost mandatory to detect diseases, at early stages of the crop. Usually, pustules, deformities or discolourations form the symptoms of diseases. They have to be sharply photographed and identified by the sensors. However, waiting for formation of larger patches of disease in the fields is not a good idea because pathogen may have already caused the damage by then.

Scouting the field crops manually for diseases that may erupt and spread is a major preoccupation in any agrarian belt and it requires relevant skills. It has to be sharp and focussed. Scouting has to be accomplished right when the disease is at early stages. At the bottom line, we have to appreciate that diseases are wide spread on crops. Therefore, crops have to be scouted periodically and sometimes frequently. So far, farm technicians and their skills have been applied to detect and map such maladies. We obviously spend a great deal of time, incur costs repeatedly and suffer large-scale human drudgery in fields. These could be avoided, if drones are adopted on a large scale, during crop scouting. It helps to detect diseases and pests efficiently. A few reports suggest that we can even offer advanced information about a disease. We can forecast the spread of a disease in a geographic location, by employing drones, to detect disease propagules (or vectors) in the atmosphere. We have to corroborate propagules of disease causing agent with the occurrence of a particular disease in the vicinity. We should also take note of wind and sand drifts across the farms, mainly because they could induce the spread of disease propagules. This is a situation comparable to advance warning about weather.

Reports by Aylor et al. (2011) suggest that sporangia of the fungal causal agent *Phytophthora* sp could be detected in the atmosphere, using drones fitted with traps. Fungal propagules were actually estimated and compared, using two different techniques—first, using drone and second using ground tests. Reports suggest that drones could be effectively used to assess the density of propagules of Asian Rust fungus (Schmale, 2012). A different report emanating from Virginia Technological University at Blacksburg states that drones are excellent aides during aerobiological sampling for propagules of plant pathogens. They say, propagules of fungal causal agents such as *Phytophthora* and *Fusarium* could be detected, by analysing their spore densities in the atmosphere above cereal crops (Winkle, 2013). In fact, first, aerobiological samplings are done using drones. Then, rapidly, microbial species (disease causing agent) is detected and identified. Such a procedure helps in forecasting disease spread. Prophylactic sprays could be adopted using the same drones. Drone models with pesticide tanks are utilized to spray plant-protection chemicals.

Aerobiological sampling for propagules (sporangia, spores) of plant pathogens that are dreaded to cause severe damage on crops is practiced. It is confined to only some geographic locations. No doubt, drones are among the most recent and best bets to capture the propagules in the atmosphere. Propagules just above the crop canopy are trapped (Aylor and Fernandino, 2008; West et al., 2010). West et al. (2010) have stated that drone-aided sampling of atmosphere allows research agencies and farmers to obtain information regarding disease build-up. We can also assess the spread of disease to neighbouring districts. The disease severity could also be gauged using the density of propagules captured by the drones with traps. The density of propagules is actually utilized as a guide. Once the spatial distribution of disease for yesteryears is known, farmers may adopt suitable prophylactic measures of control. Farmers could also overlayer imagery of crop yield, disease maps and aerobiological data, and then ascertain the reasons for disease manifestation in a particular zone and consequent reduction of grain yield (West et al., 2009, 2010; Aylor and Fernandino, 2008). In any agrarian region, detection of crop diseases using drones fitted with multispectral sensors and facility to capture disease causing propagules is a clear possibility. It should help us in understanding the epidemiology of crop diseases better. Such a method warns us about spread of maladies (Brown and Hovmoller, 2002).

Let us consider few more examples. In India, groundnut is an important legume grown under dryland conditions. Groundnut crop is exposed to several diseases. Diseases may appear at various stages of growth and pod maturation. They need to be detected rapidly, mapped and control measures have to be devised accordingly. This procedure is usually done by the agricultural extension agencies. Drones could actually allow the extension agents to conduct survey for disease, quickly and without tedious sampling of entire plants, its leaves or pods. Recent reports from Gujarat and Rajasthan, in India, suggest that drones flown over the crops could help the farm agents, to map the diseases and insects that have afflicted the crop. They could obtain high-resolution images at short intervals and map them. Actually, they could map and determine the extent of area covered by healthy crop and the portion affected by diseases (The Economic Times, 2015). The data and maps derived were useful, in devising disease/pest control measures for the whole area (districts), including all the small farms.

7.2.2   DRONES IN CROP DISEASE CONTROL: A FEW EXAMPLES

Regarding the role of drones in practical farming, particularly in disease control, we may note that there are several companies that provide aerial imagery, of course at a cost to farmers. For example, HSE Inc. is a provider of aerial images that shows up diseases, if any, that occur in crop fields. They provide well-processed images, so that farmers can adopt precision techniques (HSE, 2015; Trimble, 2015; Zhu et al., 2014). There are several drone/satellite-related agencies and companies that offer such services. They usually offer aerial imagery, prescription of plant-protection chemicals and cultural procedures (Trimble, 2015).

Rush (2014) states that wheat crop in the Texan High Plains is exposed to a few different types of diseases. The diseases deter full expression of grain yield potential. Periodic droughts also reduce yield. Wheat crop here is vulnerable to mite-vectored virus diseases and a few others. These are actually mild diseases. Wheat diseases considered here are transmitted by wheat curl mite. The consequence is felt as reduction in root and aboveground parts. It reduces the quality of forage. Remote-sensing techniques have also been applied to study viral disease, such as Wheat Streak Mosaic (Rush et al., 2008). Farmers in Texas are currently exposed to use of copter drones. Such drones detect damage by the wheat streak virus and others transmitted by mites. They have obtained images of disease-affected crop fields derived using handheld multispectral sensors. Then, they have compared it with those got using drones and satellites. Each method has its advantages and constraints. A drone with facility for close-up shots and multispectral imagery seems better than others. Handheld sensors operated by skilled technicians are often tedious. Human-fatigue-related factors reduce the accuracy. In addition, human scouts are exposed to drudgery. Labour costs could escalate if skilled farm labor is hired for long durations. Satellite images are less accurate and resolution is much lower than that offered, by sensors on drones. Satellite images could be affected by cloudiness and haze in the atmosphere. Overall, wheat farmers changing over to drone-aided disease detection and accrual of digital data may be at an advantage. Drones seem more efficient (see Lacewell and Harrington, 2015).

Traditional techniques, such as active scouting of fields for yellow rust (*Puccinia recondita*) affected plants and patches, are indeed tedious and time consuming. Again, such methods are prone to inconsistencies related to human fatigue and skills. Reports suggest that, handheld multispectral sensors are effective in detecting the fungus affected plants, in a field. The spectral reflectance of disease affected crop is higher, if measured at 560–670-nm bandwidth. Next, the reflectance of yellow-rust-disease-affected canopy was lower at infrared band widths. The chlorophyll traits are also affected, if wheat is afflicted by yellow rust disease. The depth of absorption by chlorophyll at red band width gets lessened. Reflectance at green band width is also reduced, in disease affected plants (Huang et al., 2008b). These traits of yellow-rust-affected wheat crop could be utilized effectively. Drones with appropriate sensors could help us in detecting the wheat disease and map its spread. Early detection of yellow wheat rust could help farmers in adopting suitable plant-protection procedures. In Europe, studies on wheat afflicted simultaneously with powdery mildew (*Blumeria graminis*) and yellow rust (*P. recondita*) have shown that high-resolution multispectral sensors are effective. Multispectral images can actually detect a bunch of different diseases and pest species. Therefore, drones carrying such sensors offer farmers with immensely valuable information, about the actual status of wheat crop in the field.

Maize is an important cereal crop in Eastern Africa. Its cultivation necessitates periodic surveillance for diseases and pests. Maize rust caused by *Puccinia sorghi* affects the crop, by reducing photosynthetic efficiency of leaves. The rust pustules are thick and red to black in colour. They interfere with photosynthesis. Therefore, it leads to reduction in biomass accumulation and grain productivity. Farmers adopt a range of measures such as planting rust-resistant varieties, minimizing rust inoculum by culling affected seedlings right at early stage, adopting cropping systems that create a break in rust propagule build up, destroying the alternate host—*Oxalis* species—and so forth. Whatever be the agronomic measures adopted, at the bottom line, procedures such as surveillance of the crop, detection of maize rust quickly and adopting appropriate spray schedules are necessary. Recent reports suggest that it is possible to use agricultural drones, to monitor the rust disease. We can detect its occurrence and build-up in the maize fields (Palilo, 2014). Drones fitted with multispectral sensors capture spectral signatures of healthy and rust-affected maize canopy. Drone's imagery supplied to farmers could be effectively used, to control maize rust. Often, a few other maladies too could be traced, using the same imagery. Further, it has been stated that drones could become popular. Drone could be a useful gadget in the farmland, particularly if cooperatives and agencies that can afford drones could become active and offer help to farmers.

Potato is an important crop in Idaho, Washington and other northwestern states of the United States of America. Researchers at Washington State University, in particular, have evaluated the use of unmanned aerial systems. They have used drones to scout and monitor potato crop for various maladies emanating due to pathogens, pests and environmental vagaries. For example, they have tried to standardize drone-aided aerial imagery, to assess necrosis that affects potato crop's quality and productivity. They have reported that green normalized difference vegetation index data for about 64 fields were assessed using drones. The correlation with ground truth data provided by farm workers was high. The correlation was linear and it reached a value $r^2 = 0.91$ (Khot et al., 2014).

In Louisiana, USA, agricultural engineers have expressed that drones are apt to scout tall crops, such as maize or sugarcane. In the case of sugarcane, drones have performed well to scout and obtain imagery of frost damage, to a mature crop. Plant protection specialists opine that drones are better. They could be adopted to scout for diseases or even pests and weeds that affect taller crops, such as sugarcane (Schultz, 2013).

There are reports from Brazil suggesting that drones could be effectively adopted to scout the strawberry (*Frageria x ananassa* Duschesne) creepers. Drones could scout for general growth traits, diseases and pests that afflict strawberry. Strawberry is generally grown because it offers better economic gains to farmers in Brazil (Rieder et al., 2014). Therefore, strawberry farmers are stringent and take disease control measures without fail. However, we ought to realize that early detection of diseases, if any, and mapping them are essential. Farmers adopting traditional plant protection measures do use large quantities of pesticides. Pesticide application increases the cost of production. Hence, the fruits are accordingly priced high in the market. Recently, drones have been tested to survey and map strawberry crop area. Drones are used for surveillance of disease patches in the field, right at an early stage. Farmers are in fact warned early, about the impending spread of diseases on their strawberry crop. This is highly helpful to small farm holders, so that they could rationalize the use of plant protection chemicals. Farmers are also provided with 3D maps of strawberry farm. It helps them to know the phenology of entire strawberry field. Of course, individual spots too could be studied in detail. This is done by focussing sensors correctly on the region.

Apple orchards in northeast United States of America are affected by a few different diseases. Such diseases reduce productivity and economic value. Apple scab has been an endemic disease caused by the fungus *Venturia inaequalis*. It affects apple orchards, by causing large dark scabby lesions on foliage and fruits. The fungus does not affect the flavour or sweetness of the fruits. But, scabby damage of skin is not relished. Hence, fruits are valued low or even rejected by consumers. Researchers at the University of New Hampshire have opted to use drones, to rapidly detect scab infection on the foliage and adopt control measures (Kara, 2013). Farmers are being guided to take up sprays. This is to ensure that fungal life cycle is affected and its progress into different regions of trees, including fruits is totally avoided. The data from drones is assessed using a few predictive models (e.g. Dutch program RIMpro). Then, chemical spray schedules are decided (Modern Farmer, 2013).

Now, let us consider a fruit plantation wherein drones seem to be getting popular and economically more efficient, than traditional management techniques. Citrus belt in Florida is exposed to several diseases. At present, Citrus Greening disease also known as 'HLB disease' is occurring in devastating proportions. It is transmitted through a vector named *Diaphorina citri*, a psyllid insect. Florida region supports citrus on 220,000 ha. The tree plantation's economic worth and seasonal productivity could be severely reduced, if the orchard is afflicted with HLB disease. Garcia-Ruiz et al. (2013) state that first step in restricting HLB disease and obtaining its control is to detect the disease. Detection should be done at the earliest possible stage of infection. Symptoms that HLB disease produces have to be identified and detected by farmers. Morphologically, symptoms related to HLB disease are yellowing (etiolation), also chlorosis of leaf surface and veins. Trees die if HLB infection is severe and has spread all through the canopy. Here, drones have an important role in periodic monitoring of each tree and its canopy, leaves and twigs for HLB disease. The high-resolution sensors focused on each tree can provide the farmer, with details about health and HLB symptoms, if any. Human scouts find it very difficult to monitor the diseases from above the canopy. Obtaining a bird's eye view of a particular tree or entire orchard is not easy. Researchers at Lake Alfred, in Florida, USA, have examined the utility of low flying drones. Small drones are fitted with high-resolution multispectral sensors. Researchers are testing drones for their ability to offer detailed images of citrus canopy and leaves. Drones were also compared with manned aircrafts and their efficacy in detecting HLB disease. Garcia-Ruiz et al. (2013) state that drones fitted with sensors that operate at six different band widths could offer excellent details about HLB disease on citrus trees. Accuracies of positive detection of HLB compared with ground realities showed that 68–85% of aerial imagery was correct. Spectral bandwidths between 500 and 800 nm are recommended to detect HLB disease on citrus (Ehsani and Sankaran, 2010; Sankaran and Ehsani, 2012).

We have to note that a small drone needs low investment. They are highly cost effective compared with skilled plantation workers. These drones fetch information on the status of trees in a matter minutes. Drones cover several hundreds of acres in an hour. Drones could also be fitted with chemical tanks, to spray the citrus orchard immediately. Further, if variable-rate techniques are adopted using special nozzles, then, requirement of plant-protection chemical is reduced immensely. Sometimes,

80–90% reduction in chemical usage is possible. Compared with the above advantages of drones, a best effort by human scouts will involve several hours of scouting each tree. Further, human labor necessitates high cost. Biochemical tests that confirm HLB need careful sampling and processing. Further, citrus orchards under traditional scouting and management are now demanding higher investments. It is attributed to cost escalation. It has reached around 4331 US$·ha^{-1} in Florida. Forecasts, based on several advantages attributable to drones, suggest that, soon, most citrus orchards in Florida will adopt drones. Drones will be used to monitor HLB disease and even others. There are companies that employ drones. This is in addition to satellite and aircrafts (manned) images. They help farmers in detecting HLB and other citrus diseases rapidly. Farmers may have to just buy services of drone companies, to become economically efficient.

Avocados are subtropical fruits cultivated in Florida, USA. In this region, avocados are second in economic importance to citrus, if horticultural prowess of Florida is considered. Laurel wilt that gets spread by Ambrosia beetle is common. The Laurel wilt has affected avocados in Everglades region of Florida. They say, eradicating Ambrosia beetle is tedious, costly and it is difficult to achieve total control. The causal agent is actually a fungal complex. It is said that, by the time farmer observes the fungal filaments protruding out of tree trunks, the tree would have been damaged irreversibly. Hence, farmers use dogs that sniff the beetles and fungus-affected trees. A recent report states that drones fitted with specialized thermal and infrared cameras could detect affected avocado trees. Therefore, farmers now intend to use dogs and drones together, to detect laurel wilt in its early stages. Then, they adopt fungicidal sprays (Associated Press, 2015b).

Olive (*Olivia europa*) plantations are prominent in parts of Cordoba in Spain. They are exposed to several diseases. But during recent years, Verticillium wilt has become the most limiting biotic factor, affecting olive orchard's productivity. It seems traditional Spanish olive zones are experiencing severe reduction in oil bearing fruit yield (Calderon et al., 2013). The fungal pathogen *Verticillium dahliae* is a soil-borne organism. It affects the vascular system of the trees. Therefore, it blocks translocation of water. Trees develop Verticillium wilt disease in 18–24 months after planting. The disease development is dependent on propagule density encountered in the soil. Eventually, trees show up water stress and perish. Detection of fungal disease in the entire orchard is a necessity. Scouting, as usual, involves costs that could be prohibitive. Farm workers (scouts) do not get aerial view of entire canopy, unless plantations are visualized from vantage points. Scouting could be a slow process because a careful study of each tree is required. Calderon et al. (2013, 2015) have reported use of aerial observation using drones. Drones fitted with visual, infrared, near-infrared and thermal sensors were used. Drones derive data on water stress of trees, chlorophyll fluorescence and wilt damage. The canopy-level chlorophyll fluorescence drops as severity of Verticillium fungus increases. They say, early detection of Verticillium wilt is definitely possible, using aerial imagery. Actually, tree crown temperature and chlorophyll fluorescence measured using sensors are useful. It helps in detecting and mapping Verticillium wilt. Disease-control measures involving spray of plant-protection chemicals and culling are adopted, using drone imagery as a guide.

Banana plantations are important source of nourishment and exchequer to farmers in Philippines. Banana plantations are affected by several diseases. Among them, Black Sigoteka caused by *Mycosphaerella fijiensis var. difformis* and Panama disease caused by *Fusarium oxysporum* f. sp. *cubense* are endemic. They result in wide spread damage to the crop. They affect the foliage and fruit productivity alike. During recent years, devastation due to panama disease is on the rise (Triple20, 2015). Triple20 is a Dutch-based group that has introduced drones, to detect Panama-disease-affected patches in banana plantations. Drones are also being utilized to spray plantations with suitable plant-protection chemicals (Triple20, 2015).

Drone technology has been applied to detect several other crop diseases. It is based on symptoms induced by pathogens. Calderon et al. (2013) state that most studies have focussed on comparing leaf colour changes using hyperspectral imagery. Then, its relationship to disease severity has been evaluated. Multispectral analysis using remote sensing (drones) has proved useful, in detecting head blight of wheat. Soybean root rot and its effects on canopy have also been detected using drones (Wang et al., 2004). A few other diseases detected and mapped using aerial imagery are *Rhizoctonia* blight of grasses, Crown rot of sugar beet caused by *Rhizoctonia, Fusarium* caused head blight of winter wheat and effect of parasitic nematodes on crops (Calderon et al., 2013). Studying changes in canopy temperature of infected and healthy crops has also led us, to useful applications of drones. There are indeed several diseases caused by bacterial, fungal and viral agents that affect crops worldwide. Many of them are amenable to investigation, using drone technology. The resolution of multispectral sensors and vantage points that drones take above the crop canopy is crucial. It decides the accuracy and usefulness of digital data accrued. Clearly, drone techniques have a long distance to traverse. They have a great future in helping farmers, particularly, in detecting diseases and their spread, also in spraying plant protection chemicals efficiently and swiftly.

7.3 DRONES TO CONTROL INSECT PESTS ON AGRICULTURAL CROPS

7.3.1 DRONES TO AID DETECTION AND MAPPING OF INSECT PESTS ON CROPS

Initial reactions for use of drones to detect insect pests that attack crops are very encouraging. For example, United Soybean Board has mentioned that drones are excellent tools in the hands of farm technicians dealing with precision techniques. In particular, drones that fly past the soybean crop swiftly and spot the insect population could revolutionize the soybean production procedures. Farmers say that imagery from the sensors on drones helps them. They can look for exact spots afflicted by insect population. Problem areas can be marked using Global Positioning System (GPS) coordinates on the digital map, offered by drones. Drones could detect pests at a very early stage. Sensors could be focused even on a single plant or few leaves in a canopy. Then, the area could be studied in detail, for the kind of insect damage that has occurred. Mapping insect damage in detail is a clear possibility. Such details help us in evolving most appropriate control measures (United Soybean Board, 2013).

Reports from Louisiana Agricultural Centre, Louisiana, USA clearly shows that drones are helpful in scouting and then mapping insect pest attack on crops. Therefore, in future, pesticide spray could be very well focused. In addition to mapping and conducting pesticide spray, these drones could also help in applying herbicides and so forth (LSU Ag Centre, 2013).

A report by Zhang et al. (2015) states, there are insect pests such as *Spodoptera frugiperda* and related species. They are collectively called 'army worms'. They cause large devastations on maize cultivated in different continents. They are particularly severe in the North-American and Chinese maize production zones. At present, agricultural extension officers use satellite-aided multispectral imagery, to map the damage caused by larval instars of army worms. The satellite imagery lacks in resolution. Detection of the pest is possible only when damaged zones are large enough to be picked by sensors. Instead, drones with high-resolution multispectral imaging devices are perhaps better. They provide data on insect damage both in a small field and in a large maize production zone. Accurate digital data and maps could be obtained using drones. Mapping insect damage at an early stage has its advantages. Of course, drones seem to be well suited to perform such a task.

Reports from Pennsylvania, USA, suggest that a range of drone models, both flat-winged and copter types, are being tested for their utility in scouting soybean fields. It is aimed specifically to map the pest attack on crop. Soybean plots are scouted and monitored right from planting stage. Drones are used specifically to surveil crops for insect attack and help the farmers with digital maps. Such maps accurately depict locations affected by insect damage. Drones literally lead farmers to focal points in the field. Such foci of insect attack could otherwise initiate rapid spread of the pest in soybean fields (Vogel, 2014). These drones are remotely controlled. However, they could be programmed ahead, for flight routes and imagery. Incidentally, copter drones with facility for holding pesticides and equipped with variable-rate nozzles are in use. They are used to spray the crop with plant-protection chemicals. Farmers in these areas are prone to use drones, to obtain maps of entire fields and insect damaged regions, if any. The same drone could also be used to conduct pesticide sprays. Reports suggest that soybean fields of Minnesota, USA, are often attacked by a few insect species. They affect chlorophyll pigment formation on leaves and cause discolouration. Drones that fly above the pest-attacked areas clearly pick these zones. They have used the high-resolution camera to offer digital data and pictures. Farmers may then take appropriate control measures (Tigue, 2014).

During practical precision farming, having a map showing insect damage accurately is a prerequisite. Accuracy regarding species or subspecies of insect, type of damage and its extent needs emphasis. Let us consider an example. Wheat production is wide spread in the Central Plains of North America. This crop is attacked by several species of insects at different intensities. They cause loss of grain yield. Detecting insect infestations at an early stage, mapping the crop damage and adopting pesticide sprays employing drones seem mostly appropriate. In Kansas, over 9 million acres of wheat is cultivated. The whole stretch needs to be rapidly scouted and mapped, and digital data has to be accrued. A large section of wheat farmers in Kansas still adopt human scouts to detect the insect attack in their farms. However, at present, drones are among the best bets, to conduct such an operation swiftly and in time, particularly to restrict insect infestation in the wheat belt (Associated Press, 2015a). Obtaining digital maps for variable-rate pesticide sprayers is important. The variable-rate sprayers could be located on ground vehicles or on the drone itself. Since, the Kansas wheat belt is infested by several different species of insect pests, it seems that farmers and technicians are being trained and guided, to map different insect species. Knowledge about diversity of pest species and extent of damage that potentially occurs due to each insect species is essential. Pesticide formulations that are most effective on specific species need to be sprayed. To be effective, drones are actually fitted with high-resolution multispectral sensors, so that maps derived are accurate, detailed and show up infestation by different species of insect. It is believed that such an effort should be very effective in controlling insect pests, right at an early stage. Actually, reports suggest that wheat crop in Australia and Kansas, USA do suffer similar detriments from insect attacks. Hence, projects are examining drones and their ability to offer rapid imagery of wheat crop and insect infestations, if any. Mapping each insect species and subspecies separately for the damage that they cause requires more sophisticated sensors such as high-resolution cameras and matching computer-based decision support that depends on an excellent data bank. Data bank should have information about insects that affect wheat crop. Such techniques could be later replicated and tested at different locations, all over the globe. Overall, rapid and accurate mapping of insect species (pest) and their damaging effects on crops is important.

7.3.2 DRONES TO CONTROL PESTS AFFECTING CROPS: A FEW EXAMPLES

A few reports about crop-production procedures that we generally adopt suggest that improper decisions and spraying sprees of plant-protection chemicals could occur. If extrapolated to an agrarian belt, it leads to massive loss of material and exchequer. Sometimes, the loss increases to a tune of few billion US$ each year. Adoption of precision farming techniques seems apt. Drones have the ability to offer us with accurate digital data. Drone technology should be explored and utilized (Rosenstock, 2013). Incidentally, pesticides are defined as agents used as biocides. The purpose is to control insects and other species that cause damage on food crops. Farmers use a range of chemicals and spray them on to crops. Pesticides utilized commonly are classified as organophosphates, carbamates, organochlorines, pyrethroids, sulphonylurea and microbial pesticides (e.g. *Bacillus thuringiensis*). Physically, these pesticides could be a liquid formulation, powder, granules or flakes. Most of these have side effects on farm workers and their health, if they get exposed to it consistently. We should note that farm worker fatalities due to mishandling of pesticides are not uncommon in agrarian belts. A drone keeps farmers and workers away from pesticide contamination, specifically during spray operation that covers large areas of crop field.

Reports suggest that drones are being evaluated in North America, to scout the crop field for seed germination, general growth characteristics such as plant height, foliage, canopy size and normalized difference vegetation index. In addition, farmers are using drones to detect insect attack, if any, and study its progression. Drones are efficient because they help us in applying pesticides, only at locations affected by insect attack and not on the entire field (Norman, 2014). Further, reports from North American farm belt shows that multi-rotor agricultural drones could have a great future in performing tasks, which regulate insect pests on field crops. Such drones could be used to monitor crops, obtain digital data about insect attack and, then, spray the crop with pesticides from a close range. Drones actually spray pesticides from just above the crop canopy (MMCUAV, 2015; Ehmke, 2013). The pest control operation could be videoed and images reexamined, if needed, at any time. Consequently, pesticide spray rates, drone's navigation pattern and speed could be altered.

Whatever be the advantages listed in support of drone technology, we have to note that a versatile drone, with ability to fly low over the crop canopy and cover the field area as stipulated, is a prerequisite. More important is the spray system fitted to the drone. The spray system should release the plant protection chemicals, at rates fixed by computer decision-support systems or that determined manually. At present, a small range of drone models fitted with pesticide release systems are available in the market.

Prominent examples are Yintong's copters and Yamaha's RMAX copter. They take up flight over the crop as directed by a remote control or as fixed by computer signals (see Chapter 1; Yintong Aviation Supplies Company, 2012; RMAX, 2015; Szondy, 2013; Ehmke, 2013). Now, let us consider one example of a drone and its pesticide spray system. It has been practically evaluated in a grape vine yard of the Napa Valley of California, USA. As stated earlier, Yamaha's RMAX is a popular drone tested in a few agrarian belts, particularly for its utility as a 'sprayer drone'. Giles and Billing (2014a, 2014b) have evaluated this drone (RMAX) and its sprayer system with nozzles, for performance on grapes. The primary test area in the Napa Valley was 0.61-ha block that supported Cabernet Sauvignon grape vines. There were 61 rows spaced each at 2.4 m apart from the other. The spray system on the RMAX drone included an electrically powered small pump. It supplied liquid pesticide to flat nozzles (Teejet XR 11002 by Spraying Systems Inc.). The flow rate of nozzles was 1.3–2 $L \cdot min^{-1}$. We should note that aspects such as droplet size, spray width and consistency of pesticide fluid are also important. In this case, based on nozzles and compressor pressure, the droplets released by the drone were categorized as fine to medium. The RMAX drone can saddle two tanks holding 8 L of pesticide each.

Giles and Billing (2014a, 2014b) further report that, pesticide spray deposition and its performance depended on the spray method employed, particularly the swath width and flight pattern of the drone. The drone could spray pesticide at a rate of 2.0–4.5 $ha \cdot h^{-1}$. The volume of pesticide dispensed from the tanks was 14–39 $L \cdot ha^{-1}$, as decided by computer decision-support system. The spray pattern and droplets were similar to that achieved, using ground sprayers or those fitted to a manned aircraft sprayer. There are several variations of spray equipment, nozzle types and computer software that regulate pesticide sprays. Appropriate systems could be fitted and adopted on the drone (Zhu et al., 2014). Atomizers are also used in some types of drones. Results from experiments in the Napa Valley showed that drones can release 10–40 L of pesticide per hectare. Drones have an effective work rate of 2–5 $ha \cdot h^{-1}$. Drone operator's skills are also important. Overall, drones with efficient sprayers systems are safer to human population surrounding the farms and to farm workers. At the same time, it offers significantly high economic returns. Drones utilize very low quantities of pesticides, if precision techniques and variable-rate nozzles are used. Therefore, drones reduce cost on the pesticide material. It is said that drones and precision agriculture are a natural fit. However, more important fact is that, drones are excellent bets, when we have to handle hazardous chemicals such as pesticides. As drones fly very close to crop canopy, pesticides released reach the leaves. Usually, very little quantity of pesticides trickles to soil/ground. Drones apply pesticides only on areas that are affected by pests. They do not wastefully apply everywhere in the field. Hence, drones are efficient.

In the Southern plains, efforts by United States Department of Agriculture (USDA) are directed towards adoption of drones to detect pests on crops and map them and also to standardize procedures to spray pesticides at variable rates. Sensors that rapidly detect different pests are being developed and evaluated. More important is their effort to develop decision-support systems of help, to operate variable-rate applicators. Crop consultancies try to supply aerial maps of fields with pest attacked zones marked on them. They also offer appropriate advice on pesticide sprays. But, these agencies depend strongly on sharp sensors. During aerial spray, drift of pesticide is a problem. It needs to be overcome. Hence, under this project, Bradley et al. (2015) aimed at noting pesticide drifts and their effect on crops and neighboring fields. Importantly, we have to ensure that nontarget areas and crop species (or fields) are kept, at a safe distance away from drone-aided sprays of pesticides. Drone's flight path and selection of days with least wind distraction are crucial. A closely related project, again conducted by researchers at USDA, is about standardizing drone technology, to remotely image the crop for pest damage, then spray pesticide by adopting site-specific management systems. Sensors with very high resolution and accurate pesticide spray systems will be developed eventually. It is said that drones are cheaper and can focus on small fields, whereas a manned aircraft is less affordable. It flies at rapid pace but imagery of small regions is not easy. In most cropping belts, drones with sensors for accurate detection and mapping insect damage is preferable. Further, drones that immediately apply site-specific methods (variable rates) to distribute pesticides are preferred (Huang et al., 2010).

The emphasis here in this section of the chapter is predominantly on insect pests that affect crops and plantations. Also, on the way, we can employ drones to control such detrimental biotic factors. At present, high input farms in the developed nations employ large dosages of fertilizers and crop protection chemicals. Pesticide sprays are done, mainly, to keep the crop free of insect attacks. Usually, high-intensity blanket sprays of pesticide on entire field are done. Huang et al. (2009) state that accurate sprays done using sophisticated variable-rate applicators has become a necessity, particularly if site-specific techniques are opted. Therefore, currently, drones with ability for scouting and variable-rate pesticide sprays are being opted more frequently in North America. It seems drones need relatively less capital. They are cost effective during operation and lessen need for pesticides. Further, Huang et al. (2008a, 2009) report that they have developed a low volume sprayer set that could be fitted to agricultural drones. For example, two drone models such as SR20 and SR200 Rotomation are fitted with low-volume sprayers. Drones could be fully autonomous or semi-autonomous and still perform well, during pesticide spray. Sprayer activation and volume could be controlled accurately, using GPS coordinates. Such sprayers are apt for use during site-specific farming. Quantity of pesticides released could be controlled, by varying flow-rate as well as number and size of nozzles. Nozzles become operative based on electronic signals. Overall, drones have excellent future for routine use during pest control (Miller, 2005; Huang et al., 2009, 2010).

We may note that crop duster airplanes too could be fitted with hyperspectral cameras and used just like the drone (Anderson, 2014). However, cost of operating such crop duster airplanes could be more than employing drones. Manned crop duster airplanes are less versatile in their flight path. They cannot fly too close to crop's canopy. Wind caused drifts to spray pattern could be prominent, if airplanes are used.

Farmers have been questioned about drones with an intention, to obtain a feedback. One of the reactions suggests that farmers are already equipped with ground vehicles and variable-rate sprayers. The nozzles in the sprayers automatically open or close, based on problem areas located. Nozzles function based on digital data contained in a chip. However, imagery from drones will be helpful in pin-pointing the problem on a map, more accurately. It provides an overview of the entire field, the crop growth and pest infested zones (Lyseng, 2006). Incidentally, drones are also utilized to oversee spray schedules by ground vehicles (Plate 7.1).

Taylor (2014) reports about a few case studies involving drones that scout for insects and diseases using high-resolution imagery. He points out that farmers in Ohio currently sample small pockets of their crop land. They try to distribute sampling spots as evenly as possible and try to derive recommendations. However, drones scout the crop fields rapidly and in entirety. Therefore, any of the recommendations on pest control will be that much more accurate. For example, *E384* is a drone model used by farmers

PLATE 7.1 A flat-winged drone helping farmers in scouting and overseeing pesticide spray operation in a large farm in Kansas, USA.

Source: Tom Nicholson, Ag Eagle Inc. Neodisha, Kansas.

in Ohio. It helps them to trace pests and related maladies inflicted on cereal crops. The drone flies 1000 m above the crop. It captures digital images of 2-cm resolution. At 400 m above the crop, it really scouts rapidly. It offers 140 pictures of 3 cm·pixel^{-1} in 15 min. The above drone covered a 1000-ac farm in 15 min of flight. It is really a rapid effort in finding spots affected by pests. In case, farmers require greater details about insect species (or any malady), then sensors have to be upgraded. High-resolution cameras need to be fitted in the payload area. Drone-derived pictures generally offer an overall image of the large area of field.

It is interesting to note that in Germany, farmers have used a low-flying drone, to distract and ward bird pests and other animals off from farms that have mature panicles (Microdrones GMBH, 2014). In Minnesota, USA, drones have been examined for their efficiency in detecting and quantifying animal caused damage to maize crops (University of Minnesota Extension Services, 2014).

7.3.2.1   DRONES AND BIOLOGICAL CONTROL OF INSECTS

Farmers are also conversant with pest control methods that adopt 'biological control'. Biological control regulates the insect pest population and damage that ensues. There are several examples of successful biological control of insect pests. However, there are factors such as accurate application and establishment of biological control agent, which need attention. Lepidopterous pests such as cutworms, pod borers and stem borers are often controlled using vasps that oviposit in the lepidopterous larva and reduce the pest population. Eggs of biological control agent are placed in cards at different spots in the field, so that parasite gets well distributed. A recent idea that has been tested successfully deals with spray of adult *Trichogramma* (parasite) to eradicate the European corn borer (*Lepidoptera*), by using low flying drones (Dronologista, 2015). This drone is manufactured by Height-Tech Inc. located at Blakefield, in Germany. Drones drop *Trichogramma* species at several points in the field. It ensures that the biological control agent is uniformly distributed. It can then lead to establishment of *Trichogramma*, thoroughly in the entire field. Incidentally, European corn borer is a major pest on corn. Its regulation at threshold levels is essential. Hence, we may integrate use of drones in the pest control procedures, in addition to other methods.

This trend to use agricultural drones, to spray, parasites involved in biological control of pests is gaining in popularity. Let us consider another example. In Queensland, Australia, UAVs are employed to spread the beneficial parasites namely mites (Californicus mite) (Schiller, 2015). The primary aim is to reduce pests on field crops. They say dusting with mites using drones helps in reducing chemical sprays. It avoids environmental contamination. Light weight drones of 5.5 lb and powered by batteries are apt for spraying mites. The mites are actually stored in glass cylinders with vermiculite as a carrier medium. The drone takes hardly a few minutes, about 15 min, to cover a large area of field crop with biological agent (mite). The efficacy of drone-sprayed mites is being evaluated in several farms, using cameras and noting crop growth pattern. Drones reduce cost for scouting and skilled labour required, to apply mites in fields.

7.3.2.2   DRONES TO SPRAY TREE PLANTATIONS

Drones are perhaps apt to be used in scouting and dusting tree plantations. They may perform better than a less versatile manned airplane. Winslow (2014) reports that UAVs are perfect for tree crop dusting because each and every tree could be imaged and appropriate levels of pesticides could be sprayed. Rampant use of pesticides that occurs during blanket sprays is totally avoided. The cost on pest control depreciates. In addition, drones keep humans away from contact with pesticides. Indeed, tree crop dusting using a few or a swarm of drones is an apt idea, particularly, in large plantations. Drones actually hasten up the dusting process. Regarding scouting, it has been suggested that drones that fly very close to trees can provide excellent image of the canopy, leaves, extent of damage by insects, if any. Most importantly, drones can show us exact spots that need dusting. Drones fitted with hyper-resolution sensors can pick images of trees affected by pests, fungal diseases, leaf fall and drought, simultaneously. Insect species affecting individual trees can be identified, because images by drones are usually sharp and clear. Forecasts suggest that drone usage could improve pest control, lessen yield depreciation and therefore enhance profits for farmers.

7.3.2.3   AGRICULTURAL DRONES IN CHINA

Pesticide sprays involve costs to farmers. Often, farmers may have to arrange five to six sprays within a crop season. Prior to sprays, farmers also incur costs on scouting large fields. However, one of the recent reports from farmland of China suggests that drones really offer efficient alternative in terms of scouting, pesticide consumption and spray time. For example, farmers near the city of Changsha, in Hunan province of China, are using a drone named '*32D–10A*'. It is an extremely low altitude copter drone with capability to spray insecticide rapidly (Crienglish News, 2014). Further, they stated that pesticide consumption got reduced by half, if they used drones. Pesticide sprays were accurate and evenly well distributed leading to a good control of pests. Drone usage definitely reduced the cost of pesticide spray. Costs incurred reduced to just 0.5 yuan for gasoline. Drones covered 1334 m² for every yuan spent. In all, labour and material cost got reduced by 70% of traditional level. Often, we can use drones to scout for a few different parameters, not just for pests or disease monitoring. In fact, these farmers in China state that drone pictures also helped them in arranging irrigation properly. Therefore, water consumption too got reduced enormously. Water consumption reduced by almost 90%, compared with traditional systems (Crienglish News, 2014).

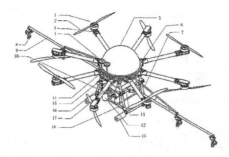

FIGURE 7.1   Diagram of a drone used to dust or spray crops.

Note: 1 Motor; 2 Propeller; 3 Motor mount, 4 GPS antenna; 5 Dome; 6 Folding arms; 7 Support rod; 8 Nozzle; 9 Sprayer pipe; 10 Pump interface; 11 Tank; 12 Pump; 13 Landing gear; 14 GPS holder; 15 Top plate; 16 On board electronics; 17 Bottom plate and 18 Landing gear vertical support.

Source: Mr. Adam Najberg, DJI Inc. Shanzen, China; Department of Intellectual Property, DJI IInc. Shenzen, China.

China supports a very large expanse of agricultural enterprise. Crop-production strategists have to be alert to several detrimental factors, to maximize productivity. Hence, usage of pesticides and other plant protection chemicals is relatively high. Pesticide usage is pronounced in the intensive farming belts of south and northeast China. Farmers utilize over 100 million skilled farm workers. They all deal with several dangerous chemicals. They are all exposed to possible ill effects of pesticides. Hence, Chinese farms are prone to adopt drone technology eagerly. They may use drones to survey, map and apply pesticides to control insects. Pesticide applicators attached to drones release chemicals without affecting any human labourer. Drones are also apt for use during precision agriculture. Drones fitted with variable-rate pesticide appli-cators could reduce pesticide consumption (Yintong Aviation Supplies Co, 2012). For example, Yintong's agricultural drones are highly useful. They spray pesticide covering an area of 2.8-m perimeter. This drone covers an area of 2.25 ac·min^{-1}. The variable-rate sprayers release about 2.5 L pesticide per min. Hence, they are also quick in accomplishing plant protection tasks.

Agricultural drones have also been utilized to detect diseases/pests that occur on horticultural crops cultivated in China. They are trying to scout and map diseases that occur on fruit crops such as apple or peach. The same small drone is also used to spray plant protection chemicals (Cao et al., 2010; Figure 7.1; Plates 7.2 and 7.3).

PLATE 7.2   'DJI GH-1 Agras' is a pesticide spraying drone.
Note: DJI GH-1 Agras has facility to hold plant-protection chemicals and to spray them on crops. Drones with decision support computers and GPS guidance can dispense liquid formulations, using digital maps and precision techniques.
Source: Mr. Adam Najberg, DJI Inc. Shanzen, China; Department of Intellectual Property, DJI IInc. Shenzen, China.

7.4   AN OVERVIEW

There is no doubt that diseases and pests affect a very wide range of agricultural crops. Agricultural regions experience major losses due to such maladies. On an average, they say at least 10% of food grain generation potential is reduced due to diseases/pests. Further, in an effort to control diseases/pest, farmers use excessive plant-protection chemicals. So, they run into environmental problems. Irrigation and ground water contamination is encountered frequently. Satellite imagery and airborne campaigns have been used since long. They help to detect crop damage (Huang et al., 2008a). Satellite imagery has its limitations of resolution. The images that it captures could be hazy. Also, farmers will have to wait for its transit path over their fields. On the other hand, airborne images are costly. They also involve higher investment. To quote a few examples of airborne campaigns, it seems citrus blackfly, citrus scale and whitefly damages could be monitored and imaged,

PLATE 7.3 A spray nozzle and accessories attached to drones to spray pesticides.

Note: Drones used during plant protection may often be attached with nozzles that are electronically controlled and operated at variable rates, based on instructions from computer-based decision systems or chips.

Specifications of the above sprayer are as follows: spray droplet 60–180 μm; optimal flying speed 1–10 m·s^{-1}; spraying swath 3–5 mm (even atomization); pesticide/liquid load 10–20 kg; optimal spraying height 1–5 m; control remotely controlled or automatic; discharge six spray nozzle; outflow rate 0.5–2.4 min^{-1}.

Source: Mr. Adam Najberg, DJI Inc. Shanzen, China; Department of Intellectual Property, DJI Inc. Shenzen, China; and Mrs. Lincy Prasanna, Vinveli Unmanned Systems Inc. Cedar Rapids, Iowa, USA.

using manned-aircraft-aided remote sensing (Everitt et al., 1991, 1994, 1996; Hart and Myers, 1968; Hart et al., 1973). Airborne campaigns have been successful in helping farmers with requisite images and information. However, if we need greater details and wish to survey crop fields at low cost, then close-up photography using drones attached with multispectral sensors seems apt. There are indeed innumerable pests and diseases whose detection and mapping could be effectively done, using drones. Drones have a long way to traverse regarding number of diseases/pests, yet to be tackled. *Foremost, data banks with spectral signatures of disease affected and healthy crops (seedlings) need to be developed.* Drones could eventually replace airborne campaigns to monitor crop damage.

Adoption of drones to control disease or pests is still rudimentary. However, the potential to use them is too great as almost every agrarian region will need the help of drones to regulate insect and microbe-related maladies. Such drones could collectively help farmers in raising better crops. It could lead us to enhanced grain and biomass yield per unit area. *Drones actually offer to reduce human drudgery immensely, during plant protection.* At present, there are only a few examples (see Plates 7.4 and 7.5). Agricultural drones are being used by personnel at Agricultural

PLATE 7.4   Vero 8A—A sprayer drone used in Agricultural fields.
Source: Mrs Lincy Prasanna, Vinveli Unmanned Systems Inc. Cedar Rapids, Iowa.

PLATE 7.5   An agricultural drone spraying pesticides on to a rice crop.
Source: Mrs Lincy Prasanna, Vinveli Unmanned Systems Inc. Cedar Rapids, Iowa.

TABLE 7.1 Agricultural Drones Applied to Rescue Crops from Disease and Pest Attack: A Few Examples.

Crop/Location	Disease/Pest	Reference
Diseases		
Wheat/Central Plains, Kansas, USA	Yellow Rust of Wheat (*Puccinia striformis*)	Kansas State University (2015)
Wheat/Amarillo, Texas, USA	Wheat leaf streak and Mite-vectored virus diseases	Rush (2014)
Maize/Southern Highlands, Morogoro, Tanzania	Maize Rust (*Puccinia sorghi*)	Palilo (2014)
Soybean/Northern Great Plains, USA	Cyst Nematode (*Heterodera glycines*)	United Soybean Board (2013)
Strawberry/Passo Fundo, Brazil	*Botrytis cinerea* fruit rot	Rieder et al. (2014)
Poppy/Central Highlands, Tasmania, Australia	Systemic Downy Mildew (*Peronospora arboscens*)	Dakis (2015)
Citrus/CREC, Lake Alfred, USA	Citrus Greening (Huanglongbing)	Garcia-Ruiz et al. (2013) and Ehsani and Sankaran (2010)
Olive/Cordoba, Spain	*Verticillium* wilt *Verticillium dehliae*	Calderon et al. (2013)
Pears/Washington State, USA	Fungal complex, several species	Khot et al. (2014)
Pests		
Wheat/Central Plains, Kansas, USA	Russian Wheat Aphid (*Diuraphis noxia*)	Kansas State University (2015)
Rice/Central Japan	Aphids, Shoot flies, Cutworms and borers	Giles and Ryan (2014a, 2014b)
Maize/Blake field, Germany	European corn borer (*Ostrinia nubilalis*)	Dronologista (2015)
Grapes/Napa Valley, California, USA	European Grape vine Moth (*Lobesia botrana*); Western Grape leaf Skeletonizer (*Harrisina metalica*)	Giles and Ryan (2014a, 2014b)

Note: Several reports about drones and their utility during disease/insect monitoring in crop fields do not identify the names of maladies, organism, species, location nor the extent of damage reduced, due to usage of drones. Drone models used and their specifications too are missing in some reports. Details on these aspects are indeed essential, in future, as drone technology becomes more popular in agricultural regions. Often, a particular drone model may suit best to survey a type of terrain, a crop species and identify specific disease/insects based on sensors and so forth. Standardizations regarding terrain, crop species, disease/insect pest malady, drone model and sensors are needed. Matching drone models with a specific task or set of agronomic procedures seems a good idea for now.

Experimental Stations and in farmer's fields, to control/regulate crop disease and pest (see Table 7.1).

7.5 DRONES TO SAFEGUARD AGRARIAN BELTS WORLDWIDE

Managing epidemics of plant pathogens and ensuing crop damage involves early detection and warning systems. Farms worldwide are exposed to periodic onslaughts (epidemics) of crop diseases, insects and other pests. Many of these are also endemic to certain crops and locations. They damage crops in variety of ways and reduce forage, grain and fruit productivity. Regular surveillance of each crop field is essential. Plant protection measures and their efficacy largely depend on the rapidity and accuracy with which, emerging insect pests and diseases are tracked, and their potential spread is judged by farmers. According to Agricultural Researchers at Kansas Agricultural Experimental Station, Manhattan, USA, there is a strong need to standardize protocols for active surveillance of wheat and other major grain crops. Procedures for using aerial detection of crop diseases and pests need to be streamlined. Drones (Unmanned Aerial Systems) are apt to be used, to achieve regular surveillance of large areas of Kansas wheat belt (over 9 million ac). Drones accomplish surveillance quickly and accurately, to conduct, timely sprays of pesticides on crops (Kansas State University, 2015). At present, there is a collaborative project involving Agricultural Drone Technologists of Kansas State, a Consortium of Queensland University of Technology and New Zealand's Agricultural Department. They are actually operative under a consortium known as 'Plant Biosecurity Cooperative Research Centre'. Together, they intend to optimize surveillance protocols for various disease and pests that attack crops worldwide. At present, the aims are confined to wheat crop cultivated in North America and the Australian continent. Incidentally, wheat-cropping zones in Central Plains of North America and Australia do share several common maladies. On a wider horizon, we can summarize saying that, drone-based surveillance techniques should be adopted worldwide. We can monitor agricultural zones and report about emergence of any disease or insect pest attack, promptly and quickly. *Global crop security using drone and sensor techniques is the intended theme.* In fact, currently, farm researchers at Manhattan, Kansas State University are striving to develop easy protocols. They intend using drones and multispectral imagery, to detect diseases and insect pest early. This is to safeguard cereal crops and plantations. The idea is to detect crop maladies before they accentuate and take proportions of severe attacks or epidemics (Kansas State University, 2015). A few forecasts state that, soon, agricultural drones will fly and hover over America's Bread Baskets. They could reduce use of pesticides and plant protection chemicals (Holt, 2013). Globally, there are a few agrarian regions similar to Great Plains of North America, such as the European plains, Cerrados of Brazil, Pampas of Argentina, Gangetic plains and so forth. These are important food grain generation systems. These agricultural regions too need drones and their services. *It is now becoming easy to guess that, ultimately, drones will take guard of global crops and their productivity. They will monitor diseases and pests that attack crops. They will do so by flying periodically over the fields and informing crop protection agencies, in time, to adopt control measures.* Global crop protection using drones and drone-related agencies is in a way mandatory. It works to offer food-grain security to the large human population on earth. Drone technology is said to reduce costs on global biosecurity of crops (Lyons, 2015). Ultimately, aspects such as agricultural drones, global crop protection, removal of human drudgery, reduction in the use of plant protection chemicals and environmental protection all seem interrelated. A global crop protection agency based exclusively or predominantly, on the usage of agricultural drones, seems an excellent proposition, at this juncture. Agricultural administrators need to take note of it.

KEYWORDS

- pest control
- crop disease control
- biological control
- crop dusting
- site-specific methods
- crop protection chemicals

REFERENCES

Anderson, C. Drones vs Crop Dusters to Gather Crop Data. DIY Drones. 2014, pp. 1–5. http://diyydrones.com/profiles/blogsdrones-vs-crop-dusters-to-gather-crop-data (accessed Oct 9, 2014).

Associated Press. Kansas State University Researchers Using Drones to protect Crops. *The Topeka Capital Journal.* 2015a, pp 1–2. http://cjonline.com/news/2015-03-19/kansas-state-university-researches-using-drones (accessed May 22, 2015).

Associated Press. Drones, Dogs Deployed to Save Avocados from Deadly Fungus in Florida. The Associated Press. 2015b, pp 1–6. http://www.nydailynews.com/life-style/eats/drones-dogs-deployed-save-avocados-deadly-fungus-in-floirda (accessed Oct 18, 2015).

Aylor, D. E.; Fernandino, F. J. Prospects for Precision Agriculture to Manage Aerially Dispersed Pathogens in Patchy Landscapes. *Proceedings of 9th International congress of Plant pathology, healthy and safe food for everybody,* Torino, Italy. *Plant Pathol.* **2008,** *90,* 59.

Aylor, D. E.; Schmale, D. G.; Shields, E. J.; Newcombe, M.; Nappo, C. J. Tracking the Potato Late Blight Pathogen in the Atmosphere Using Unmanned Aerial Vehicles and Lagrangian modelling. *Agric. Forest Meteorol.* **2011,** *151,* 251–260.

Bradley, C. F.; Martin, D.; Westbrook, J.; Yang, C. *Aerial Application Research for Efficient Crop Production;* United States Department of Agriculture: Beltsville, USA, 2015; pp 1–2. http://www.ars.usda.gov/research/projects.htm?ACCN_NO-414825 (accessed Oct 9, 2015).

Brown, J. K. M.; Hovmoller, M. S. Epidemiology: Aerial Dispersal of Plant Pathogens on the Global and Continental Scales and its impact on Plant Disease. *Science* **2002,** *297,* 537–541.

Calderon, R.; Naves, H. J. A.; Lucena, C.; Zarco-Tajeda, P. J. High-resolution Airborne Hyperspectral and Thermal Imagery for Early Detection of Verticillium Wilt of Olive Using Fluorescence, Temperature and Narrow-band Spectral Indices. *Remote Sens. Environ.* **2013,** *139,* 231–245.

Calderon, R.; Naves, J. A.; Zarco-Tejada, P. J. Early Detection and Quantification of Verticillium Wilt in Olive Using Hyperspectral and Thermal Imagery Over Large Areas. *Remote Sens.* **2015,** *7,* 5584–5610.

Cao, H.; Yang, Y.; Pei, Z.; Zhang, W.; Ge, D.; Sha, Y.; Zhang, W.; Fu, K.; Liu, Y.; Chen, Y.; Dai, H.; Zhang, H. Intellectualized Identifying and Precision Control System for Horticultural Crop Diseases Based on Small Unmanned Aerial Vehicles. *Comput. Comput. Technol. Agric.* **2010,** *VI,* *393,*199–202.

Crienglish News. Farmer Uses 'Helicopter' in Pest Control. 2014, pp 1–3. http://englsih. cri.cn/12394/2014/10/16/2361s848132.htm (accessed Oct 9, 2014).

Dakis, S. Drones Helping Tasmania's Poppy Industry Combat Serious Systemic Mildew Disease. Rural. 2015, pp 1–4. http://abc.net.u/news/2015-04-16/agricultural-drones-poppy-mildew-precision/ (accessed Oct 18, 2015).

Dronologista. Drones for Pest Control. 2015, pp 1–3. http://dronologista.wordpress.com/2014/04/09 (accessed Oct 9, 2015).

Ehmke, T. Unmanned Aerial Systems. American Society of Agronomy, CSA News. 2013, pp 1–7.

Ehsani, R.; Sankaran, S. Sensors and Sensing Technologies for Disease Detection. Citrus Industry. 2010, pp 1–5. http://www.crec.ifas.ufl.edu/.../2010junesensoringtechnology.pdf (accessed Oct 14, 2015).

Ehsani, R. Applications of Small Unmanned Aerial Systems in Agriculture. Symposium on Agricultural Remote sensing with UAVs: Challenges and Opportunities. American Society of Agronomy Madison, Wisconsin, USA. 2015, pp 1–28. http://dl.science societies.org/publications/meetings/2014am/13358/87813 (accessed Oct 1, 2015).

Everitt, J. H.; Escobar, D. E.; Villareal, R.; Noriega, J. R.; Davis, M. Airborne Video Systems for Agricultural Assessment. *Remote Sens. Environ.* **1991,** *35,* 231–242.

Everitt, J. H.; Escobar, D. E.; Summy, K. R.; Davis, M. R. Using Airborne Video, Global Positioning System and Geographical Information System Technologies for Detecting and Mapping Citrus Blackfly Infestations. *Southwest. Entomol.* **1994,** *19,* 129–138.

Everitt, J. H.; Escobar, D. E.; Summy, K. R.; Alaniz, M. A.; Davis, M. R. Using Spatial Information Technologies for Detecting and Mapping Whitefly and Harvester and Infestations in South Texas. *Southwest. Entomol.* **1996,** *21,* 178–182.

Garcia-Ruiz, F.; Sankaran, S.; Maja, J. M.; Lee, W. S.; Rasmussen, J.; Ehsani, R. Comparison of Two Aerial Imaging Platforms for Identification of Huanglongbing Infected Citrus Trees. *Comput. Electron. Agric.* **2013,** *91,* 106–115. DOI: http://dx.doi.org/10.1016/j.compag.2012.12.002 (accessed Sept 7, 2016).

Giles, D. K.; Billing, R. Unmanned Aerial Platforms for Spraying: Deployment and Performance. *Aspects Appl. Biol.* **2014a,** *122,* 63–69.

Giles, D. K.; Billing, R. Deployment and Performance of an Unmanned Arial Vehicle for Spraying of Speciality Crops. *Proceedings of International Conference on Agricultural Engineering,* Zurich, g.eu pp Switzerland, 2014b, pp 1–7. http://www.eurageng (accessed Oct. 10, 2015).

Hart, W. G.; Myers, V. I. Infrared Aerial Colour Photography for Detection of Populations of Brown Soft Scale in Citrus Groves. *J. Econ. Entomol.* **1968,** *61,* 617–624.

Hart, W. G.; Ingle, S. H.; Davis, M. R.; Mangum, C. Aerial Photography with Infrared Colour Film as a Method of Surveying for Citrus Black Fly. *J. Econ. Entomol.* **1973,** *66,* 190–194.

Holt, S. How Will Drones be Used on American Farms? 2013, pp 1–7. http://www.takepart.com/article/2013/05/24/how-will-drones-be-used-amricans-farms (accessed Oct 9, 2015).

HSE. Intelligent Imaging. 2015, pp 1–3. http://www.uavcropdustersprayers.com/agriculture_delta_fw_70_fixed_wing_uav.htm (accessed May 19, 2015).

Huang, Y. *Development of Unmanned Arial Vehicles for Crop Production and Production.* Symposium on Agricultural Remote sensing with UAVs: Challenges and Opportunities; American society of Agronomy Madison: Wisconsin, USA, 2014; pp 1–12. http://dl.science societies.org/publications/meetings/2014am/13358/88127 (accessed Sept 6, 2016).

Huang, Y.; Thomson, S. J. Remote Sensing for Cotton Farming. American Society of Agronomy. *Agron Monogr.* **2015,** *57,* 183. DOI: 10.2134/agronomymonogr57.2013.0030 (accessed July 4, 2015).

Huang, Y.; Lan, Y.; Hoffmann, W. C. Use of Airborne Multispectral Imagery for Area Wide Pest Management. *Agric. Eng. Int.: The CIGR Ejournal Manuscript IT07 010.* **2008a,** *10,* 1–14.

Huang, W.; Lin, L.; Haung, M.; Wang, J.; Wan, H. Monitoring Wheat Yellow Rust with Dynamic Hyperspectral Data. 2008b, pp 1–12. http://www.researchgate.net/c/nveacx/javascript/lib/pdfjs/web/viewer.html?file=http (accessed Sept 29, 2015).

Huang, Y.; Hoffmann, W. C.; Lan, Y.; Wu. W.; Fritz, B. K. Development of a Spray for an Unmanned Aerial Vehicle Platform. *Appl. Eng. Agric.* **2009,** *25,* 803–809.

Huang, Y.; Hoffman, W. C.; Lan, Y.; Thomson, S.; Bradley, F. *Development of Unmanned Aerial Vehicles for Site Specific Crop Production Management*; United States Department of Agriculture: Beltsville, Maryland, USA, 2010; p 1. http://ars.usda.gov/research/publications.htm?SEQ_NO_115=253718 (accessed March, 20, 2014).

Kansas State University. Project Using Drones to Detect Emerging Pest Insects, Diseases in Crops. Ag Professional. 2015, pp 1–2. http://www.agprofessional.com/news/project-using-drones-detect-emerging-pest-insects-dieases-in-crops (accessed Oct 9, 2015).

Kara, A. Low Cost UAV Disease Devastating Apple Crops. Aris Plex, 2013, pp 1–3.

Khot, L.; Sankaran, S.; Cummings, T.; Johnson, D.; Carter, A.; Serra, S.; Musacchi, S. Unmanned Aerial Systems Applications in Washington State Agriculture. *Proceedings of 12th International Conference on Precision Agriculture,* Sacramento, California, USA, 2014, p 16.

Lacewell, R. D.; Harrington, P. Potential Cropping Benefits of Unmanned Aerial Vehicles (UAVs) Applications. Texas A and M Agrlife, College Station, Texas, Texas Water Resource Institute Technical Report 477; 2015, pp 1–7.

LSU Ag Centre. *Ag Centre Researchers Study Use of Drones in Crop Monitoring.* Louisiana State University Agricultural Centre: Baton Rouge, Louisiana, USA, 2013; pp 1–2. http://www.Lsuagcenter.com/news_archive/2013/december/headline_news/AgCentre-researchers-study-use-of-drones-in-crop-monitoring (accessed Oct 9, 2015).

Lyons, P. Drone Tech to Reduce Biosecurity Cost in Australia. 2015, pp 1–3. http://www.freshplaza.com/araticle/141449/Dronne-tech-to-reduce-biosecurity-costs-in-australia. (accessed Oct 18, 2015).

Lyseng, R. Ag Drones: Farm Tools or Expensive Toys. Western Producer. 2006, pp 1–6. http://producer.com/2006/02/ag-drones-farm-tools-or-expensive-toys/ (accessed Nov 3, 2014).

Miller, J. W. Report on the Development and Operation of an UAV for an Experiment on Unmanned Application of Pesticides. AFRL., US Air Force, Youngstown, Ohio. 2005, pp 1–10.

Microdrones GMBH. Flying Farmers-Future Agriculture: Precision Farming with Microdrones. 2014, pp 1–5. (accessed June 25, 2015).

MMCUAV. Best Drones/multi Rotor UAVs Application for Agriculture Pesticide Spray test. 2015, pp 1–2. http://www.yooutube.comwatch?v=fDPOzmMcsuO (accessed Oct 9, 2015).

Modern Farmer. Using Drones in the Fight Against Apple scab. 2013, p 1. http://huffingtonpost.com/modern-farmer/using-drones-in-the-fight_b_4171110 (accessed Oct 16, 2015).

Norman, K. Drones Over Farm Fields: Unmanned Aircraft Systems may be Used to Monitor Crops. Forum News Services. 2014, pp 1–4. http://twincities.com/localnews/ci_24966038/drones-over-farm-fields-unmanned aircrafts (accessed Oct 9, 2015).

Palilo, A. *Monitoring and Management of Maize Rust (Puccinia Sorghi) by a Drone Proto-type in Southern Highlands, Tanzania.* Sokoine University of Agriculture: Morogoro, Tanzania, 2014; pp 1–15. http://www.academia.edu/8063999/MONITORING_AND_MANAGEMENTOFMAIZE-RUSTDISEASE (accessed May 30, 2015).

RMAX. RMAX Specifications. Yamaha Motor Company, Japan. 2015, pp. 1–4. http://www.max.yamaha-motor.Drone.au/specifications (accessed Sept 8, 2015).

Rieder, R.; Pavan, W.; Carro Maciello, J. M.; Cunha Fernandes, J. M.; Pinho, M. S. A Virtual Reality System to Monitor and Control Diseases in Strawberry with Drones: A Project. International Environmental Modelling and Software Society. *Proceedings of 7 International Congress on Environmental Modelling and Software*; Ames, D. P.; Nigel, W. T.; Rizzoli, A. E., Eds.; San Diego: California, USA, 2014; p. 108.

Rosenstock, S. How Drones are Helping Farmers Produce Healthier Crops. DIY Drones. 2013, pp 1–5. http://diydrones.com/profiles/aha-how-dronesare-helping-farmers-produce-healthier-crops. (accessed Oct 14, 2015).

Rush, C. Scientist Uses Helicopter Drone to Detect Wheat Disease Progression. Iowa Farmer Today. 2014, pp 1–5. http://www.iowafarmertoday.com/news/crop/scien-tist-uses-helicopter-drone-to-detect-wheat-disease-progression/article_9d38c4f8-78a5-11e3-bb2963f4.html (accessed Oct 14, 2015).

Rush, C. M.; Workneh, F.; price, F. Application of remote sensing technologies for study of wheat steak mosaic virus. *Proceedings of 9th International Congress of Plant Pathology: healthy and Safe Food for Everybody,* Torino, Italy. *Plant Pathol.* 2008, *90S*, 60–61.

Sankaran, S.; Ehsani, R. A Detection of Huanglongbing Disease in Citrus Using Fluores-cence Spectroscopy. *Trans. ASABE.* 2012, *55*, 313–320.

Sankaran, S.; Mishra, A.; Ehsani, R.; Davis, C. A. Review of Advance Techniques for Detecting Plant Disease. *Comput. Electron. Agric.* 2010, 1–13. http://www.science direct.com/science/article/pii/S0168169910000438 (accessed July 7, 2015).

Schiller, B. *Drones are Now Delivering Bugs to Farms to Help Crops;* University of Queensland: Brisbane, Australia, 2015; pp 1–2. http://www.fastcoexist.com/3045339/drones-are-now-delivering-bugs-to-farms-tohelp-crops (accessed Oct 11, 2015).

Schmale, D. G. Drone Planes Take Aerial Imaging to a New Level. In *Tiny Planes Coming to Scout Crops*; Ruen, J., Ed.; Corn and Soybean Digest; 2012; pp 1–3. http://cornand soybeandigest.com/corn/tiny-planes-coming-to-scout-crops (accessed Aug 16, 2014).

Schultz, B. Louisiana: Researchers Study Use of Drones in Crop Monitoring. Agfax.com. 2013, pp 1–4. http://www.lsuagcenter.com/news_archive/2013/deceber/headline_news/AgCenter-researchers-study-use-of-drones-in-crop-monitoring (accessed Jan 1, 2016).

Szondy, D. UC Davis Investigates Using Helicopter Drones for Crop Dusting. 2013, pp 1–8. http://www.gizmag.com/uav-crop-dusting/27974 (accessed May 23, 2015).

Taylor, J. Crop Scouting with Drones—A Case Study in Precision Agriculture. Drone Yard. 2014, pp 1–5. http://droneyard.com/2014/08/18/crop-scouting-agriculture/ (accessed Oct 10, 2015).

The Economic Times. Drones to Help Rajasthan, Gujarat Farmers Detect Crop Diseases. 2015, pp 1–2. http://articles.econnmictimes.indiatimes.com/201502-19/news/59305367_1_groundnut (accessed May 19, 2015).

Tigue, K. University of Minnesota Research Group Pushes for Ag. Minnesota Daily. Precision Farming Dealer. 2014, pp 1–3. http://www.mndaily.com/news/campus/2014/04/29/u-research-pushes-agriculture-drones.

Trimble. Trimble UX5 Aerial Imaging Solution for Agriculture. 2015, pp 1–3. http://www.trimble.com/Agriculture/UX5.aspx (accessed May 20, 2015).

Triple20. Using Drones to Detect Crop Diseases. 2015, pp 1–2. http://www.foodvalleyupdate.com/news/using-drones-to-detect-crop-diseases/ (accessed Oct 18, 2015).

United Soybean Board. Farming's Newest Precision Agriculture Tool Takes Data to on New Horizon. 2013, pp 1–5. http://www.unitedsoybean.org/article/new-precision-agriculture-could-revoluitonize-farming (accessed Oct 26, 2015).

University of Minnesota Extension Services. *Camera Drones in Agriculture: Using the Required Engineering Principles;* University of Minnesota, Extension Services: St Paul, USA, 2014; pp 1–8. (accessed Nov 1, 2015).

Vogel, J. Ready to Fly a UAV Drone Over Your Fields. Prairie Farmer Magazine. 2014, pp 1–3. http://farmprogress.com/story-ready-fly-uav-drone-fields-9-120813 (accessed Oct 9, 2015).

Wang, D.; Kurle, J. E.; Estevez Jensen, C.; Percich, J. A. Radiometric Assessment of Tillage and Seed Treatment Effect on Soybean Root Rot Caused by *Fusarium* sp in Central Minnesota. *Plant Soil.* **2004,** *258,* 319–331.

West, J. S.; Bravo, C.; Oberti, R. The Potential of Optical Canopy Measurement for Targeted Control of Crop Diseases. *Annu. Rev. Phytopathol.* **2003,** *41,* 593–614.

West, J. S.; Atkins, S. D.; Fitt, B. D. L. Detection of Airborne Plant Pathogens: Halting Epidemics Before they Start. *Outlooks Pest Manage.* **2009,** *20,* 111–113.

West, J. S.; Bravo, C.; Moshou, D.; Ramon, H.; Alastair McCartney, H. Detection of Fungal Diseases Optically and Pathogen Inoculum by Air Sampling. In: *Precision Crop Production: The Challenges and Use of heterogeneity.* 2010; pp 135–149. http://lnk.springer.com/chapter/10.1007/978-90-481-9277-9_9?no-a (accessed Oct 16, 2015).

Winkle, E. Crop Scouting Drones. Hymark High Spots. 2013, pp 1–3. http://hymark.blogspot.in/2013/05/crop-scouting-drones.html (accessed May 29, 2015).

Winslow, L. UAVs for Spraying Fruit Trees and Crops? EzineArticles. 2013, pp 1–2. http://ezinearticles.com/?UAVs-for-Spraying-Fruit-Trees-and-Crops?id=7816365 (accessed Oct, 9, 2015).

Yintong Aviation Supplies Company. Precision Agriculture UAV. 2012, http://www.made-in-china.com/showroom/zhythkqc/product-detailZbKnaseBoPNVy/ (accessed Oct 9, 2015).

Zhang, J.; Huang, Y.; Yuan, L.; Yang, G.; Chen, L.; Zhao, C. Using Satellite Multispectral Imagery for Damage Mapping of Armyworm (*Spodaptera frugidera*) in Maize at a Regional Scale. *Pest Manage. Sci.* **2015,** 1–3. DOI: 10.1002/ps.4003 (accessed Oct 18, 2015).

Zhu, H.; Lan, Y.; Wu, W.; Hoffman, C.; Huang, Y.; Xue, X.; Liang, J.; Fritz, B. Development of a Precision Spraying Controller for Unmanned Aerial Vehicles. 2014, pp 1–8. DOI: 10.1016/S1672-66529(10)60251-X (accessed May 25, 2015).

CHAPTER 8

DRONES IN CROP YIELD ESTIMATION AND FORECASTING

CONTENTS

8.1 INTRODUCTION

In near future, agricultural drones may have a major say in accomplishing several of the agronomic procedures needed to raise a crop, that is, from seeding till harvest of grains. Forecasting crop yield is an important task that drones could perform efficiently and with greater ease. At present, yield forecasting involves tedious collection of crop growth data and evaluation along with that available for previous years/seasons. It also involves use of established crop growth models. Latest trend in large farms is to adopt combine harvesters with electronic yield monitors and Global Positioning System (GPS) tagging facility. The yield maps generated are used by superimposing and comparing soil fertility data.

The idea explored in this chapter is the use of drones to provide imagery of crop at various stages of growth from seedling till seed set. Then devise statistical methods (equations/models) that help to forecast yield. In this chapter, the first half explains terminologies and methods already available. They are related to combine harvesters and crop yield mapping methods adopted with in these farm vehicles. The usage of drones in yield forecasting is still rudimentary. It is not at all used in most agrarian belts. However, there are a few recent examples, where in, drone's role in yield estimation using red, green and blue optical sensors has been experimented and explained. Such studies are discussed in the later portion of this chapter.

First, let us define and explain a few terminologies relevant to crop yield, its harvest, estimation and its connotation.

8.1.1 CROP YIELD

Crop yield occurs in different forms based on crop species and its portion harvested and utilized. Crop yield most commonly refers to grain/forage harvests from field crops such as cereals, legumes and oil seeds. Crop yield also refers to leaves, flowers, bolls, stem, tubers, bulbs and fruits. Crop yield is often a measurement of grains harvested per unit area of the field if it is a grain crop. In the literature, crop yield is also considered and referred as 'agricultural output'. It has been stated that despite most frequent reference to crop yield in the agricultural literature, its definition and accuracy with which it is used to explain crop performance is still not very satisfactory (ACIAR, 2013). The crop yield data for entire farm, a district/county or region could be arrived at by collecting and compiling the grain harvest data for individual fields. Usually, such yield data is depicted in tons per hectare. Crop yield per farm tells us about the influence of soil and crop management practices adopted on the grain harvests in a specific farm. They also indicate the effect of cropping intensity on grain/forage yield. In the present context, we may note that drones with multispectral sensors could be utilized to closely monitor the crop all through the season as it develops tillers, panicles and then grains get filled. The imagery could also be indicative of the development of panicles and grain. Further, grain yield or crop yield could be forecasted using appropriate statistical procedures and models. The imagery from drone's sensors could be used to obtain an idea about crop yield by measuring normalized difference vegetative index (NDVI) for an entire farm or county.

8.1.2 POTENTIAL YIELD

Potential yield is the expected grain yield with the best suited variety, adopting best soil, water and crop management procedures, and in the absence of abiotic and biotic stresses. Potential yield is usually measured for field plots, but it could be extrapolated to larger areas such as county, district, state or an agrarian belt. Field plots, environmental parameters, inputs and agronomic procedures adopted by famers in the large belts have to be similar or uniform. Uniformity should be ensured, particularly if extrapolations are made. *Potential yield of a crop, in any field is a good yard stick using which we can compare the performance of a genotype/crop.* Usually, potential yields are obtained from field trials by comparing different cultivars or a single cultivar exposed to different inputs and crop husbandry procedures. Regular location trials of different cultivars also provide data on potential yields possible for a given genotype (ACIAR, 2013). Potential yields can also be obtained by careful experimentation. Appropriate inputs such as fertilizers, irrigation and pesticides have to be supplied. There are also crop simulation models that allow farmers to know the potential yield of a variety (genotype). Simulations could be used to assess the consequences if genotypes get exposed to different seasons, planting densities, fertilizer input levels and irrigation. In future, drones may have a major role in obtaining data that is required for identification of apt crop cultivars. Drone technology could be used to suggest treatments that lead us to potentially higher yield. Periodic flights by drones over crop fields at different stages, beginning with seedlings until grain fill are essential. Drones may collect data pertaining to leaf number, leaf area index (LAI), leaf chlorophyll, leaf-N, plant water status, panicle number, maturity and so forth. These parameters could then be utilized in the most recent and updated crop model to derive values for potential yield.

Incidentally, we can assess performance of a crop genotype using theoretical values and trace the limits for low and high yield potentials. It is called 'theoretical yield'. Another bench mark yield is termed the 'attainable yield' in a given location and environmental conditions. 'Attainable yield' refers to one that is obtained by farmers rather routinely. This is an important bench mark yield. It is useful to compare the current crop and its performance at a given stage of growth. It is said, 'attainable yield' is usually 20–30% lower than best yield or potential yield derived from a high-input field (ACIAR, 2013). Drones could help us in obtaining data about crops grown by farmers with routine and economically optimum inputs. Farmers use economically optimal levels of inputs. They cultivate crops under normal weather conditions prevailing in a farm. Such data could be fed to computer simulation models. This way, in-season modifications could be envisaged if farmers wish to obtain higher yield levels and reach potential yields.

8.1.3   YIELD POTENTIAL OF A CROP GENOTYPE

Yield potential of crop genotype is the grain/forage yield harvested when the genotype is grown in an area to which it is adapted well, and when nutrients and water are non-limiting, there is no lodging, pests or diseases affecting the crop growth and weather pattern is normal (Evans and Fischer, 1999). This term 'yield potential' refers more to the crop species and its specific genotype grown. So far, we have enhanced yield potential of various crops by breeding for higher yield, say, by dwarfing and improving harvest index in case of wheat. In case of maize, the yield potential was enhanced by making genotypes perform well even in high-density planting situations. In many other species, widening genetic base has helped in resisting pest and diseases pressure. Therefore, it has led to better yield potential. In the experimental fields, drones could be excellent in collecting data about several genotypes of a crop. We can compare genotypes at various stages of growth and the final yield potential. Such comparisons could be done using NDVI and panicle reflectance if it is a cereal like wheat, rice or pearl millet.

8.1.4   YIELD RESPONSE

Crop's response to usual but important inputs such as nutrients (fertilizer), water and other amendments varies immensely. There are established databases that list crop species, location that is agrarian region, soil type and major environmental parameters as background data. They emphasise the actual yield responses obtained by farmers for a given input level. For example, response of rice genotypes to fertilizer (N, P, K) supply in Alfisols of Southeast Asia. Farmers may consult such data along with images drawn periodically using drones. Then, they can appropriately modify the inputs. FAO-FERTI-BASE is one such database that could be consulted if farmers aim at closing yield gaps in a given agrarian region. The data bank provides crop yield maps for a given location. A particular crop, say maize and its response to a range of nutrient levels and timings could be referred. Such maps also depict crop's response to in-season fertilizer application (see IFPRI, 2016).

8.1.5 YIELD GOAL

Yield goals are usually set by farmers on the basis of several factors, particularly those related to topographic conditions, soil and its fertility status, and fertilizer supply. A few other factors considered are seasonal weather, particularly, precipitation quantity and pattern, crop species and its genotype preferred and economic gains. Often, most appropriate levels of nutrients and water are applied to achieve the set yield goals. The way nutrients and irrigation are channelled depends on the farming systems adopted. The nutrient and irrigation inputs are applied based on grid or management zone methods (USDA, 2010). The variable applicator and its sophistication have its impact on yield goals set for each and every grid or management zone. The number of yield goals set decreases if only few management zones are made out of a big field. Drone imagery could be adopted effectively to decide about the number of management blocks or grids formed in a field. Drone images about topography, soil type, crop growth trends and NDVI could be used to form management blocks. Often, 3–5 different yield goals are set in a field. It matches the number of management blocks marked using drone imagery. They say such yield goals set for each block also helps farmers in fixing appropriate nutrient budgets. It aids in record keeping about in-season growth and final yield. Yield goals and previous yield data are some of the best guides to design nutrient/water supply trends in a farm. Drone's imagery can be of great help in monitoring the crop growth changes in relation to yield goals set by farmers.

8.1.6 YIELD GAP AND ITS ANALYSIS

Crop yield gaps are studied with the hope to reduce it and attain higher productivity. Yield gaps could be defined both in economic terms and as biological yield, such as forage/grains. However, here, we are concerned more with biological yield gaps, particularly the grain yield that could be enhanced to close the gaps (Beddow et al., 2015; Lobell, 2013; Lobell et al., 2009). As stated earlier, 'attainable yield' or 'farm yield' are obtained by adopting economically optimum inputs and procedures that allow profitable levels of grain yield. Such yield is usually much less than potential grain yield known for a genotype. The gap between 'attainable' or 'farm yield' and 'potential yield' is known as 'yield gap'. A review of relevant literature suggests that most yield gaps are about 23–30% of farm yield. Farmers try to reduce this yield gap by adopting improved genotypes, higher inputs, better timing of inputs, pesticide sprays and so forth. Grain/forage yield difference between potentially higher yield possible in a field location versus that obtained during the particular season/year is important. There may be a range of factors related to soil fertility, soil maladies such as improper pH, alkalinity, high Al and Mn levels, low organic matter, shallow profile depth, excessive gravely conditions, low nutrient content and availability, and depreciated moisture content that affect yield. These factors widen the yield gap.

Yield gap analysis is an important preoccupation for farmers and farm consultancy companies. Yield gap analysis is conducted under an assumption or notion that potential yield in that location and under prevalent environment is higher than that reaped, in that season or consistently for a few seasons. Drone images are used by few companies, to identify the causes and the extent of yield gaps. A few others adopt 'crop surface models (CSMs)' or regular crop growth models, to ascertain factors that cause the depletion of grain/forage yield. Such information helps farmers to rectify the situation in the following season and try to close that yield gap. Drones, with their ability for detection of spatial and temporal variations in growth pattern and yield formation, may offer extra and more useful information to farmers. Farmers can compare data with established crop surface and growth models. Then, decide on appropriate agronomic procedures and input levels (IFPRI, 2016).

Yield gaps could be identified between different yield types. For example, Pasuquin et al. (2007) identify yield potential as the theoretical high yield obtainable, when inputs are high and management is commensurately efficient. The difference between yield potential and maximum yield attainable under favourable condition is termed *yield gap-1*. The difference between maximum attainable yield and attainable yield, under average farmer's condition, is termed *yield gap-2*. In addition, the difference between average attainable yield and actual yield realized by farmers is termed *yield gap-3*. Obviously, the yield gap between yield potential and actual yield realized by farmers is wider. Therefore, new techniques may have to be searched to reduce the yield gap.

The yield potential, attainable yield and yield gaps can also be traced at farm, county, state or national level. The yield gap quantified for larger regions may help policy makers to arrive at appropriate decisions regarding inputs and their timely distribution. Usually, in a field, a farm or even county/state, the yield attainable reaches a plateau, once it reaches 75–80% of potential yield. Policy makers may have to be alert to such realities. They have to routinely monitor crop growth at various stages, and at yield formation stages all through the county/state. Drones can be adopted to monitor large areas of the cropping belt in order to obtain information on growth rates and yield formation (Van Wan et al., 2013). Drones offer data sets, based on which in-season changes of agronomic procedures could be effected.

8.1.7 YIELD FORECAST

Knowledge about possible final yield, in other words, yield forecasting is important. Yield forecasts are usually based on systematic analysis of preliminary data about several factors. Major factors considered are the soil and its fertility, seasonal weather parameters and previous yield data. In addition, factors such as the extent of inputs, agronomic procedures, their intensity and accuracy are also computed. Crop growth parameters and grain yield attained in several instances are compared, for a given location. There are established computer-based crop growth forecasting models. Such models could be consulted. We can also consider drone images at various stages of crop growth during a season and compare it with the data accrued. Parameters such as NDVI, leaf chlorophyll content and nitrogen content are considered. CSM could also be compared to forecast possible grain/forage yield, at the final instance of harvest. There are established crop growth models such as ORYZA, CERES-Maize, CERES-wheat, APSIM and so forth that could be utilized to forecast yield. However, these models are meant for data that is collected methodically using farm scouts. They are not amenable for drone-derived aerial imagery and digital data. Modifications should be possible, but they need to be investigated and developed. Amalgamation of crop growth models with 'CSM' developed using drone imagery should be possible.

8.2 DRONES IN YIELD MONITORING AND MAPPING

Routine yield monitoring and mapping became possible after combine harvesters produced were fitted with GPS tagging facility. Grain yield monitoring and mapping using combine harvesters has several advantages. First, we can instantly have an idea about total grain yield and variations encountered within the field. We can guess about the genotype used in different locations or management blocks and its impact on grain yield. Yield maps allow accurate comparison of different genotypes harvested by the combines. Yield maps obtained could be superimposed with soil moisture maps. Then, we can compare it for relevance of irrigation to the grain yield harvested. Farmers can make a better judgement about the genotype to be sown in the next season especially, after studying and analysing the yield maps derived from combines. While storing harvested grains, knowledge about moisture content of seeds is important. The yield maps that show moisture content of different genotypes can help farmers. They can modify the grain storage conditions accordingly. Yield maps can also be used by insurance companies to assess the value of harvests. A step further, if drone-derived imagery of crop fields (in-season) is available, then, such maps could also be consulted or superimposed with grain yield maps. It provides better insights about performance of genotypes.

There are several types of combine harvesters and yield mapping systems adopted by farmers. Each one may have some definite advantage in terms of GPS connectivity, grain flow measurement, groundspeed control or cutting width and so forth. The accuracy of yield maps too may differ based on combine harvester model. In order to select best suited combines and yield mapping systems, there are facilities for simulation. There are virtual fields with crops at grain harvest stage. The suitable combines (virtual) could be driven past the virtual field and studied. It is done prior to actual harvest in real fields (Maertens et al., 2004).

Measuring crop yield is an important aspect of farming. Farmers have generally collected grains after the crop has been harvested. The harvest is expressed using units such as tons, kilometre, bushels, or pounds per acre or hectare. During recent years, crop harvesters also called 'combine harvesters' are used regularly to harvest large fields. In a short, swift, accurate and clean operation, these combine harvesters with yield monitors remove grains. They record grain yield data per point or area. Yield monitors are of great utility to farmers. They offer an idea regarding the quantity of grain harvested at that location (GPS tagged). Yield monitors also help in identifying the variations in grain yield harvested. Such grain yield data with variations depicted on maps could be of immense value to farmers. In a practical crop field situation, farmers have to take note of the components of combine harvesters. They should carefully select grain flow sensors, volumetric and mass flow detection, ground speed sensor, harvest controls, yield data collection points and yield mapping. Originally, combine harvesters were provided with monitors that gave only final yield per plot. However, if farmers adopt site-specific or precision farming techniques, yield monitors that note the grain harvested at each point (say per 2–3 s) is utilized. It helps to generate an yield map. Yield maps provide farmers with an idea about which area of the field is poor in soil fertility, which is affected by drought or flooding or disease. Drone imagery done periodically over the same field or area can be consulted or superimposed. This is to analyse the cause of poor or higher grain yields at a particular point or zone.

Crop yield achieved is actually an end result of interaction of crop genotype, soil properties, environment and production inputs. Management blocks are often marked using yield maps. It is believed that if grain yield data (maps) for multiple years are used to mark management blocks, then it provides greater authenticity and yield stability, particularly when precision techniques are adopted (Diker et al., 2004). In other words, drone-derived imagery for multiple years should be preferred while marking management blocks.

Yield monitors are often quick and offer data instantaneously. Immediate relay of data to computer stations or computer screen in the cabin is possible. The data recorded could be summarized and offered. The data is usually tagged with GPS coordinates, so that further detailed analysis about each grid or management zone could be made. Aerial imagery by drones could augment further the data obtained by monitors placed in combine harvesters.

Shearer et al. (1997) had opined that yield data obtained using multiple combine harvesters and processing it into yield maps has indeed allowed certain advantages to farmers. It helps to analyse their fields' productivity trends and causes for depression or increases in grain yield. Yield monitors improve farmers' managerial skills for the next season. Post-processing of yield monitor's data leads to yield maps that are self-evident. Such maps show up the cause and effect relationships. In the recent years, drones with ability for close-up shots and high-resolution imagery have evolved. They possess the ability to obtain data on NDVI, growth pattern, leaf chlorophyll, canopy temperature during in-season and grain yield imagery. They can be of great help to farmers. Drone imagery could be corroborated with ground data obtained using combines with sophisticated yield monitors. Drone's data and those obtained from combine harvesters could be superimposed and verified using GPS coordinates.

8.2.1 YIELD MAP DEVELOPMENT

In the general course, a farmer can record forage/grain yield after making grids or management blocks. Farmers can map yield variations manually. It helps to understand the variable response of crops to inputs and various agronomic procedures. Farmers have been performing these procedures since many decades. Such yield maps offer a generalized idea about variation of crop growth and yield in a field. Yield maps could be obtained using drones' imagery. Images are usually processed by computer software that stiches the ortho-mosaics and shows up grain yield variations accurately. Most common software packages now available are *Drone-Mapper, AgisoftPhotoscan, Pix4D, Microsoft ICE, Visual SFM, CMVS* and *CMOVMS* (Agmapsonline, 2016).

The harvest of grain yield and its instantaneous mapping is possible with most of the recent combine harvesters (or headers). In most cases, grain yield data collected using monitors on the combine harvesters are downloaded and imported on to computer stations. Appropriate software packages are available. We may note that post-harvest processing of yield data is important. Often, yield maps with colour codes are prepared to indicate yield levels, classes and quality. The digital copy of yield data could be utilized to compare. We can superimpose grain yield data from headers with drone-derived data about NDVI, GNDVI and leaf chlorophyll content, if needed. A thoroughly prepared map with accurate interpretation of grain yield variation is generally preserved in data files and as digital images. Yield map interpretation is equally important. Drone imagery about topography, soil flooding, drought, pests, diseases and crop genotypes used could be superimposed and accurate interpretations made (USDA, 2010).

Let us consider a few examples of yield mapping. Torres-Sanchez et al. (2014) state that multi-temporal mapping of wheat in the early stages of crop is a useful procedure. Drones, in fact, offer sharper and high-resolution

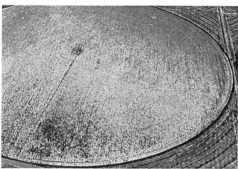

PLATE 8.1   Left: A Quadcopter flying low over a legume crop in mid-stages of growth.
Note: Such drones obtain images and offer values of vegetative indices, soil and canopy temperature. Final yield maps obtained from combine harvesters could be over-layered with those derived from drones and then analysed. Right: An aerial map of cereal crop grown using centre-pivot irrigation. The picture depicts NDVI and its variations.
Source: Mr. Sean Wagner, Spatial land Analysis, Bakersfield, California.

images. It is because they fly at low altitude over wheat crop. The ultra-high-resolution images of seedlings are used to derive a series of indices such as NDVI, CIVE, EXG, NGRDI, VEG and Woebebecke index. The vegetation fraction of wheat fields could be mapped. We can utilize images from different dates and different flight altitudes of drone. They say low-cost drones fitted with commercial cameras are good enough to provide useful data. Such data can be utilized to prepare maps depicting temporal changes of wheat crop. These maps should help farmers in obtaining an idea about crop vigour and anticipated final yield (Plate 8.1).

Yield maps are a necessity during precision farming. It involves series of steps of processing and interpretation. It is done after the data has been obtained by combine harvesters (Grisso et al., 2008). Yield maps obtained from combine harvesters repeatedly over the same field for several seasons could be compared effectively. Based on it, useful conclusions could be drawn. Field maps are merged using boundaries and specific locations on the map, for all four seasons. Yield maps obtained for the same field in subsequent seasons too could be superimposed. Then, yield at each data point could be compared and analysed (GIS Ag Maps, 2015). The data from different seasons could be collected into one file. Then, we can compare the grain yield obtained, point wise, at that particular location. It is also possible that yield data from all seasons could be associated with same master file and compared. We can compare yield data for different crops at the same point, in the same year but from different seasons. The data for each crop is usually normalized to the mean and compared. This way, farmers can compare the performance of crop as well as note down yield trend, for almost each point in the field, for different years. Drone imagery and multiple maps for the same crop field could be an added advantage. The NDVI data and disease/pest spots imaged at several stages of crop growth could be compared, along with yield data for several seasons from yield monitors. We can prepare maps depicting standard deviation and coefficient of variation. We can also mark the highest and lowest yield reaped for the same point in the field, for each of the several seasons and crops. Availability of statistical maps for multiple seasons for a single crop or different species could be useful to farmers (GIS Ag Maps, 2015).

Most importantly, yield maps and drone's imagery of vegetative growth (NDVI) helps farmers, especially in assessing the effect of inputs and agronomic procedures adopted at a given point through several seasons. For example, yield maps for multiple seasons showing effect of fertilizer-N input on wheat yield at a particular point in the field could be useful. It helps farmers to assess effect of fertilizer for several seasons. The residual effects, if any, could also be deciphered.

Anderson et al. (2014) suggest that in the near future, drones would throng the air space above crops in most of the agrarian regions. The 'big data' would be consulted and utilized by computer decision-support systems, rather, too frequently. Satellites may guide crop production systems. Yet our ability, right now, to map the crops or entire cropping systems adopted in a field is not perfect. There are lacunae in our ability to show variations and pin-point them accurately. Drones should come to the rescue of farmers who adopt precision techniques, by offering the crop growth and yield maps, periodically. Such maps may have to provide great details about variations in vegetative growth and yield formation. Such an effort will help farmers to understand the extent of yield gaps. They can also trace the factors that caused the yield gaps. Yield depressions and high productivity zones could be treated with appropriate remedies. For example, if uncongenial pH is the cause, after identifying variations in soil pH, these farmers could appropriately apply gypsum. Similarly, variation in drought effect could be corrected, using variable-rate irrigators. The basic requirement is that drones should image the variation as accurately as possible. It then allows farmers to react to the situation. Digital maps of both large agrarian regions and single fields are a necessity, in future.

PLATE 8.2   An ultra-high resolution Crop Surface Model of corn grown in California, USA
Note: The map depicts NDVI values and thermal indices for a crop, which is irrigated using centre-pivot irrigation system. The variation in Vegetative Indices and moisture is visible.
Source: Mr. Sean Wagner, Spatial and Land Analysis LLC, Bakersfield, California, USA, www.spatiallandanalysis.com

At this juncture, we may note that, drone imagery depicting variations in NDVI, GNDVI and leaf chlorophyll could be consulted or over-layered and yield maps could be analysed appropriately. It is generally said that, biomass and/or grain yield maps, derived using drones may after all, be a photo depiction of the crop from above. We should be able to analyse, interpret and utilize the data effectively. In the future years, farmers may be in a position to make effective decisions on crop species, genotypes, soil management techniques, irrigation and harvesting procedures.

In a field kept under precision farming, data from yield monitors, yield maps with GPS tags, location of management blocks and soil fertility or fertilizer input data can be carefully studied. Superimposed maps could help farmers in deciding about the fertilizer dosages for the next season. The NDVI data from drone imagery will be of immense help in matching growth variation with fertilizer and water supply at variable rates (see Plate 8.2 and 8.3).

Crop researchers at the University of Nebraska, Lincoln, have been striving to spread precision techniques during maize and wheat production.

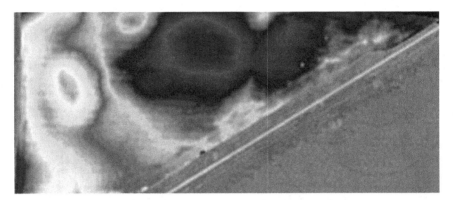

PLATE 8.3 A drone-derived image showing soybean crop over-layered with a map of topography.

Note: Map depicting field topography is at the top left of the diagonal separation. A NDVI map showing center-pivot irrigated field is at the bottom right of the superimposed maps. Light grey color indicates depressions and dark patches show hill tops. (in the color picture (e-book) red indicates depression and purple shows hill tops).

Source: Mr. Michael Dunn, Anez Consulting LLC, Little Falls, Minnesota.

Yield maps are most valuable sources as they are often referred to by farmers, while adopting variable-rate techniques. Yield maps are essential to know the soil fertility status and productivity attained at different locations in the field. It is always useful to know crop yield history of a field. Typically, 3–5 year grain yield maps are averaged and utilized, while adopting variable-rate techniques (UNL, 2015). Yield map smoothening helps in the interpretation of maps better. Agricultural consultancy agencies utilize processed yield maps to investigate the cause for yield depression, if any. Further, if the causes for yield variation are known unequivocally, then they could be eliminated through variable-rate techniques. This way, entire field could be brought to a uniform yield potential. Some examples where yield maps can help in removing variations are: (a) soil pH variation that can be removed by variable-rate application of gypsum on to soil; (b) if the variation in yield is due to soil compaction at different spots in the field, then deep ploughing with disc is helpful (UNL, 2015). It is previous years' yield maps that help in decision-making and removing variations. Drones could play a vital role by offering maps of the crop during the current season. Again, accurate overlayering of previous yield maps and current imagery derived via drones is crucial.

In the Canadian prairies, wheat, canola, soybean and corn are main crops. Here, farmers have utilized high-input precision farming techniques on the basis of aerial imagery derived from drones/satellites. Farm companies offer images of crops at different stages of crop along with yield forecasts. Farmers can access yield data of previous years and overlayer them to derive exact prescriptions, particularly those related to seeding density, fertilizer application, water supply and pesticides (ProSeeds Inc., 2015).

Without doubt, soil fertility and yield maps of yester years are the basic data on which the precision techniques are based. Soil fertility maps depicting soil physicochemical characters (texture, soil tilth, bulk density, soil pH, CEC, redox potential and so forth.) and availability of macro- and micronutrients are tedious to prepare. However, yield maps that indirectly show variations in soil productivity could be obtained using GPS-connected combine harvests. Yearly variation in crop growth response is common. Hence, decisions based on immediate one or two previous seasons may be improper. Yield map for several years should be consulted, if available, to understand the variations in soil fertility and productivity. Several factors such as variations in soil moisture, pH, nutrient availability, may affect crop yield differently. Hence, it is often advisable to overlayer yield maps and maps depicting various soil factors for which the data is available. Then, allow a good software to decide the necessary procedures in the next season. No doubt, yield maps, NDVI and soil data, when amalgamated and analysed, will lead us to better decisions (Country Guide Canada, 2015).

8.3 DRONES IN YIELD FORECASTING

Yield forecasting is an important aspect of farming, irrespective of geographic location, crop species cultivated, whether it is an individual small farm or large farm or a high-input commercial agricultural company. Obviously, there are advantages in forecasting grain yields and working to achieve it. Farmers have adopted several different techniques to arrive at most accurate grain yield forecast. Previous data and current measurements of soil, crop and weather parameters have served well in yield forecasting. Basso et al. (2015) have recently reviewed the methods adopted to forecast yield. They have also explained methods by which farmers could be provided with early warning of yield depressions, if any. It depends on detrimental factors that afflict the crop, such as pests, disease, uncongenial pH, drought or floods and so forth.

Most common traditional method of yield forecasting involves a careful scouting of the entire crop field(s), by crop production experts.

Observations made throughout the crop season prior to grain maturity are considered. Crop experts usually rely on parameters such as number of tillers, panicles, their size and grain-fill percentage, grain weight and moisture content. Grain damage due to insect attack, diseases such as grain smut, discolourations and so forth are also considered, while forecasting yield. It is common to encounter gaps between yield forecasted and that actually realized after harvest. Forecasting or an authentic idea about achievable grain yield is a necessity for farmers or crop production companies. It helps farmers to invest and supply inputs accordingly and to achieve the yield goals set.

Crop yields are also forecasted using tedious data collection, pertaining mainly to growth and yield factors. A series of crop traits such as plant height, number of leaves, chlorophyll content, number of tillers, LAI, panicles, grain number, weight and moisture percentage are analysed using appropriate programs. Simple or even complex statistical models are applied to arrive at tangible yield forecasts (National Agricultural Statistics Service-NASS, 2006; Lobell, 2013). A few of the statistical models are applicable to single farms or small patches of crops. Several other statistical models could be extrapolated to large area or even an agrarian zone. The accuracy of yield forecasting is affected by the statistical model adopted, soil fertility variation, weather patterns and crop traits considered.

Basso et al. (2015) state that agrarian zones that support several different crop species are actually complex. The final yield achieved by each crop or in a field is the culmination of series of interactions of several factors. The interaction of crops with natural factors such as soil, atmosphere, weather pattern and input trends all need consideration. The statistical methods usually consider each of them separately. In nature, however, the influence of soil-plant-atmosphere continuum, all interacting simultaneously seems important. Crop experts try to adopt 'crop simulation models' and assess the influence of several factors, simultaneously. Several crop simulation models are available and many of them are crop-specific (Asseng et al., 2013; APSIM, SALUS (Basso et al., 2012, 2010); Cammarano, 2010). There are also constraints encountered, while adopting crop simulation models and statistical methods. For example, the requirement of minimum data sets. In some cases, excessive parameterization may be required. Also, it involves costs to procure data for several parameters. That too, all through the crop season until grain-fill stage.

Remote sensing using satellites has offered images that could be utilized to forecast grain harvest. But, they are apt for a large area or an entire agrarian belt. The resolution of satellite imagery does not allow small farms to be focussed and grain harvests analysed. There are actually several models developed, using spectral data. They are used to distinguish crop species and their growth stage. Thermal imagery is adopted to judge water distribution in the crop belt (Basso et al., 2015). Regression equations between satellite imagery (NDVI) and actual ground data on samples have been widely used, to ascertain crop yield. Yield forecasting is also possible using imagery from aeroplanes, but it is costly. The current trend is to adopt low cost, low flying drones with multi-spectral cameras and software, for processing the images. The yield forecasting could be done using NDVI, GNDVI, CSM and so forth.

8.3.1 CROP SURFACE MODELS

There are currently several drone companies and agricultural service providers that deal with assessing crop health, growth rate and forecast the final grain yield. They base it on a new method known as 'Digital Surface Models (DSM)'. For example, Khumalo (2014) reported that, Anez Consulting Inc. an agricultural service provider situated in Little Falls, Minnesota, USA, helps farmers in forecasting crop yield. They analyse the crop using the usual NDVI, GNDVI, Leaf chlorophyll content. However, they transform the drone-derived data into 'digital surface models', to estimate biomass accumulation. The drone's imagery is also used to measure plant height, using DSM at full canopy and the topography data (land surface). Actually, using DSM of corn, they have developed authentic forecasts of possible final grain yield. There are several private drone agencies that develop geo-referenced images of fields. They prepare DSMs, offer NDVI values and help in conducting precision farming. Mainly, they supply basic data and DSM that could be utilized by farmers to forecast grain yield (Dronemapper, 2016; see Plate 8.4).

Reports suggest that NDVI data is being utilized with some reservations to develop a yield map. Therefore, Anez Consulting Inc. (Minnesota, USA) has resorted to modification of NDVI data and transforming it, through a computer program (e.g. *Pix4Dmapper*), into DSM. It is done to visualize variations in crop yield, at full canopy (Waypoint, 2015).

PLATE 8.4   Top: An aerial image of maize obtained using red, green and blue band width cameras mounted on a SenseFly's drone. Bottom: A high resolution Digital Crop Surface Model of the same field obtained, using a low flying drone.

Note for Top: It is a conventional NDVI map. The image was processed using Pix4Dmapper software. The conventional NDVI map is skewed and hazy due to clouds, sun-glare and shadows. Also, note that the crop and centre-pivot appear slightly hazy and unclear.

Note for Bottom: Crop surface and even centre-pivot are seen relatively clear, without glare. Such data could be used in variable-rate applicators.

Source: Mr. Michael Dunn, Anez Consulting LLC, Little Falls, Minnesota.

Further, they state that, mapping relative biomass instead of just NDVI values offers better idea about crop yield and its variations. Developing relative biomass maps and comparing it, with ground reality data is also a useful method for farmers. Specifically, it is helpful, to those who wish to understand the potential of their fields and crop genotype, in terms of productivity. Relative biomass mapping and DSM offer clearer picture of crop growth and yield variations in the field.

Agribotix Inc. is another drone-based agricultural consultancy company situated in Colorado, USA. It helps farmers with NDVI data, DSMs and imagery depicting spatial variation in growth and grain yield. They help in forecasting grain yield. They also suggest about fertilizer and irrigation inputs to match the variation in soil fertility and moisture (Agribotix Inc. 2015).

Geipel et al. (2014) have evaluated information from spectral imagery depicting spatial variations of growth parameters at various stages of maize crop. They have assessed utility of such data during yield prediction. They have actually obtained spectral data using drones fitted with red, green, blue (R, G and B) bandwidth cameras. The data from R, G and B bandwidth sensors were computed to develop a crop surface model and to detect plant height. The data from R, G and B and vegetative indices (VI) calculated were utilized to develop regression equations/models to forecast the final grain yield. They have reported that correlation between R, G, and B and VI data collected during mid-season and the final grain yield predicted was high, at $R^2 = 0.74$. The crop stage at which the data was collected and CSM developed using such data had its impact on yield forecasting. Drone imagery done at early stages of seedling development should be of high resolution. Similarly, yield predictions were more accurate when imagery and plant height estimation was conducted at later stages of plant growth, that is, when canopy closure occurred. Plant height measured using drone images and done at early or mid-season of maize crop had its impact on yield forecasts (Yin et al., 2011a, 2011b). It is said that once canopy closure occurs, the need for high resolution imagery is lessened and disadvantages due to noise also reduce. Yet, it is always preferable to develop CSM, using high resolution drone imagery (Bendig et al., 2013a, 2013b; Bendig et al., 2014). Integrating CSM with known crop growth/yield models such as CERES-maize has its advantages (Fang et al., 2008). Over all, Geipel et al. (2014) have suggested that, drone imagery done at mid-season (growth stages Z39-Z58), and adopted using a combination of spectral data and CSM, proves suitable to forecast maize grain yield. Such methods could be explored for other cereals, legumes and oil seeds. This way, yield forecasts for several crops could be developed, at a fast pace. Farmers may then adopt suitable strategies to maximize yield.

Recent discussions with experts from satellite-based farm agencies suggest that extremes of weather and fluctuations in economic value of food grains become conspicuous. This necessitates accurate yield forecast. The crop yield forecasting systems have also gained in sophistication. They try to match the environmental vagaries. Remote sensing using satellites is an established method. It adopts sensors that assess and measure reflectance of crops/vegetation. In an agrarian region, factors such as soil type and its known fertility levels and input range are considered. Using these, it has been possible to make long-range forecast about yield levels. Canopy reflectance maps too could be constructed and used effectively during grain-fill stage of crops such as corn, soybean, wheat and rapeseed.

In case of rice (*Oryza sativa*), Chang et al. (2005) have attempted to study the spectral reflectance of crop at booting stage. They aimed to trace quantitative relationships with final rice grain yield. They utilized fields sown to rice genotype, namely Tainung-67 at two different fertilizer-N levels. The canopy reflectance (NIR/RED) and (NIR/GRN) were estimated along with ground data for rice grain yield. The regression equations clearly proved that reflectance measured at booting stage is directly related to final rice grain yield. Hence, drone imagery at booting stage could be of great utility to farmers, particularly to those who may wish to have some idea about final yield. It has been reported that severe drought or disease could cause considerable noise to reflectance data. Appropriate corrections/filters are needed.

The drone-based methods to image crop's canopy growth, leaf area, plant height, assess its nutrient status (N) and water status (thermal imagery) are available. These have been used along with other parameters and crop growth models to forecast grain yield. Drone-based methods to assess effects of inputs such as fertilizers and water on crop growth and yield are also in place. Drones fitted with suitable cameras that can fly low on crop canopy at panicle development and grain-fill stage are needed. Such a system allows imaging the panicle length, grain number, moisture content and weight. Spectral data that differentiates panicles/grains from leaves is necessary. In addition, statistical methods to arrive at most accurate grain yield levels using drone-derived data are essential.

KEYWORDS

- crop yield
- potential yield
- yield gap analysis
- crop surface models
- yield forecast
- yield monitoring
- yield map development

REFERENCES

ACIAR. ACIAR Crop Yields and Global Food Security; Australian Centre for International Agricultural Research: Canberra, Australia. Internal Report; 2013; pp 1–22.

Agmapsonline. Unmanned Aerial Vehicles (UAV) in Precision Agriculture. 2016, pp 1–18. https://agmapsonline.com/2014/01/06/unmanned-aerial-vehicles-uav-in-precision-agriculture (accessed May 4, 2016).

Agribotix. Agribotix Launches First on-demand Data Processing Platform for Agricultural Drones. Drones.org. 2015, pp 1–6. http://drones.org/agricultural/agribotix-first-on-demand-data-processing-platform-for (accessed May 3, 2016).

Anderson, W.; You, L.; Anisimova, E. Mapping Crops to Improve Food Security. International Food Policy Research Institute, Washington D.C. 2014, pp 1-5. http://www.ifpri.org/blog/mapping-crops-improve-food-security/ (accessed May 5, 2016).

Asseng, S.; Ewert, F.; Roensweig, C.; Jones, J. W.; Hatfield, J. L.; Ruane, A. C.; Boote, K. J.; Thorburn, P. J.; Rotter, R. P.; Cammarano, D.; Brisson, D.; Bassso, B.; Martre, P.; Aggarwal, P. K.; Angulo, C.; Bertuzzi, P.; Biermath, C.; Challinor, A. J.; Doltra, J.; Gayloor, S.; Goldberg, R.; Grant, R.; Heng, L.; Hooker, J.; Hunt, L. A. Uncertainty in Simulating Wheat Yields Under Climate Change. *Nat. Clim. Change* **2013**. DOI: 10.1038/nclimate1916 (accessed Aug. 31, 2016).

Basso, B.; Cammarano, D.; Troccoli, A.; Chen, D.; Ritchie, J. T. Long-term Wheat Response to Nitrogen in a Rain Fed Mediterranean Environment: Field Data and Simulation Analysis. *Eur. J. Agron.* **2010**, *33*, 132–138.

Basso, B.; Fiorentino, C.; Cammarano, D.; Cafiero, G.; Dardanelli, J. Analysis of Rainfall Distribution on Spatial and Temporal Patterns of Wheat Yield in Mediterranean Environment. *Eur. J. Agron.* **2012**, *41*, 52–65

Basso, B.; Cammarano, D.; Carfagna, E. Review of Crop Yield Methods and Early Warning Systems. *Proceedings of the First Meeting of the Scientific Advisory Committee of the Global Strategy to Improve Agricultural and Rural Statistics*, FAO Headquarters, Rome, Italy 2015, pp 1–58.

Beddow, J.; Hutley, J. M.; Pardey, P. G.; Alston, J. Rethinking Yield Gaps. Department of Applied Economics; University of Minnesota: St Paul, Minnesota, USA, 2015; Staff Paper No 15, pp 1–17.

Bendig, J.; Bolten, A.; Bareth, G. UAV Based Imaging for Multi-Temporal Very High Resolution Crop Surface Models to Monitor Crop Growth Variability. *Photogrammetry Fernerkund. Geoinformation* **2013a**, *13*, 551–562.

Bendig, J.; Willkomm, M.; Tilly, N.; Gnyp, M. L.; Bennertz, S.; Qiang, C.; Miao, Y.; Lenz-Wiedmann, V. I. S.; Bareth, G. Very High Resolution Crop Surface Models (CSMs) from UAV-based Stereo Images for Rice Growth Monitoring in Northeast China. *International Archives of Photogrammetry and Remote Sensing and Spatial Information Science.* 4–6 Sept; 2013b, 40, 45–50.

Bendig, J.; Bolten, A.; Bennertz, S.; Broscheit, J.; Eichfuss, S.; Bareth, G. Estimating Biomass of Barley Using Crop Surface Models (CSMs) Derived from UAV—based RGB Imaging. *Remote Sens.* **2014**, *6*, 10395–10412.

Cammarano, D. Spatial Integration of Remote Sensing and Crop Simulation Modelling for Wheat Nitrogen Management. Ph.D. Thesis, University of Melbourne: Victoria, Australia, January 2010; p 148.

Chang, K.; Shen, Y.; Lo, J. C. Predicting Rice Yield Using Canopy Reflectance Measured at Booting Stage. *Agron. J.* **2005**, *97*, 872–878.

Country Guide Canada. Improving Nitrogen Efficiency with Precision Farming. Precision farming Dealer. 2015, pp 1–4. http://www.Precisionfarmingdealer.com/content/improving-nitroegen-efficncy-precision-farming. (accessed Jan 2, 2016).

Diker, K; Heermann, D. F.; Bradahl, M. K. Frequency Analysis of Yield for Delineating Yield Response Zone. *Precis. Agric.* **2004**, *5*, 435–444.

Dronemapper. Geo-referenced Ortho-mosaic, DEM, DSM, NDVI and Point Cloud Generation. 2016, pp 1–5. http://dronemapper.com (accessed May 4, 2016).

Evans, L. T.; Fischer, A. Yield Potential: Its Definition, Measurement and Significance. *Crop Sci.* **1999**, *39*, 156–159.

Fang, H., Liang, S., Hoogenboom, G. Teasdale, J.; Cavigelli, M. M. Corn Yield Estimation Through Assimilation of Remotely Sensed Data into the CSM-CERES-maize Model. *Int. J. Remote Sens.* **2008**, *29*, 3011–3032.

Geipel, J.; Link, J.; Claupein, W. W. Combined Spectral and Spatial of Corn Yield Based on Aerial Images and Crop Surface Model Acquired with an Unmanned Air Craft System. *Remote Sens.* **2014**, *11*, 103335–10355; DOI: 10.3390/r861110335.

GIS AG. Maps Yield Management and Statistical Mapping Raw Yield Map Points. GIS Maps.com. 2015, pp 1–15. http://www.gismaps.com/yield-map-management/ (accessed July 6, 2015).

Grisso, R. B.; Abey, M.; Phillips, S. Interpreting Yield Maps—I Gotta a Yield Map—Now What. Virginia Technological University Extension Service. 2008, pp 1–6. http://pubs.ext.vt.edu/442/442-509/442-509.html (accessed June 30, 2015).

IFPRI. Yield Analysis. Harvest Choice. 2016, pp 1–2. http://harvestchoice.org/topics/yield-analysis. (accessed May 3, 2016).

Khumalo, V. Using Drone Technology to Estimate Crop Yields and Assess Plant Health. 2015, pp 1–4. http://www.rocketmine.com/using-drone-technology-to-estimate-crop-yields-assess-plant health/ (accessed May 3, 2016).

Lobell, D. B. Cassman, K. G.; Field, C. B. Crop Yield Gaps: Their Importance, Magnitudes, and Causes. Annu. Rev. Environ. Resour. 2009, *34*, 179–204.

Lobell, D. B. The Use of Satellite Data for Crop Yield Gap Analysis. *Field Crop Res.* 2013, *143*, 56–64.

Maertens, K.; Reyniers, M.; De Baerdmaker, J. Using a Virtual Combine Harvester as an Evaluation Tool for Yield Mapping Systems. *Precis. Agric.* 2004, *5*, 179–195.

National Agricultural Statistics Service (NASS). The Yield Forecasting Program of NASS by the Statistical Methods Branch, Estimates Division, National Agricultural Statistics Service. NASS Staff Report No. SMB 06-01; U.S. Department of Agriculture: Washington, D.C., May 2006.

Pasuquin, J. M. C. A.; Witt, C. *Yield Potential and Yield Gaps.* International Plant Nutrition Institute: Norcross, Georgia USA, 2007; pp 1–4. http://www.ipipotash.org/en/eifc/2007/14/4/english (accessed May 13, 2016).

ProSeeds Inc. See Regional Data for Corn, Canola and Soybean. Farmers Yield Data Centre. 2015, pp 1–3. http://yielddata.farms.com (accessed May 14, 2015).

Shearer, S. A.; Higgins, S. G.; McNeill, S. G.; Watkins, R. I.; Barnishel, J. C.; Doyle, J. H.; Leach, J. H.; Fulton, J. P. Data Filtering and Correction Techniques for Generating Yield Maps from Multiple-Combine Harvesting Systems. *Proceedings of ASAE Annual International Meeting,* Minneapolis, Minnesota, USA, Paper No: 971034, 1997, pp 1–7.

Torres-Sanchez, J.; Pena, J. M.; De Castro, A. I.; Lopez-Granados, F. *Comput. Electron. Agric.* 2014, *103*, 104–113.

UNL. Listening to the Story Told by Yield Maps. 2015, pp 1–14. http://cropwatch.unl.edu/ssm/mapping (accessed May 5, 2016).

USDA. *Precision Nutrient Management Planning;* Agronomy Technical Note No 3; United States Department of Agriculture: Beltsville, USA, 2010; pp 1–10.

Van Wan, J.; Kersbaum, K. C.; Peng, S.; Miner, M; Cassman, K. Estimating Crop Yield at Regional to National Scales. *Field Crops Res.* 2013, *143*, 34–43.

Waypoint. Using Drone's Surface Model to Estimate Crop Yields and Assess Plant Health. 2015, pp 1–10. http://waypoint.sensfly.com/drone-surface-model-estimate-crop-yield-plant -health/ (accessed Nov 25, 2015).

Yin, X.; Jaja, N.; McClure, M. A.; Hayes, R. M. Comparison of Models in Assessing Relationship of Corn Yield with Plant Height Measured During Early- to Mid-season. *J. Agric. Sci.* 2011a, *3*, 14–24.

Yin, X.; McClure, M. A.; Jaja, N.; Tyler, D. D.; Hayes, R. M. In-season Prediction of Corn Yield Plant Height Under Major Production Systems. *Agron. J.* 2011b, *103*, 923–929.

CHAPTER 9

ECONOMIC AND REGULATORY ASPECTS OF AGRICULTURAL DRONES

CONTENTS

9.1 INTRODUCTION

Drones presently available to farmers have been conjured as a technology that can transform agricultural technology. It is presumed to enhance the efficiency of farm management and productivity (Elmquist, 2015). We should note that drone technology that is rapidly being imbibed into the farming sector has to face a certain degree of competition from other technologies. Drone technology has to pass the test for economic feasibility. The introduction of these 'agricultural drones' could affect the economic aspects of a small farmer, a large farmer, an agricultural company with vast acreage, a district/county or even a nation. Regardless of the scale of usage, firstly, drones have to be economically beneficial. The gains from using this technology should outweigh its disadvantages, if any. Further, Nicole et al. (2015) have rightly pointed out that worldwide, drone techniques have to negotiate a constantly changing economic scenario during the current period. Economic aspects and governmental subsidies to the farming sector sometimes could fluctuate drastically in agrarian regions. This means that investment on drones and net gains from using drones have to be significant, despite such fluctuations in economic parameters. Experts dealing with drones (Association for Unmanned Aerial Vehicles, AUVSI) opine that agriculture would be the sector that adopts drones massively. Agriculture is expected to account for 80% of total usage of drones in all aspects. In USA, forecasts indicate bullish purchases of agricultural drones. It is actually based on the trend noted with the usage of drones in Japanese farms during 1990s (Fernholz, 2013). Lacewell and Harrington (2015) evaluated the potential of drones in the agrarian belts of Texas. They suggested that a careful assessment of crops, their economic value and profits that accrue need to be assessed prior to touting drones. As stated earlier, there is 'Drone Flight Calculator' that quantifies the economic benefits of using drones as a service for any type of scouting, spraying and so forth Drone Flight Calculator offers accurate figures of profits per acre. It also shows the amount of inputs such as fertilizers, pesticides and herbicides saved through the usage of drones (Torres-Declet, 2016).

Reports suggest that farms earning a gross income at the rate of 900–1100 US$ ac^{-1} have improved their exchequer by 10%, if drone technology plus precision methods are adopted. Payback period for costs incurred on drones in a corn field of just 250 acer is small. The other advantages attributable to drones are the ease of operation and collection of data required during adoption of variable-rate methods (Precision Drone, 2015).

Drones are expected to be most useful and profitable when they are used for large-scale monitoring, scouting and surveillance of crops. Drones effectively replace manned aircrafts and satellites. Moreover, they offer better economic gains for time-sensitive operation. According to King (2013), drones offer greater profits to farmers under two conditions: (1) when their farms are large and (2) when they adopt cash crops and high input crop production methods. One of the observations made relates to farm size versus drone usage and its impact on economic gains. During the past few decades, farms were small. Many of them were actually family farms. They were easily observable by climbing on to a vantage point or on to a field bund. However, during recent years, farm consolidation has created very large-sized farms. Farm cooperatives and private companies often own and manage large-sized farm units. These units are not easily observable in entirety, in one go. At present, farmers need a view of the fields from the sky and in entirety, wherever feasible. Therefore, large farm size makes drones an essential farm gadget and at the same time profitable (Stutman, 2013).

Nicole (2015) believes that although drones are apt for large or moderately sized farms, they may be equally advantageous for small family farms. They may offer both social and economic advantages to farm families. Drones could be effectively used by groups of family farms and cooperatives. Further, many of the crop production procedures that employ drones could be attended in a collective fashion in a village.

Drones are currently evaluated in many agricultural experimental stations for their efficiency in completing the tasks. The aim is to quantify consequent economic gains to farmers. Agricultural drone experts from Kansas State University have suggested that in due course, drones will be sought by many farmers. This demand is attributable to their innumerable uses in the farming sector. Among them, the ability to reduce drudgery is highly perceptible. They predict that by 2025, the drone industry may well cross 100 billion US\$ turnover per year. According to them, economic gains due to drone usage in farms will outweigh other aspects. Introducing drones into farms may also become much easier. Drones used in crop fields are small and will cost about 2000–12,000 US\$ at best. Most models could be fetched at 7000–8000 US\$. Currently, in the United States of America, people are just waiting for the finalization of Federal Aviation Agency (FAA) rules and regulations. Drone market is said to blow up, once the regulations are finalized (Johensen 2014). A forecast suggests that FAA regulations on the commercial use of drones could be finalized until the end of 2016 or early 2017. In fact, Morgan (2015) states that delay in finalization of rules for the usage of drones has indeed made many companies in the United States of America and other nations to wait. They have refrained from using these flying robots until then. This has resulted in loss of billions of sums to many industries and farmers. They could have otherwise used drones in farms and saved a share of expenditure on human skilled workers.

Farm consulting agencies are expected to depend extensively on the data that drones accrue and store. Farm technicians could use these data to further improve efficiency during fertilizer, water and pesticide application. Drone-derived data allows greater flexibility to farmers regarding farm operations. Many of these drone-aided decision-making by farm consultants ultimately lead us to higher economic advantages. The economic gains could be due to reduction in the use of chemicals and water. Drone-aided agronomic procedures too could add to economic benefits. Drones offer immense economic advantages through reduction in labour costs on technicians, particularly during mapping and layering (Labre Crop Consulting, 2014). In the general course, without drone technology, obtaining, collating and deciphering trends (variations) in crop growth are tedious and difficult. Therefore, ascertaining the factors that affect crop productivity becomes a time-consuming and costly procedure, for both consulting firms and farmers alike.

We should note that drones, on their own, would reduce cost on human labour. They improve efficiency during scouting, mapping the field and spraying fertilizers or plant protection chemicals. Consequently, this decreases the cost of crop production. Hence, profits tend to increase. In addition, when drones are used in conjunction with precision farming methods, it reduces use of inputs further, leading to enhanced profits for farmers. Drones also improve profits by enabling farmers to know about soil erosion, drought, flood or pests at early stages of crop production. Overall, drones promise to enhance returns to farmers.

9.2   INDUCTION OF DRONES REQUIRES CAPITAL INVESTMENT BY FARMERS

Drone systems range widely across different farming zones. The type of drone needed, its level of sophistication, particularly, in terms of imagery, resolution, computers, software and variable-rate sprayers, add to costs. Drones used to obtain data about normalized difference vegetative index (NDVI) and chlorophyll/N status of crops, and those needed for scouting are relatively less costly. They are sought by small- and medium-sized farmers. Otto (2014) states that currently in USA, a starter system such as quadcopter and small flat-winged cost 3000–4000 US$. However, the cost of the drone system increases as the farmer opts for more cameras, sharper imagery and endurance. Often, farmers who require only the imagery of terrain (3D), crop growth, NDVI and crop-N status prefer flat-winged drone versions (4000–8000 US$). Farmers requiring NDVI data sometimes hire drone services. They may only need to pay for imagery and the processing done by the agencies.

Farmers surveyed in Idaho, USA, state that drones are highly useful. Particularly, while deciding on replanting at various spots in a large farm of say 900–10,000 ha. Drone imagery helps them to decide about the remedial measures for erosion control. Farm drones' cost ranges from 2000 to 60,000 US$. However, farmers spending 30,000–40,000 US$ for a drone report that the economic advantages due to drones outweigh their initial costs. Drones detect erratic seed germination and gaps in crop stand. They could reduce on scouting costs perceptibly. Considering only this aspect simply breaks even the costs incurred on drones within few seasons (The Des Moines Register, 2014; Vanac, 2013).

In the past 5 years, several different types of drone models, both flat-winged and copters, have been designed, developed and produced in large numbers in industries. Many of them are specific for the purpose. Farmers are selective regarding drones that they purchase. Usually, drone of a certain size, compactness, cameras and their resolution, remote control equipment and computer decision-support are preferred for specific activities. Important farm activities considered are crop monitoring, disease/insect surveillance, detection and mapping soil erosion and so forth. There may not be a best drone for all locations and crops. The drone model, agricultural operations and crops have to match. In addition, there is an important factor that should be considered. To many farmers, the cost of the drone model is an important constraint, particularly for small farm holders. Yet, there are popular models of drones preferred more often by farmers. One such list that claims 7 best drones in the market in 2014 is shown in Table 9.1.

Let us consider few aspects about drones and their impact at a larger scale, say a crop belt or an agroecosystem. In case of 'Corn Belt of USA', it is said that in each season, farmers and farming companies inevitably scout, collect samples and analyse soils, and monitor and assess cropping systems. Then, they map them using computer programs, or else, skilled technicians prepare the maps. Several types of field maps, particularly those depicting terrain, topography, soil type, fertility status and grain yield productivity, are prepared. These procedures involve costs to farmers. At this juncture, we should note that the greatest economic advantage of drones is their ability to rapidly scout and offer computer-generated maps showing variations in field. Economic advantage actually occurs due to reduced costs on scouting and mapping field variations. Now, consider the 170 million acres of corn belt scouted each season by

TABLE 9.1 Top Few Drone Models, Their Endurance, Camera, Cost of the Equipment and the Manufacturing Company (*Source:* Amato, 2014)

Drone model	Battery life (min)	Camera (Pixels)	Cost (US$)	Company
Ag Drone UAS	31–60	1080 p HD	9,995	HoneyComb Inc. Wilsonville, OR, USA
eBee Ag	31–60	1080 p HD	25,000	SenseFly Inc. Lausanne, Switzerland
Lancaster Hawk Eye -III	31–60	1080 p HD	25,000	Precision Hawk Inc. Raleigh, NC state, USA
Crop Mapper DT-18	> 60	1080 p HD	37,700	Delair-Tech, Toulouse, France
AG550	< 30	1080 p HD	3,000	Aerial Technology, Portland, OR, USA
Quad Indago	31–60	1080 p HD	25,000	Lockheed-Martin Inc. Bethesda, MD, USA
Ag Eagle	31–60	1080 p HD	13,500	AgEagle Inc. Neodisha, KS, USA

Note: Cost of drones based on 2014 market prices. *HD:* high definition.

using human skilled farm workers versus the ease with which drones and computer programmers offer the digital maps (Paul, 2015). If we apply drones on a large scale, they may offer proportionately higher produce (grains/forage). They can alter the economics of large-scale production of crops and grains. Now, let us consider other large agrarian belts such as European cereal belt, Indo-Gangetic plains, Northeast China or even Guinean savannas of West Africa. Here, drones could be of great service to agrarian pursuits plus could alter the economics to better profits. There is a strong reason to spread the usage of drones in such crop belts. It helps both to reduce human drudgery and to attain higher profits. At most, skyline may have to accommodate periodic flights, one time a week or fortnight, by drones. Therefore, it does not seem to be a cumbersome idea to make efforts for large-scale usage of drones.

Aspects such as economic gains, rapid collection of data and the accuracy and ease with which drones could be operated by farmers seem to outweigh the other methods. For instance, a single manned flight (airplanes) over the large farm needed 9000 US$. This is only to obtain aerial imagery. By using the images, farmers could identify regions that require fertilizer and water. However, a 9.0-lb drone could accomplish similar task with greater versatility and clarity of images. Drones could be flown as many times and at any instance, whenever the farmer desired. They provided clear images of 1500 acres of land that was otherwise difficult to obtain using farm labour. The economic advantages derived were impressive. Therefore, some farmers tended to invest and obtain more number of drones. Swarms of drones inducted into fields could indeed be efficient (Shinn, 2014). They could finish the tasks in a matter of minutes.

9.2.1 COST OF DRONES AND THEIR USAGE IN AGRICULTURAL FARMS

Farmers are currently exposed to a wide range of drone models with an array of sophistication. The cost of drones available in the market shows a very wide range from 500 US$ for a simple one to 100,000–120,000 US$ for complex models. The costs are based on size, flight endurance, durability, operations possible, software, image quality, flight pattern and tasks (e.g. scouting, spraying) possible (Dobberstein et al., 2013). A popular and versatile model such as Yamaha's RMAX along with accessories costs 150,000–160,000 US$. A simple flat-winged drone such as *eBee* or a local copter drone may be obtained, at 7000–10,000 US$. Such drones allow rapid imagery. They possess great versatility of flight and flight path, and easy processing of images.

Drone technology has attracted large number of investors. They have funded many start-up companies. Such industrial units are primarily geared at producing agricultural drones. Many types of drones that focus on surveillance of natural resources, industries, and mines are also popular. However, recently, agricultural drones have gained importance. For example, at a recent contest on entrepreneurship, 'Raptor Maps'—a project dealing with mapping agriculture zones—was preferred and a prize of 100,000 US$ was awarded to the start-up. Agricultural drones were preferred from among contestants dealing with medical and other industrial uses of drones. Agricultural drone start-ups were preferred on the basis of their immediate utility and applications. Other aspects considered were business turnover and economic gains (MIT Sloan Management, 2015).

Drones are among the best bets for crop surveillance. They focus on variety of aspects such as crop growth, insect/disease damage, damage due to storms and inclement weather, panicle initiation, seed filling and grain maturity, lodging if any, and so forth Reports suggest that drones are highly efficient in accomplishing these above tasks, particularly in terms of accuracy and economics. A skilled farm worker may have to be hired at 2–4 US\$·ac^{-1} to perform visual inspection of the crop field. The accuracy depends on his skills, particularly, his ability to mark and describe the kind of damage the crop has undergone. In contrast, drones provide accurate imagery that could be repeatedly seen on a computer screen. Then, appropriate remedial measures could be considered. The return on investments for the drone could be achieved in a crop season or even less. Furthermore, farmers can own drones and use them for several years (Precision Drone, 2013). Reports state that drone-aided (e.g. Precision Hawk) imagery, data storage and processing costs only about 10–20 cents·ac^{-1} (Modern Farmer, 2015).

Several aspects of drone, such as its mechanics, operation control and purposes such as surveillance of crops affect the economic benefits derived. Often, it is the time needed or saved by using drones that weighs positively for drone usage. A report by Impey (2014) states that it is the rapidity and ease of operation of drones that attracts farmers. Drones save on tedious and costlier activity of farm workers. Indeed, drones take just 30 min to cover an area of 150 ac of natural vegetation or cropped fields. They provide detailed images along with digital data. Drone flights above crops could be repeated frequently and in short notice. Drones, no doubt, cost less than human scouts. Digital data could be analysed and re-analysed, enabling farmers to have a few options. Farmers may shrewdly select agronomic procedures required to improve the crop. They could actually weigh out economic advantages and opt for a particular procedure. Many of the agronomic procedures become less costly to adopt, if drones are used in farms. For example, pesticide spraying costs less, if drones are used. Rus (2014) believes that with the advent of precision farming techniques, imagery through the sensors becomes a necessity. Frequent need for images of crop fields means greater dependence on drones. A low-cost drone that could be purchased at 2,000 US\$ may become an effective tool during precision farming. For example, a small lightweight drone such as 'Skyhunter', a fixed-wing drone, bought at 2000 US\$ is easily a very good bet to farmers. This drone covers an area of 150 ha in less than 30 min. It can fly over the fields repeatedly in short intervals, thus allowing farmers to have a good idea about crop growth variation. Lopes (2015) states that drones offer one of the most advanced technologies through multispectral imagery. They may finally outclass both manned aircrafts and human labour. While a manned aircraft costs 1000 US$ per hour to fly past crop fields, we can buy a drone entirely for ourselves at 1000 US$. Then, the requisite information about crop fields could be obtained as many times as required in a day.

We may also have to consider the cost of the drone, keeping in view, the fiscal status of farmer/farms. Additionally, we should consider the farm enterprise and its product. For example, is it low input crop that offers commensurately low returns or is it a highly profitable cash crop. Farms that are large and profitable may not find it difficult to buy a few drones. A small farmer may need easy payment schedules. Factors such as payment schedules should also be given weightage, if a farming enterprise buys a series of drones. For example, a citrus grove in Florida may easily afford an agricultural drone model costing 7000–10,000 US$. It serves the citrus farmer in several ways. The drone reduces the cost of labour for scouting, spraying pesticides and conducting general surveillance of citrus orchards. It saves on such expenditure and offers better crop (Zimmcomm New Media LLC, 2011). Bechman (2014) opines that in USA, there are drones meant for two different classes of clients. One set involves farmers who want to fly the drones and obtain pictures of their farm and survey it. Such drones may cost 1000–10,000 US$. The second group comprises farmers with large units or farm consultancy firms that need very detailed and accurate pictures. They opt for ortho-mosaics and computer-processing facilities, so that prescriptions could be developed. These clients use slightly sophisticated drones and cameras that cost more than 10,000 US$. A certain set of drones with specialized advantages related to visual and red edge bandwidth imagery is also available. For example, Lehman Aviation Inc. has released a flat-winged drone with multispectral sensors. It focuses on red edge bandwidth photography (Precision Farming Dealer, 2015c).

Reviews by MIT (USA) engineers suggest that most unmanned aerial vehicles used in the farming sector are of cheaper version. Yet, they offer excellent returns because they provide close-up images of crops to farmers. Images of crops are offered at a very low cost. As stated earlier, a manned aircraft costs about 1000 US$·h⁻¹ to image the crop fields. Compared with it, a low-cost drone model that fetches similar or often better images than aircrafts costs just 1000 US$ to buy the instrument outright (Anderson, 2014). Such a drone lasts for several years and provides high resolution imagery. On a wider horizon, the advent and use of such low-cost drones in farming may improve crop production. The impact of drone technology on farm economics could be felt, perhaps, in almost all farming belts of the world. In due course, if farm co-operatives are effectively used, drones could be cheaper by many folds. They may become affordable even to small farmers, particularly those practising subsistence farming.

King (2013) states that many of the agronomic procedures are time-sensitive. Therefore, drones that are small, cost less and are versatile to be commissioned at any time are apt. Further, she states that in Oregon, farmers are trying to adopt smaller versions of drones in agriculture. Farmers producing potato spend about 4000 US$·ac^{-1}. An average-sized crop field requires about 50,000 US$ if aerial images from manned aircrafts are used or if human scouts are adopted. Hence, switching to drone-aided aerial imagery to survey and scout the potato crop is preferred. Drones cost much less than manned aircrafts.

Agricultural drones are still an instrument of curiosity. However, they are efficient in reducing farm drudgery. Drones promise higher accuracy during scouting and identifying areas with problem (soil maladies, pests/diseases) in the crop fields. Many farm companies and farmers with large units are still experimenting with drones. They are testing and ascertaining profits out of drones. So, they are purchasing drones at a wide range of costs to them. Such purchases of drones depend on the enterprise, its size, crop species cultivated and the expected benefits. For example, an *eBee* flat-winged drone costing 5,000–7,000 US$ has been procured in many farms. A farmer with a unit of size 6000 ac has preferred, a slightly low-priced *eBee* drone. An expensive and a bit highly sophisticated version of the same *eBee* drone costing 23,000 US$ has been opted by larger, high input farms of USA (Heacox, 2014). Green (2013) has opined that average-sized farms found in USA have all opted for drones costing around 9,000 US$. They are thought to drastically reduce the cost of land survey, scouting and detecting diseases in crop fields. Farmers have stated that a drone is affordable because, it is a one-time expenditure. Tallying farm size, purpose, cost of drones and profits is the crux if one wants to deploy drones in their farms. Presently, drone production is increasing due to inquisitiveness of farming companies. Farm consultancy agencies need them utmost. Drone prices are decreasing because several models are being flooded into the market. However, we have to be cautious about the quality of drone machines. They should also be apt for the purpose and should possess full repertoire of sophisticated cameras, computer chips and digital data storage facility (American Farm Bureau Federation, 2014). In practical farming, a major challenge to farmers is to buy drones at low cost. Drones should also be efficient in procuring pertinent data.

Drones built by established companies located in developed nations are being imported by some nations in Latin America, Africa and Asia. Although drones are not costly items to import, most nations have allowed the establishment of several drone companies. This helps to satiate the forecasted local demand, particularly in the farming sector. Incidentally, the farming sector drones are usually small and their technology is still simple. They only need a few sensors and software to process the ortho-images. Drones with variable-rate applicators and software to regulate them are being produced by several aviation companies in North America, Europe, China and Japan. Generally, large farms and farm companies are at ease fiscally while procuring such agricultural drones. Farm companies are assured of return on investment, particularly, in terms of reduction in labour costs and inputs. Small farmers, however, may need drones that are cheaper. They may have to develop their own indigenous models by using local material and technical knowhow. Such drone models may indeed be equally effective, when it concerns reducing costs on labour and inputs. Let us consider an example. The Chinese Agricultural Ministry has invested in developing indigenous models of drones. Home-made platforms using bamboos, local wood material or other tough items has been fitted with sensors. Such unmanned aerial vehicles have been effectively used, to scout for diseases and pests in vegetable fields. These low cost drones have also been fitted with sprayers, so that, pesticides could be sprayed on to crop fields. No doubt, cost incurred by farmers on such indigenous small drones is paid up, in a matter of seasons. Again, reports by China's Agricultural Ministry states that, such local models of drones are effective in controlling aphis and powdery mildew attacks on vegetable crops (Ministry of Agriculture-Beijing, 2013). Over all, we may have to appreciate that drone technology is apt for high-input intensive farming zones where in, farmers/companies could invest more to procure high technology studded drones. At the same time, drone-based techniques could easily penetrate low-input farming zones traced in different regions of the world.

9.3 JOB LOSS AND CREATION DUE TO AGRICULTURAL DRONES

Jobs are an integral portion of economic aspects of any small- or large-scale farm enterprise. In the present context, drones are gadgets that could improve agricultural crop production efficiency, particularly in terms of net energy required to conduct certain agronomic operations. They reduce need for skilled labour. Therefore, drones have direct impact on cost of crop production. Certain sections of farm jobs may get reduced. Farm workers may face retrenchment and job loss, if drones are adopted. At the same time, drone technology is a new aspect introduced into farming sector. It needs skilled farm technicians and labour to operate them, fly them, maintain them and decipher the images using specialized computer programs. Computer skills related to processing drone imagery, storing digital data and using them during variable-rate application of chemicals is almost mandatory. Therefore, several drone technologists could be employed in farms. The exact effect of introduction of drones in farm sector is difficult to guess right now. Yet, there are innumerable reports, forecasts and calculations reported from across different agrarian belts. Such reports deal with drones and their impact on farm labour and economics. Let us examine a few reports from different regions. A survey by Association for Unmanned Aerial Vehicles, USA, indicates that in a single state such as Kansas, in USA, agriculture related spending on drone technology would reach 75 million US$/year. It may create over 1000 jobs each year in the state. Further, in Kansas, out of a total 82 million US$ expenditure on drones, agricultural drones consume 75 million US$. The rest of the aspects, such as general surveillance and public safety may account for a paltry 3.5 million US$ year^{-1}. In California, expenditure on agriculture drones is expected to cross 2.3 billion US$, in 2015 (Nicole, 2015). It could create over 12,000 jobs in the state. In the entire United States of America, economic impact of agricultural drones is expected to be 65 billion US$/year, for a period of 2015–2017. It is expected to gradually increase to 75 billion US$/year in the next decade. Eventually, by 2025, drone techniques may offer an economic turnover of 112 billion US$/year (Nicole, 2015) and create jobs proportionately.

Yet another report suggests that in the United States of America, 70,000 jobs involving farm drones will be created in the first 3 years from 2012. The number could increase to 1000,000 as drones gain popularity. In California alone, over 12,500 jobs related to drone production, software development and usage will be created in first 3 years. Drone usage could inject a further 2.39 billion US $ into California's farm business. The forecasts about entire United States of America state that over 1 billon US$ could be introduced into agricultural drone market in this decade (The UAV-Belize Ltd., 2013). It means several thousand jobs will be created in the subsectors such as drone production, computer processing, farm technicians and so forth. In the Caribbean, drones are expected to be deployed in farm sector. They are expected to perform tasks related to crop dusting, scouting for disease, crop growth and yield monitoring, fertilizer management and livestock tracking. Therefore, job loss, if any, due to agricultural drones are to be expected within these farm functions (The UAV-Belize Ltd., 2013). At the same time, all of these functions need well-trained farm technicians, drone operators, computer specialists and skilled workers. Therefore, several new jobs are expected within the same area of farm operations.

In Great Britain, House of Lords formed a committee that conducted a discussion about drones and their usage in farm sector. They stressed on the types of rules and stringent safety measures required, if drones are to become common in farms. However, they opined that very strict regulations and restricting use of drones has consequences on crop productivity. It also affects job creation trends in farm sector and drone industry. In the near future, drones are to be more popular so more than 150,000 jobs could be easily created. These jobs are required to just feed the farm sector with drones and to operate, service and maintain these gadgets (Crop Site, 2015). A few British farmers have expressed that as precision farming becomes more common, need for drones and drone technologists (jobs) will be perceived more than at present. Therefore, training in drone technology in farms should be imparted.

Drone usage could spread rapidly into different agrarian regions. This means rapid alterations in job loss versus creation trends due to introduction of drone technology. However, Paul (2015), a drone company chief, believes that much depends on the FAA and the rules that get promulgated. Most farmers are waiting for the FAA guidelines. Hence, in the near future, that is, 2015–2016 and next few years, drones could spread creating new jobs for skilled computer technicians and flight masters. It depends on the regulations and actual advantages reaped by farmers. There also a few set-backs to be considered, if and when they occur. For example, in Saskatchewan (Canada), a few crop dusters experienced disasters, because proficiency and handling experience was insufficient. Such events could slow down drone usage and jobs that they bring with them.

We may note that jobs that involve drone operations are not difficult to obtain. There are several states in the United States of America that have devised rules and regulations that permit use of drone for commercial purposes. For example, FAA has issued permits to agricultural service companies and realtors to monitor crops/properties using drones. The conditions stipulate that ground pilot and an observer should at least hold a private pilot certificate and a medical certificate. It also stipulates that the operator should stay within the sight of the drone's flight path (AG-UAV News, 2015). At this juncture, we have to note that on a wider scale, teaching and training farm workers about drone technology also costs. Sometimes, individual large farm company may train their own personnel. There are also training courses developed by Universities and Aircraft Management Training Schools. Drone companies too train farmers in flying their drone models.

9.4 AGRICULTURAL DRONES IN DIFFERENT AGRARIAN REGIONS: ECONOMIC ASPECTS

Economic analysis about drone usage in agriculture is required. Aspects that need attention are its consequences on: (a) reduction in farm worker requirements, (b) reduction in time needed to scout field of a specific size, (c) reduction in cost of scouting, (d) savings due to rapid accrual of aerial imagery, (e) reduction in costs for assessment of fertilizer needs and its application using variable-rate applicators, (f) reduction in the use of plant protection chemicals, (g) enhanced efficiency of agronomic procedures, in general and (h) overall increase in exchequer attributable to drones. Extensive analysis of drone technology is required, specifically, to authenticate economic gains due to them. Drones have to be tested in different agrarian regions, climatic conditions and at different input regimes in order to decipher economic advantages. Drone models are being churned out rapidly and in large numbers by different companies. Some of them could be highly efficient in certain agrarian settings, and on specific crops. They may accomplish specific agronomic procedures efficiently. Hence, profits due to drone usage also depend on specific drone model that the farmer uses. No doubt, there is a need for series of experimental evaluation of drones, specifically to compare and ascertain economic gains accurately. So far, there are only very few studies that even venture to estimate accurate financial costs and fiscal advantages per unit area, due to the usage of drones. We have to essentially match the drone model with specific tasks and economic advantages that accrue.

While judging the overall economic advantages due to adoption of drones, we have to be ready with and compare facts about costs and gains that occur due to drones versus human scouts. Firstly, accurate data about drones are essential. Some of them are as follows: (a) drone models and their cost, depreciation of value of drones through time, life of drones; (b)cost of fuel and fuel efficiency; (c) sensors and their cost for maintenance, repair and replacement, if any; (d) cost of software used to obtain images and to process them; (e) cost of technicians who operate and fly the drone; (f) cost on software experts who convert ortho-images to digital data and images useful to farmers; (g) cost of variable-rate applicators, if drones are used in precision farming and (h) cost of garage/hanger if the drone is a larger model.

Advantages due to drone usage that need consideration are: (a) reduction in farm labour costs to scout for obtaining digital images and prepare maps of fields/crops; (b) reduction in the requirements of fertilizers, plant protection chemicals and irrigation due to adoption of drones and precision farming concepts; (c) elimination of cost escalations due to scarcity of farm workers in the area during certain seasons and (d) avoidance of costs on upkeep of health and efficiency of farm workers. The quality of digital maps prepared by skilled technicians may often fall short, whereas excellent accuracies could be obtained using sensors on drones. We should note that higher resolution and accuracy has an economic value, that is, higher quality costs.

There is also another angle to compare drones with human skilled labour. Drones indeed just do not replace human farm workers. Apart from absolute economic gains due to usage of drones in farms, we may have to realise the fact that drones remove human drudgery in fields. This fact just outweighs and induces farmers to adopt the drones. We have adopted several farm vehicles, instruments and gadgets just because, they reduce physical toil by farmers, avoid fatigue and make things easy and efficient for farmers, on an overall basis. Yet, as a thumb rule, perhaps, farmers and farming companies will first weigh and compare quickly, the cost of drones, fuel costs, technician charges and image processing charges. Then, they account for the number of farm workers that this activity replaces and reduces costs on labour. In addition, they may make a quick estimate of fiscal gains due to reduction of inputs such as fertilizers, plant protection chemicals and irrigation. An immediate requirement is a series of trials that compare economics of drones versus human workers in fields. Such comparison within agricultural experimental stations, farmer's fields and if possible on large scale, say, at county or district level are required. Long term field trials to evaluate economic advantages of drones need to be envisaged.

Farmers and drone purchasers need computer-based software that compares a wide range of salient features of drones. Such software should include initial cost, depreciation, flight-related information, mechanical aspects, scouting details, reduction in farm worker requirement and cost versus benefits to farmer. The reduced need for plant protection chemicals, irrigation and farm labour also needs to be computed accurately. Some of these benefits due to drones need to be evaluated, using accurate data and computer programs. Based on it, farmers can buy drones that are apt for the purposes envisaged. A computer-based ready reckoner with software that simulates and provides a well-evaluated comparison and options, prior to drone purchase is almost mandatory. A recent report states that a computer software closely matching the above statements has been released into the agricultural drone market. 'Measure'—a drone operator has developed and released first ever 'Drone Flight Calculator'. It helps farmers to assess the economic value of using the drone. The 'Drone Flight Calculator' allows farmers to determine, if drone technology can improve the farm operations (Torres-Decelet, 2016). The Drone flight calculator actually quantifies the economic advantages of a drone, if used to scout the field or for spraying or both. This software has been tested on crops such as soybean, corn and grapes. Farmers will have to enter the field data and the calculator provides the profits per acre. Farmers are informed about the quantity of inputs such as fertilizer and pesticides saved due to drone technology (Precision farming Dealer, 2016).

In the United States of America, drone technology is being touted by private agencies, drone production industries, agricultural agencies and experts in crop production. They have all forecasted a rapid spread of drones into the farm sector. The potential economic benefits from drones are expected to be high. It is expected to be good enough, to pay up for its introduction. Additionally, it could offer farmers with extra exchequer. Extension agencies in several states have offered to help the farmers with necessary training. For example, Munson (2013) states that in Nevada, discussions and training about drones are aimed at improving their role in several aspects of farming.

Several researchers from different agrarian zones within Spain have tested drones for use in aerial imagery, acquisition of digital data for variable-rate applicators of fertilizers, pesticides and herbicides. No doubt, further testing and evaluation of efficiency of drones is needed. There are reports that, drones could be effectively used for eradication of weeds and pests. However, economic aspects of drone usage and its consequences, once it is adopted on a larger scale needs evaluation. Genik (2015) has reported that, wild oat is an important weed that infests wheat, barley and canola fields in North America. Using drones, to detect wild oat affected regions and spray them with herbicides, actually costs much less, than scouting and spraying by humans. For example, for a field of 52 ac, about 900 US$ could be saved, if drones are used. Lacewell and Harrington (2015) state that, drones show a great potential in avoiding insect/disease caused losses. At an advantage of 5.0 US$ ac^{-1} due to drones, the gross potential of drones applied to 340 million ha of crop land is 1.7 billion US$ and at an advantage of 25 US$ ac^{-1} it is 8.5 billion US$.

We may have to note that, readiness to invest on manufacture of drones in high numbers, depends on improved production of grains/fruits with simultaneous reduction of inputs. At the same time, quality of products such as fruits has to be maintained at higher level. In other words, farm profitability has to no doubt increase. In Germany, for example, vine growers are adopting precision techniques, so that, large scale blanket application of harmful pesticides is reduced. They are using drones to obtain accurate aerial imagery, prior to pesticide sprays. It seems quality of grapes is also held high, if drone images and precision techniques are used. The German Ministry of Agriculture has invested 800,000 €, to buy drones and supply them to vineyards (Microdrones GMBH, 2014).

In China, millions of farmers use pesticides. They often use human skilled workers to spray the crop. Pesticide spraying is tedious. It is dangerous if one gets exposed to mists/fumes. It becomes costly if repeated pesticide applications are needed. Most importantly, if the farm is large, scouting and identifying regions that exactly need pesticide application is difficult and costly. During recent years, several drone models have been deployed for military purposes. But, a few companies have already identified the advantages of using drone in the Chinese farms. For example, Yintong Aviation Supplies Ltd. (2014) has produced two agricultural models. They could be adopted to scout for pests/diseases and spray chemicals. Models using electric batteries are light and efficient. Most interesting is the fact that, using 'Yintong Drone' has helped farmers, to reduce pesticide requirement by 50% compared to manual sprays. Drones could after all spread rapidly in agrarian regions of China. Drones are sought, particularly, if the farm is large and managing fields, using large number of human skilled workers is costly. Therefore, on a large scale, say a county or state or an agrarian region, reduction in inputs such as fertilizers, pesticides, herbicides and water could be immense. Such reductions occur during each crop season. Over all, economic advantages due to drones could be highly perceptible.

Let us consider yet another example. Agricultural crop production is an important aspect of self-reliance and economics of Kenya. Newer methods and technology that improves crop productivity are always desired. Maize is among the major cereal food crops. Hence, there have been attempts to standardize drone usage, to scout and spray chemicals to the maize crop. So far, tests have shown that, drones provide farmers with greater details, through aerial imagery (Okune, 2014). Practically, in farmer's fields, there are few tests that are yet to be conducted, to establish economic advantages due to drones. It may be few years before Kenyan farmers realize the advantages of agricultural drones. Farm agencies that supply digital information to farmers could make headway, since they are less costly. Timely application of inputs and reduction in farm worker needs are other main economic gains, for a Kenyan farmer.

One of the reports about drones in Southeast Asia states that, this region spent about 590 US$ million on drones. It included both military and civilian models. Forecast suggests it could rise to 1.4 billion US$ by 2017. A large share of increase in investment is to be attributed to agricultural drones mainly because, farm drones are gaining in popularity at present (Lamb, 2013).

9.5   REGULATORY ASPECTS OF DRONE USAGE IN AGRICULTURE

The ultimate purpose of guidelines for drone usage is to see that farmers, farming company professionals and farm extension service personnel are well informed and trained to handle drones. Drones should be safely and efficiently operated to one's own advantage during crop production. Hence, in the United States of America, FAA is aiming at providing set of guidelines called 'B4UFLy' (Huerta, 2015). This website offers detailed, updated information to farmers regarding federal regulations. It deals with what they should do with drones and what they should not. It also guides farmers about restrictions for drone usage, if there is any, in that region. There is also a 'know before you fly' campaign that provides unmanned aerial systems (UAS) operators with information for safe adoption of drone technology in the farms (Huerta, 2015). The FAA regulations and information needed to fly drones are expected to be available on mobiles in due course. Forecasts suggest that when the regulations are finally announced, they may transform the drone industry and farming practices. It could usher in a revolution in terms of farm drudgery. Farm worker's drudgery may get reduced or altogether removed. They may also create jobs in great numbers, particularly, those needing aircraft engineering skills, software development and knowledge about agricultural practices.

9.5.1   REGULATIONS BY FEDERAL AVIATION AGENCY (FAA) OF UNITED STATES OF AMERICA

Periodic releases of information from FAA suggest that rules and regulatory instructions do vary among the states (Dorr, 2014). A similar situation exists among European nations, wherein, categorization of drones into different classes, such as agricultural drones, commercial drones to monitor mines, installations and civil surveillance is prevalent. Rules and regulations for use of agricultural drones with in Europe too, differ between regions/nations (The Drone Log, 2014).

In the United States of America, effective December 21, 2015, the FAA requires that all drone users, or those who possess it should register their drone equipment. They have to provide details on models, year of manufacture, fitness test certificates and purpose. They have also to pay a US$ 5/- as fees (Precision Farming Dealer, 2015a). Those who do not register their drones are liable for penalties. Unregistered drones will attract a fine of US$ 27,500/-, to be paid to FAA Drone Regulatory Department (Vandermause, 2015; Precision Farming Dealer, 2015a).

Otto (2014) states that drones are getting popular in many regions within the United States of America. However, farmers are actually waiting for rules and stipulations that are to be finalized, by the FAA. Most commonly understood rules concern the fact that agricultural drones should not be allowed to fly above 400–500 ft. and should not drift into airspace of neighbours' plots, urban areas or airports.

One of the forecasts by a major drone company is that announcement of FAA regulations and finalization for their promulgation has a direct impact. It may affect the way business community related to agriculture, invests funds in drone technology. Generally, drone production is expected to increase exponentially, across high-input agrarian regions. There will be several other new farm companies that would initiate drone production. Hence, Dean (2015) states that across farming regions in North America, there is a strong need to establish proper work flows for drone production and related computer accessories. It may also require training of larger workforce in drone technology, as applied to crop production. Training and spread of FAA regulations induces more farmers to adopt drone technology.

Regulations announced regarding drone usage in agricultural settings around February 2015 suggest a series of restrictions and limits to their use. Yet, many of the rules are flexible to a certain extent. One of the flexible rules suggests that drones could be small and light in weight. The weight may range from couple of lbs to heavy ones up to 55 lb. Drones could be used to take digital pictures of crops, even sample leaves and water from irrigation canals. One other rule suggests that the drone operator has to be in sight of the flying machine. Preprogrammed flight paths that are out of sight of the operator are not allowed. Drones are to be flown at 500 ft. altitude not above. The previous two regulations seem tough to a group of farmers and drone experts managing, crop production enterprises (Star Journal, 2015). It is opined that, some of the rules released are apt to urban settings. But, in rural and agrarian zones, a few of them could be modified. For example, drones may fly low on the crop's horizon in a farming belt. A drone company in USA has stated that, regulations for drone usage could be summerized as follows: In general, drone user (i.e. farmer) must take a written test covering aspects of drone technology; farmers should use the drone only in the daylight; farmers should keep the drone below 500 ft.; drone should be flown at speed less than 100 kmph and ensure that drones stay within the sight of the operator (Ag Armour, 2015).

A recent report states that FAA has also initiated testing drones (machines). They have involved three major drone producing companies and several universities in USA. They intend to examine consequences of flying drones in locations, out of sight from the operator by using predetermined flight (Precision Farming Dealer, 2015b). The aim is to list the problems that may erupt, if drones are flown without the presence of operator. The idea is to anticipate difficulties, devise remedies and list the precautions. It will be useful, in case, a farmer really wanted to send drones out of sight of the operator, to survey the fields with crops. The results are expected to affect drone usage extensively, in the agrarian regions of different continents. Software for drone flight path needs to be modified, by including several safety items and procedures to retract drones, if they ever went astray. Finally, the idea is to devise rules and regulations for agricultural drones that could be flown, outside the visible range of operators. There is no doubt that, stringent tests and appropriate remedies are key, to flying drones on predetermined pathways or adopting remote controller, away from operator's visible range.

Drones that fly out of sight of the operator could stray into airspace used by manned aircrafts, airports and other locations. They could be distractive and at times dangerous. Safety of drones and manned aircrafts both are equally important. One of the ways that FAA is evaluating is to track the drones that fly above the crop fields and civilian zone, via satellite guidance. Preliminary plans are to employ ADS-B technology. It is a system which air planes use, to determine and broadcast their location via satellites. Drones that fly close to airports and airspace of others or manned aircrafts could detect them. It avoids any collision course. Another suggestion made is to restrict and segregate drones used, for civilian purposes and agriculture. This suggestion keeps drones away from manned aircrafts and paths of commercial airline or military fighter jets (Seeking Alpha, 2015).

Let us consider a few abridged points related to guidelines issued by FAA about registration of drones, including those used in farm sector. Firstly, any drone operator is an aviator with specific responsibilities. Drone is an equipment that needs to be registered with transportation (aviation) authorities. This essentially helps them to track the drone vehicle, to its legal owner. Registration of drones could be affected by filing the details of models and purpose, on a paper pro forma. It could also be done electronically, so that, records are intact. Drones are marked with specific registration based on location and a serial number, identifiable by National Aerospace System (NAS). Actually, by late 2016, drone companies may have sold over 1.6 million drones that will find their way, into farms and other civilian locations in USA. All of these have to be registered and certain rules have to be followed, if they have to be used. Drone registration has to be renewed every 3 years. Its renewal depends on fitness tests applied on the instrument. Safety of drones and human workers in the field seems to be a central theme, based on which the rules are framed. Persons aged 13 and above are allowed, to use the registered drones (Vandermause, 2015; Precision Farming Dealer, 2015a).

9.5.2   REGULATIONS FOR USE OF DRONES IN CANADA

Drones are in use in Canada for purposes such as delivering small packets, carrying luggage and few other items. Drones are also in use to map the natural resources, image topography, terrain, soils, water bodies, oil pipe lines, civilian movement and other activities in urban areas (Fitzpatrick and Burnett, 2014). Drones are now finding their way into agrarian regions of Canada. Let us now consider rules and regulations related to use of drones in the skies above Canadian farming zones. A report by Redmond (2014) states that during 2013, a few pilots with expertise about drones first demonstrated the way drones could be flown above the crop fields and how aerial imagery could be obtained. Such images were later processed using appropriate computer programs. However, in general, there are certain stipulations regarding drone usage. A pilot who wishes to operate drones has to first register himself with the Transport Canada—a government agency. He has to be a well-trained drone pilot with a certificate called 'Special Flight Operations Certificate (SFOC)'. In addition, there are two important criteria. The drone used should be mechanically fit and safe. Next, pilot must demonstrate minimum competence in managing drones, above crop fields. Pilots should have handled both copter and flat-winged drones (Fitzpatrick and Burnett, 2014).

Transport Canada limits drone usage to locations away from airports and military airstrips. Agricultural and commercial drones should be kept, at least 10 nautical miles away from airport control towers. Drones flown closer or in the region where airports are situated may do so. But, they have to inform the concerned airport managers 24–48 h, prior to using their drones, in that area. Drone operator should also call the airports and report to ground supervisors, whenever the drones are close to airstrips, particularly when they reach within 1–2 nautical miles. They should use radio signals to communicate. All drone instruments should have software that will allow them, to get back to the place where they started the flight (see Redmond, 2014). In Canada, it is slowly but surely becoming clear that, drones are not exclusive to military. In the recent past, drones have actually evolved to be part of civilian administration. More recently, they are becoming a sought after farm vehicle. Farmers have developed an inquisitiveness to use this gadget. However, rules and regulations that are prevalent in Canada do not mention much, about the privacy issues. Drones are going to be common and all farmers would own, a piece or couple of them and fly them. They may do so, at will, anywhere above their farms in the sky. Such drones could trespass into other's farms and collect data (Privacy Commissioner of Canada, 2013). In the United Kingdom too, similar observations have been made about shifting of drones from primarily military usage, to civilian surveillance and farms. They believe, drones may induce a kind of competition among drone users. It may induce a certain degree of easy flight paths and trespass into others' properties, farms and daily life (Doward, 2012). Privacy questions about drones, at least in commercial farms, need to be addressed within the rules and regulations framed for drone usage.

One general observation about rules and stipulations are that, once the process of certification and usage of drones gets standardized, market for agricultural drones would increase enormously. The basic fact considered is that, drones remove farm drudgery. Drones improve crop production efficiency by reducing on farm labour costs. It also reduced pesticide and other chemical usage. As a consequence, farm exchequer is expected to increase.

9.5.3 REGULATIONS FOR USE OF DRONES IN OTHER NATIONS

Drones have intruded the farming regions of Latin American continent. Over 14 nations in the Caribbean and South America are known to have already used drones in crop fields. There are no well-designed rules and regulations to the use of agricultural drones in any of the nations, excepting in Brazil (RT News Team, 2014). In Brazil, drone usage is controlled by the military agency and it also regulates civilian use of drones. Drones are currently used to monitor Amazonian forests and farming stretches. Brazilian Air force that operates drones, it seems, helps neighbouring nations such as Bolivia to conduct aerial surveys. It offers digital data about Cocoa and Coffee plantations and other crops. No special rules are in effect in Bolivia to survey crops using drone. Chilean government has its own drone producing companies. Drones are yet to be inducted into its air force. Drones could be used in the Chilean farming sector in due course. Regulatory aspects need to be standardized. Several other nations of this region such as Colombia, Jamaica, Dominican Republic, El Salvador, Panama, Haiti and Cuba use drones (RT News Team, 2014; Glickhouse, 2014). Perhaps, they adopt rules formed by FAA of USA or in due course, they may announce formal rules and regulations for drone use, in their skies above the crop fields.

Most recent pronouncement by the South African Civil Aviation Authority suggests that rules and regulations about drone usage in the agricultural sector are not well placed. However, a lot of enthusiasm has been generated for use of drones in their farms. A policy being developed by the government agencies suggests that it is basically similar to those developed by European nations and Australia (Dronologista, 2014). Following are the salient aspects: (a) agricultural drones fly only 120 m or below; (b) agricultural drones are not to trespass into airstrips or large airports; (c) drones should be away from the boundaries by at least 4.2 nautical miles; (d) agricultural drones will be flown in sight of the operator; (e) no pre-programmed flight pathway for drones, if they are out of sight or if the drones are to be used during night times and (f) no flying of drones over public property, roads and other's farms.

Japan, as stated earlier, is among the earliest users of drones within crop production zones. They have successfully used drones for two decades to dust paddy fields with pesticides. In Japan, regulatory aspects for drone usage in agricultural sector, first received attention in 2002. Farmers started using copter drones to spray pesticides on to their rice crops, around that period. The guidelines for agricultural drones has since been modified progressively and standardized by the Ministry of Economics, Trade and Industry. In 2004, they had developed a set of rules for commercial use of drones that fly at low altitudes, mainly over the crops and other zones (The Japan UAV Association, 2004). Their recent reports suggest that, new safety guidelines for agricultural drones will be developed periodically. Rules are to be modified on the basis of extent of use and whenever newer models of drones are developed.

9.5.4   RMAX—AGRICULTURAL DRONE APPROVED FOR USE IN AGRARIAN REGIONS

RMAX built by Yamaha Company Inc. located in Japan is an agricultural drone. It has been approved for use in agricultural farms of USA. Actually, this is relatively heavy (60–80 lb) equipment. It has been in use in the Japanese crop production regions for over 20 years. In USA, this drone—'Yamaha RMAX' is exempt from FAA regulations under section 333 that apply to commercial drones. It is allowed for use in the farms of USA. RMAX is an autopilot unmanned aerial vehicle that could be used effectively to obtain aerial imagery, mark the areas that need fertilizer and irrigation. We can also use it to spray the crop with pesticides and other chemicals (Gallagher, 2015; Associated Press, 2015). Similarly, this popular drone has been awarded approvals and exemptions, for use in the Australian agrarian belts. The Civil Aviation Authority of Australia has approved

RMAX, for use during scouting and spraying pesticides, on to wheat and all other crops. South Korea too has approved use of RMAX as an agricultural drone. Reports suggest that RMAX is usually flown above 150 m, that is, just around 500 ft. Therefore, it operates within the stipulations by FAA and other agencies. So far, in Japan and other countries, RMAX has been extensively tested. This drone has been typically flown between 150 and 200 m altitude above the crops. It has been used definitely far-off from populated regions. Right now, with its ability for aerial imagery, digital data collection and variable-rate spray, RMAX is among the foremost drones. It has received legal approval for wider use in agrarian regions of the world.

There are few other drone models that are almost similar with regard to technical specifications, size and efficiency for imagery and chemical spray. They may soon find approval from aviation agencies of different regions. A good example is 'Autocopter' produced by a company in North Carolina, USA. In addition, several other models, both copters and flat-winged drones may receive approval from regulatory agencies. It is a matter of time before the skyline of agrarian belts, show up larger number of drones.

Over all, as a new technology, drones should be easy to purchase, service them and maintain each machine, for a considerable length of time. Drone usage should lead us to economic benefits, reduction in human drudgery in crop fields, reduction in application of harmful chemicals and fertilizers. Therefore, drones could help us in delaying or avoiding deterioration of soils and agro-climate. Drones should be safe and fool-proof with no undue difficulties to fly them repeatedly. Farmers should be able to imbibe the technology and master it, in a short period. *In one sentence, 'drones should be profitable and safe in farmland'.*

KEYWORDS

- drone systems
- drone usage in agriculture
- job loss and creation
- drone-aided decision-making
- agronomic procedures
- FAA regulations

REFERENCES

Ag Armour. Ag Tech: Incorporating Drones to Increase Productivity. 2015, pp 1–17. http://www.agarmour.com/#/Ag-Tech-incorporating-Drones-to-increase-Productivity/cx9v/554cf95bOcf2adc1ad1cab85.htm (accessed June 19, 2015).

AG-UAV News. FAA grants UAV Permits for Agriculture and Real Estate Companies. 2015, pp 1–2. http://www.agricultureuavs.com/faa_grants_uav_permits_for_agriculture_real_estate_companies. htm (accessed May 23, 2015).

Amato, A. The 7 Best Agricultural Drones on the Market Today. 2014, pp 1–4. Dronelife. com. http://2014/10/01/best-agricultural-drones/ (accessed Oct 16, 2015).

American Farm Bureau Federation. Drones Hold Greatest Promise for Ag. Ag Professional. 2014, pp 1–2. http://www.agprofessional.com/news/drones-hold-getest-promise-for-ag-23994297 (accessed March 12, 2014).

Anderson. C. Agricultural Drones: Drones in North American California. MIT Technology Review. 2014, pp 1–4. http://www.technologyreview.com/featuredstorey/52649/agriicultural-drones/ (accessed Aug 2, 2014).

Associated Press. 21st Century Crop Dusting: FAA Approves Large Drone use on Farms. CBS News. 2015, pp 1–3. http://sanfransisco.cbslocal.com/2015/05/05/21st-century-crop-dusting-large-drone-. Htm (accessed May 24, 2015).

Bechman, T. Corn Farmers see UAV Potential in Crop Production. Prairie Farmer. 2014, pp 1–5. http://www.farmproress.com/story-corn-farmrs-uav-potential-crop-production-15-115093 (accessed May 30, 2015).

Crop Site. How Can Drones Make Farming Profitable. 2015, pp 1–3. http://www.thecropsite.com/news/17406/how-can-drones-make-farming-profits/ (accessed June 28, 2015).

Dean, C. Measure and Precision Hawk Partnership to Provide Commercial UAS Services. Precision farming Dealer. 2015, pp 1–5. http://www.Precisionfarmingdealer.com/content/measure-and-precisionhawk-provide-commercial-uas-services.htm (accessed Jan 24, 2015).

Dobberstein, J.; Kanicki, D.; Zemlicka, J. The Drones are Coming—Where are the Dealers. Farm Equipment. 2013, pp 1–3. http://www.farm-equipment.com/articles/9509-from-the-october-2013-issue-the-drones-are-coming-were-are-the-dealers (accessed Jan 2, 2016).

Dorr, L. Fact Sheet: Unmanned Aircraft System. 2014, pp 1–10. http://www.faa.gov/news/fact_sheets/news_story.cfm?newsid=14153 (accessed Aug 17, 2015).

Doward, J. Rise of Drones in UK Airspace Prompts Civil Liberties Warning: A European Commission Report Predicts Hundreds of Civil Uses for Unmanned Aircraft in the Next Decade. The Guardian, 2012.

Dronologista. The Thriving Drone Community of South Africa. 2014, pp 1–3. Robohub. http://robohub.org/the-thriving-drone-community-of-south-africa.htm (accessed May 13, 2015).

Elmquist, S. Fighter Pilot Turned Farmer to Ply Drones Over Crop Land. Stars and Stripes. 2015, pp 1–4. http://www.stripes.com/news/veterans/fighter-pilot-turned-farmer-to-ply-drones-over-cropland.htm (accessed May 12, 2015).

Fernholz, T. The US Drone Economy will Create 100,000 Jobs Say Companies who Make Drones. 2013, pp 1–3. http://qz.com/61727/the-us-drone-economy-will-create-100000-jobs-say-companies-who-make-drones.htm (accessed June 25, 2015).

Fitzpatrick, S.; Burnett, K. Regulation and Use of Drones in Canada. The Canadian Bar Association. 2014. http://www.cba.org/CBA/section_airandspace/newsletters2013/drones.aspx (accessed Sept 14, 2014).

Gallagher, S. Crop Dusting-Unmanned Helicopter Gets Cleared for Commercial Flight. Technology Laboratory/Information Technology. 2015. http://arstechnica.com/information-technology/2015/crop-dusting-unmannned-helicopter (accessed May 24, 2015).

Genik, W. Case Study: Wild Oat Control Efficiency Using UAV Imagery. AgSky Technologies Inc. 2015, pp 1–5. http://agsky.ca/case-study-wild-oat-control-efficiency-using-uav-imagery (accessed June 26, 2015).

Glickhouse, R. Expaliner: Drones in Latin America. 2013, pp 1–5. http://www.as-coa.org/articles/explainer-drones-latin-america (accessed Aug 3, 2015).

Green, M. Unmanned Drones May have Their Greatest Impact on Agriculture. The Daily Beast. 2013, pp 1–7. http://www.thedailybeast.com/araticles/2013/03/26/unmanned-drones-may-have-their-greatest-impact-on-agriculture.html (accessed March 20, 2013).

Heacox, L. Real World UAV Experience in Agriculture. Crop Life. 2013, pp 1–2. http://www.croplife.com/equipment/real-world-uav-experience-in-agriculture/ (accessed March 20, 2014).

Huerta, M. FAA Unveils B4UFLYAPP for Safe and Legal UAS Operations. Precision farming Dealer. 2015, pp 1–3. http://www.Articles/1432-faa-unviels-b4ufly-app-for-safe-and-legal-uas-operations (accessed May 13, 2015).

Impey, L. Drones to have a Bigger Role in Mapping Arable Crops. Field Star Weekly. 2014, pp 1–4. http://wwwfwi.co.uk/articles/22/01/2014/142533/drones-to-have-a-bigger-role-in-mapping-arable-crops.htm (accessed Jan 29, 2014).

Johensen, J. Drones to Increase Profitability. AgWired. 2014, pp 1–2. http://www.agwired.com/2014/01/13/drones-to-increase-profitability.htm (accessed Jan 10, 2016).

King, R. Farmers Experiment with Drones. 2013, pp 1–4. http://blogs.wsj.com/cio/2013/04/18/farmers-experiment-with-drones/ (accessed April 30, 2015).

Labre Crop Consulting. Iowa Crop Consulting Firm Offers Ag Services from Drones. AG Professional. 2014, pp 1–5. http://www.agrimarketing.com/s/87203 (accessed Feb 12, 2014).

Lacewell, R. D.; Harrington, P. Potential Cropping Benefits of Unmanned Aerial Vehicles (UAVs) Applications. Texas A and M Agrlife, College Station, Texas, Texas Water Resource Institute Technical Report 2015, 477, 1–7.

Lamb, K. Indonesia Readies Mass Production of Drones. Voice of America. 2013, pp 1–3. http://www.voanews.com/content/indonesia-readies-mass-production-of-drones (accessed Jan 1, 2016).

Lopes, L. Farming Drones and Precision Agriculture. Interesting Engineering. 2015, pp 1–2. http://interestingengineering.com/farming-drones-and-precision-agriculture/ (accessed May 5, 2015).

Microdrones GMBH. Flying Farmers—Future Agriculture: Precision Farming with Micro-drones. 2014, pp 1–5 (accessed June 25, 2015).

Ministry of Agriculture-Beijing. Beijing Applies 'helicopter' in Wheat Pest Control. Ministry of Agriculture of China, Beijing. Internal Report, 2013, pp 1–4 (accessed Sept 22, 2014).

MIT Sloan Management. Drone Technology Service Proving Crop Analytics wins MIT $100 Entrepreneurship Competition. 2015, pp 1–3. http://www.prnewswire.com/

news-releases/drone-technology-service-providing-crop-analytics-wins-mit-100k-en-trpneurship-Surve300083404 (accessed May, 2015).

Modern farmer. Meet the New Drone that Could be a Farmer's Best Friend. Modern Farmer. 2015, pp 1–4. http://modernfarmer.com/2014/01/precision-hawk.html (accessed June 29, 2015).

Morgan, D. FAA Expects to Clear US Commercial Drones within a Year. Ag Professional. 2015, pp 1–3. http://www.agprofessional.com/news/faa-expects-clear-us-commercial-drones-within -a-year.htm (accessed June 25, 2015).

Munson, J. Nevada Looks at Drones for Economic Development and Natural Resource Efforts. Nevada Appeal. 2013, pp 1–4. httpp://www.nevadaappel.com/news/lahonton-valley# (accessed Jan 8, 2016).

Nicole, P. Sustainable Technology- Drone use in Agriculture. 2015, pp 1–7. https://wiki.usask.ca/display/~pdp177/Sustainable+Technology+Drone+Use+in+Agriculture (accessed July 6, 2015).

Okune, A. NRBUZZ: Drones for Agricultural Monitoring in Kenya. IHUB Research, Nairobi, Kenya. A Report, 2014, pp 1–12.

Otto, J. UAVs are Next Wave of Agricultural Technology. Agrinews. 2014, pp 1–3. http://agrinews-pubs.com/Content/News/MoneyNws/Article/UAVs-are-next-wave- (accessed May 16, 2015).

Paul, R. Drones (UAVs) and Agriculture Some of my Experiences Using UAV. 2015, pp 1–5. http://aerialfarmer.blogspot.in (accessed April 30, 2015).

Precision Drone. Drones for Agricultural Crop Surveillance. 2013, pp 1–4. http://www.precisiondrone.com/agriculture (accessed March 20, 2014).

Precision Drone. Drones for Agricultural Crop Surveillance. 2015, pp 1–4. http://www.precisiondrone.com/agriculture (accessed April 3, 2015).

Precision Farming Dealer. Drone Owners Must Register Equipment with FAA—Starting Today. 2015a, pp 1–3. Hhttp://www.precisionfarming dealer.com/1881.htm (accessed Dec 22, 2015).

Precision Farming Dealer. FAA to Test Drones that Fly Beyond Pilot's Line of Sight. 2015b, pp 1–5. http://www.precision farming dealer.com/articles/1426-faa-to-test-drones-that-fly-beyond-pilot's-line-of-sight (accessed May 11, 2015).

Precision Farming Dealer. Lehmann Aviation Unveils LA300 UAV System. 2015c, 1–8. http://www.precisionfarmingdealer.com/articles/1381-lehman-aviation-unveils-la300UAV-system (accessed Sept 8, 2016).

Precision Farming Dealer. First-ever Drone Flight Calculator Launched. 2016, 1–3. http://www.percisionfarming dealer.com/articles/1906-first-ever-drone-flight-calculator (accessed Jan 17, 2016).

Privacy Commissioner of Canada. Drones in Canada. Privacy Research Papers. 2013. http://www.privcy.ca/information/research/drones_201303_e_aspx (accessed Sept 14, 2014).

Redmond, S. The Future of UAVs for Agriculture. Hensell District Co-operative. 2014, pp 1–3. http://www.hdc.on.ca/grain-markedly/hdc-reports/29-grain-marketing/253-hdc-future-of-UAV-ag-steve-redomd.html (accessed July 21, 2014).

RT News Team. Latin American Drone use on the Rise and Unregulated. 2014, pp 1–2. http://www.rt.com/news/latin-american-drones-unregulated-216/ (accessed Aug 3, 2014).

Rus, M. In: *Actual and potential use of drones in precision agriculture*. Danovich, A. Ed.; Hot Wires. 2014, pp 1–3. http://circuitassembly.com/blog/?p=4030 (accessed May 14, 2015).

Seeking Alpha. Regulators Consider Satellite Tracking for Delivery Drones. 2015, pp 1–8. Precisionfarmingdealer.com. http://www.regulators-consider-satellite-tracking-for-delivery-drones (accessed Jan 1, 2016).

Shinn, M. Experts See Farming as Next Big Use for Drones. Precision Farming Dealer. 2014, pp 1–4. http://www.precisionfarmingdealer.com/content/experts-see-farming-next-big-use-drones.htm (accessed Dec 3, 2014).

Star Journal. Editorial, 2/15: Tweak Rules for Agricultural Drones. 2015, pp 1–4. Journal Star.com http://ournalstar.com/news/opinion/editorial/editorial-tweak-rules-for-agricultural-drones (accessed May 11, 2015).

Stutman, J. Agricultural Drone Investing: Unmanned Aircraft Industry Ready to Soar. 2013, pp 1–4. http://www.techinvestingdaily.com/articles/agricultural-drone-investing/229&title/ (accessed May 25, 2015).

The Des Moines Register. Backers Say Drones will Prove Useful for Farmers. 2014, pp 1–2. Idahostatesman.com (accessed July 23, 2014).

The Drone Log. Drone Classification and Proposals for Europe. 2014, pp 1–4. http://www.thedronelog.com/drone-classification (accessed June 17, 2015).

The Japan UAV Association. The Japan UAV Association—A report. 2014, pp 1–3. http://www.jauv.org (accessed June 30, 2014).

The UAV-Belize Ltd. Cost Effective Fertilizer Application: A Case for UAV Helicopters in the Caribbean. 2013, pp 1–15. http://UAV-bleize.com (accessed Aug 18, 2014).

Torres-Decelet, B. In: *First ever drone flight calculator launched*. Precision Farming Dealer. 2016, pp 1–3. http://www.percisionfarming dealer.com/articles/1906-first-ever-drone-flight-calculator (accessed Jan 17, 2016).

Vanac, M. Drones are that Latest Idea to Improve Farm Productivity. The Columbus Dispatch. 2013, pp 1–3. http://www.dispatch.com/content/stories/business/2013/.../eyes-in-the-skies.htm (accessed Jan 4, 2016).

Vandermause, C. FAA to Require Small UAV Registration. Precision Farm Dealer. 2015, pp 1–8. http://www.precisionfarmingdealer.com/articles/1864-faa-to-require-small-uav-registration#sthash.7Y8lmfDw.dpuf (accessed Dec 22, 2015).

Yintong Aviation Supplies Ltd. Agricultural Crop Protection UAV. 2014, pp 1–3. http://www.China-yintong.com/article/en/productshow.asp?sortid=7&id=57 (accessed July 13, 2014).

Zimmcomm New Media LLC. Drones Tested for Agriculture Use. Farm Equipment. 2011, pp 1–3. http://www.farm-equipment.com/articles/59465-drones-tested-for-agricullture-use.htm (accessed Dec 31, 2015).

CHAPTER 10

DRONES IN AGRICULTURE: SUMMARY AND FUTURE COURSE

CONTENTS

10.1 INTRODUCTION

Unmanned aerial vehicles are termed 'drones' because of the low-intensity noise they produce when flying across a field, similar to a male honeybee at work (Matese and Di Gennaro, 2015). However, the 'agricultural drone' discussed in detail in this volume is neither dull nor monotonous in the world of farming. It is perhaps the most interesting and exciting concept. It helps farmers by minimizing drudgery in open fields. Drones add greater accuracy to farm procedures and enhance crop yield.

These 'agricultural drones' work to provide humans with food grains and fruits, at a better energy and economic efficiency. They cause least disturbance to environmental parameters. In fact, agricultural drones reduce use of harmful chemicals, particularly when farmers utilize them during precision farming procedures. They do not come in contact with the crops or harm them. All the analysis is done remotely, using visual and thermal sensors.

Agricultural drones are a boon to human kind, as they promise to offer better crop production efficiency and higher grain harvests. This is much unlike the role of drones in military engagements and conquests. Historically, we are in a period when drones are migrating from military barracks/ storehouses into agricultural farms. They fly at low altitudes across the fields. This is to help farmers in wide range of activities in the farm.

10.2 MILITARY TO FARM TECHNOLOGY: A FLUENT TRANSITION OF DRONES TO PEACEFUL PURSUITS

Historically, we have been versatile in modifying several techniques, gadgets, and procedures that were originally developed for military reasons. We have ably adapted them to serve civilian day-to-day life and farm operations. Drones are among the most recent gadgets that seem to shift, from predominantly military usage to farm lands. Drones are forecasted to reduce drudgery, labor costs, and at the same time improve farm profits. Right now, it is a phenomenon that has taken roots and it is expected to gain greater momentum in the near future. Probably, drones will become common in agrarian belts in the next 5–10 years. The need of the hour is to make this transition a fluent and easy one, with utmost utility and profitability to drone production companies and farmers. Drones have actually moved fairly quickly from military barracks and garrisons into agrarian expanses. In the near future, drones could become a major factor that mold agricultural operations worldwide. There are several agronomic procedures that currently are difficult or less efficient. Agricultural drones could alleviate this situation to a certain extent.

Not only drones but also the techniques attached with them undergo changes from military stand point to farm management. Even the personnel employed hitherto to operate and deploy military drones have been opting to change to farm drones (Elmquist, 2015; King, 2013). Let us consider an example. A fighter pilot who operated military drones in Middle East and South Asia, upon return, preferred to teach drone flying and its usage in the potato farms of Idaho. In California, USA, several of the retirees from military have opted to start farm drone companies. They operate drones and offer appropriate crop production-related services (Elmquist, 2015). In Oregon, military retirees are becoming farmers and are trying to utilize drones to reduce cost of crop scouting (King, 2013).

One of the sayings that connotes transition of drone usage from military to agriculture more vividly is as follows: 'the future of unmanned aircraft in America may be much about agricultural chemical spraying not military spying'.

Green (2013) believes that drones that are controversial and dreaded as military bombing instruments may soon find great acceptability in American farms. They will become gadgets that auger peace. Further, it is said that drones may still conjure up 'Star War' kind of situations and war machines. But, very soon they could be common farm gadgets, helping farmers in food generation. This change from war machine to agricultural drone is expected to improve the economics of both drone producers' and farmers' perceptibly. In fact, some reports suggest that, if drones reduce cost of farming by 1–5% and improve grain yield by the same 1–5% surpassing the previous statistics, it would be a great achievement.

Drones have only recently gained popularity within agricultural sector. As stated earlier, they were originally military gadgets and were used for surveillance and bomb enemy positions. However, there are drone companies that were predominantly focused to serve military. But, now, they too have shifted their focus to a certain extent. Let us consider another example. It deals with a major military supplies company – Northrop Grumman Corporation, which produces the most common and effective military drone known as 'Global Hawk' (Spence, 2013). A recent report suggests that Northrop Grumman's activity related to military is getting affected due to military budget cuts, by several countries including United States of America. The demand for military drones seems to have been plummeted. A few other models of drones produced by Northrop Grumman Corporation are R-Bat and Bat. They too are *not* gaining attraction from military establishments. Perhaps, rapid and large-scale sale of drones has already saturated the military establishments with such drones. Therefore, Northrop Grumman Corporation has decided to modify the small drones such as R-bat or Bat. They are trying to apply them for use in agricultural settings. Northrop Grumman Corporation has changed its focus partly to produce agricultural drones. This is a very clear case of a military supplies producer shifting focus to agricultural purposes – *a peaceful pursuit indeed* (Spence, 2013). One of the observations states that if Northrop Grumman Corporation or even others engaged in military supplies change focus to farm drones, they need not depend exclusively on government spending and budgets. Agricultural sector is really vast, the private farming absorbs a very large number of drone equipment produced by them.

In several other countries, drone companies that hitherto produced drones for reconnaissance and bombing targets have not lagged behind. Military drone producers have opted and are already producing farm drones in place of bombers (King, 2013). For example, Bosch Precision Agriculture GMBH have been redesigning, modifying, repurposing, and deploying their drone models to monitor and take aerial images of cropland.

Yintong Aviation Supplies Ltd. located in China is known to produce drones for military purposes. However, currently, they are also engaged in producing small, light drones, useful in scouting crops for disease/pests and spraying chemicals. In China, drone production companies are initiating production of specialized agricultural drones. They may slowly shift their focus from military to agriculture, as the demand for drones in farms increases (Yintong Aviation Supplies Ltd., 2014; Ministry of Agriculture, 2013).

The shift in the purpose for which drones are used could be noticed, right at the research level. For example, engineers in Aerospace Department, Mechatronics Research Groups, and U.S. Army Engineers themselves are designing and developing drones for civilians and farmers. They have attempted modifications to hitherto military models of drones to make them farm drones. They are using drones, to monitor natural resources and crop fields (Department of Aerospace, 2014; Garland, 2014). After all, effecting modifications to old military drones and partly changing designs to suit the recent purposes must be easier to accomplish.

One of the reasons for a shift from military to civilian, particularly, agricultural purposes is the great potential for usage of drones in farms. For example, no-till fields experience weed-related problems that could be severe. It necessitates immediate scouting of large fields once seeding has been accomplished (Rowsey, 2014). Agricultural drone has become a great attraction to farmers to trace weeds in fields. However, drone's performance is yet to be tested and advantages realized.

An interesting forecast relevant to farm drone states that 'what started as military technology, may in due course end up as most important agricultural gadget that improves, farmers' efficiency and his profits.' It further suggests that general public and farming community may eventually realize the potential role of drones and use it (Farm Info News, 2015; Garland, 2014). In future, children may see farm drones almost as a natural addition to agricultural landscape, just like the previous generations did about ploughs, tractors, and combine harvesters. Pacello (2015) opines that drones are meant for agriculture in future. They may get progressively weaned out from military zones.

A recent report from United Nations Environment Programs tries to emphasise that drones were first adopted in military by many nations. The word 'drone' is more related to military usage of small unmanned aerial vehicles. However, at present, drones are being used by many environmental monitoring agencies. Drones are utilized to detect changes in the atmospheric composition. The low-cost drones are preferred by environmental managers and meteorological agencies. As these drones measure aspects relevant to soil and atmospheric conservation, such drones have been nicknamed 'Ecodorones or Conservation drones' (UNEP-GEAS, 2013; Koh and Wich, 2012; NASA, 2013).

Countries such as Indonesia have not used drones to any great extent for military purposes ever. However, they intend to use them directly for peaceful civilian and agricultural purposes (Lamb, 2013). Reports emanating from Indonesia's Research and Technology Application Agency states that drones are being developed and mass produced to accomplish civilian surveillance tasks and agricultural crop scouting. In Australia, again, drones that were hitherto part of their military, police department, and civilian surveillance have now entered agricultural farms. Farmers find them useful in monitoring pastures, wheat crop, and plantations (Houston, 2014).

The previous section offers a summary about recent trends. It shows how drones have been progressively modified to suit the agricultural enterprise, instead of being predominantly a military-oriented gadget. Such a transition has been rapid and smooth till date.

10.3   DRONES IN AGRICULTURAL CROP PRODUCTION

Most agricultural drones are relatively smaller versions of the aerial robots. They are light weight, only 3–5 kg per unit, including the payload of cameras. Currently, only copters with facility for pesticide/liquid fertilizer tank get heavier. Yet, there are few models, particularly those popular in plant protection, which are metallic and heavy at 50–80 lbs. Lighter the drone safer it is to use in crop fields. However, the drone material has to be tough if it is to be used repeatedly. At present, most agricultural drones are made of toughened plastic and graphite material. We should also explore the possibility of 'use and throw' very light weight, cardboard-like degradable and recyclable materials. Also, a small drone should be developed (e.g., Nano copter). The cameras too could be of light weight, detachable from one platform unit and then fixed to a new one. This way, cost of drone could be immensely decreased. Farmers may find ease in flying a drone any time. All they have to do is to preserve the cameras. Even the cameras could be made small and light and can be connected to ground station via GPS. If not, drone could possess an electronic chip that records the images obtained by the cameras. A 'use and throw sensor assembly' will be an excellent item. It may make agricultural drone technology very efficient. Lenses have to be of thin material and the chip that stores digital data too has to be minute and affordable (low cost). However, this is a futuristic suggestion. Drone manufacturing companies may have to bestow some attention. Such small drones are apt, if farmers intend only to get imagery of land, natural vegetation, and crops. Nano drones do not serve purposes related to spraying or carrying heavy items.

One of the major roles that agricultural drones may perform in future, particularly in large farms, is surveillance and fetching information about the ground vehicles, farm robots if any, and their activities. They may help farmers to keep a vigil and oversee the activity of ground robots. For example, drones could watch the activity of autonomous tractors fitted with ploughing equipment (e.g. Spirit). Drones can keep a vigil on robots that weed fields and plantations (e.g. Vitirover – a weeding robot in grape yard), field sprayers, fertilizer inoculator (tractors), autonomous combine harvesters, and grain transport systems (e.g. Kinze's Autonomous Combine Harvester) (see Krishna, 2016). How drones perform along with several other semi-autonomous and totally autonomous robotic vehicles, in crop field, is a question of great curiosity. *Perhaps, it will lead us towards making all farm operations totally autonomous.* Drones that interphase and quickly offer digital data to ground robots are sought. No doubt, drones would revolutionise farm production systems and impart greater efficiency, if they are interphased with ground robots (see Krishna, 2016). Farm robotics research may offer some excellent results in near future.

The thrust is to introduce agricultural drones, and then spread information about them and their purported capabilities in farms. We should realize that drones are not capable of all the activities of a farm. They are not highly versatile farm gadgets. They are restricted to perform aerial scouting, collect data on crop growth (NDVI, GNDVI), thermal indices, leaf chlorophyll content, plant-N, and indirectly soil-N (surface soil only), crop water indices, and location characteristics. Drones can help in forecasting final yield, and they also have a role in imaging land and soil characteristics of crop fields prior to sowing. Therefore, farmers adopting drones will still depend on traditional gadgets and farm equipment. Traditional farm equipment is to be adopted to accomplish several other tasks related to crop production. For example, drones despite excellent imaging ability, will not be able to get data on soil depth, subsurface soil nutrient and moisture distribution, fruit yield under the canopy of big trees, and so forth. Even among field crops, if the grains are formed below in the canopy as in maize, cobs may not be counted or imaged accurately. Field crops where panicles and grains are formed at the apex, such as wheat, rice, foxtail millet, pearl millet, finger millet, and so forth, can only be distinguished aerially, using spectral reflectance data. Again, if tillering is not uniform in height and panicle initiation/maturity extends into a long period, then again, drones may have to be flown repeatedly, at intervals, to collect data. We have to apply appropriate correction and compute total yield. Crops that provide grains, pods, or fruits in 2–5 flushes, such as vegetables, too cannot be imaged and passed by drones in one go. Multiple flights at intervals are needed. However, senescing leaves and uniformly borne panicles with filled grains in a wheat or rice field could be distinguished.

10.3.1 DRONES IN SOIL MANAGEMENT AND PRODUCTION AGRONOMY

Now, let us consider the accrued knowledge regarding drones and their current usage in managing land, soils, and their fertility. Foremost, drones are of utility to farmers and farm companies, right from the stage when they begin land clearing, levelling, and contouring the region. Drones offer some excellent 3D images of land that needs to be converted to crop fields (see Tara de Landgrafit, 2014; Draganfly Inc., 2015; SenseFly, 2015; Anez Consulting, 2016). The visual and IR data provides details about soils. Farmers can decipher spatial variations of textural classes, surface soil moisture, organic matter content (soil colour), surface features, eroded or flooded regions, and so forth. This aspect will gain attention whenever farming stretches have to be expanded into new areas. Drone images could form an excellent basis while forming contour bunds, ridges, and furrows, laying irrigation channels and pipes.

In Europe, already, drone images are utilized to decide on ploughing schedules and sequences. Drone images are consulted to know the extent of clods, gravel, silt, and clayey locations that the ploughing machines will encounter. Optimum tilth, ridging, and contouring are decided on the basis of drone images (Crop Site, 2015). It is said that, drones are the best bets, if the field is over 1000 ha and few kilometers in length and breadth. It is costly for human scouts to survey the fields on ground vehicles. Using drone-derived maps, soil mapping and formation of management blocks is accomplished easily.

Direct imagery of soil to obtain details of soil color is a possibility. Soil color may be indicative of parent material and organic matter content. This aspect has to be explored in greater detail. Soil thermal imagery is indicative of soil moisture in the surface soil (0–10 cm depth). This aspect has been tested and found useful (Quattrochi and Luval, 1999; Kaleita et al., 2005; Esfahani et al., 2014a, 2014b; 2015). Drone images that depict soil fertility variations are of great value to farmers and agricultural consultancy agencies. Without doubt, soil fertility variations dictate crop growth and productivity. Currently, crop's traits such as NDVI, GNDVI, leaf area index (LAI), biomass and yield maps of yester years are consulted. This is done to decipher soil fertility variations. Such spectral imagery of crops is utilized along with data about topography, water resources, soil maladies if any, and previous years' productivity figures. It helps to mark and designate 'management blocks'. In future, sensor data on crop vigor, soil characteristics, and combined harvester's yield maps may become mandatory, particularly to form management blocks. This will make drone usage in farming very essential.

Soil nitrogen is among the most important factors that affects crop growth rates and yield formation. Soil-N status is computed indirectly using crop-N status data collected by drone's sensors. Drone imagery that offers data on leaf chlorophyll content is utilized to calculate leaf-N, plant-N, and to map the soil-N variations in the root zone of crops. Leaf-N is directly related to plant-N and grain yield. In fact, there are efforts to assess even grain protein using drone imagery.

Drones may be used most frequently to judge soil fertility, particularly soil-N. Drones could record crop's response to basal and in-season split-N supply. Drones with ability to spray could be utilized both to map the crop vigor and to assess nitrogen requirement. Drones could supply split-N via foliar sprays. In particular, drone-aided aerial sprays of foliar fertilizer-N may gain popularity. At present, drones are in vogue to conduct foliar sprays on rice fields in Japan (RMAX, 2015; Yamaha Inc., 2014). Agricultural drones may also distribute *Bradyrhizobium*-treated tree seeds in the agroforestry zones. The basic idea is to enhance soil-N fertility using symbiosis between leguminous tree and *Bradyrhizobium*. Organic slurry has been sprayed on soil surface using drones. This practice too adds to soil quality and soil organic matter content.

Drones, in future, will offer digital data and maps about soil fertility and moisture. Such data could be utilized effectively in the variable-rate applicators. Hence, precision farming may largely depend on how quickly and efficiently we procure data from the drone's sensors and process and use them on the vehicles that supply fertilizer. Currently, there are several drone companies and drone-based agricultural consultancy agencies. They offer soil fertility and moisture maps for a fee. The digital data that private agencies supply could be utilized for precision farming. *This aspect is perhaps the centrepiece of agricultural drone technology* (SenseFly, 2015, 2016; Anez Consultancy, 2015). One of the suggestions is to initiate training centers and suitable short courses. The aim is to train farmers in drone technology and precision farming techniques (Green, 2015).

Monitoring soil moisture in the soil surface is not so easy if the field is large, over 1000 ha. Farm crew may take several days and toil hard on soils to collect surface (0–15 cm depth) and subsurface samples. However, drones have ability to conduct thermal imagery of soils. They can relay data processed by in-built computers or those in a ground station. It is tedious for human scouts to repeatedly surveil fields for soil moisture content and decide on irrigation options. Drones, on the other hand, routinely provide images of water status of crops using IR sensors.

Soil erosion is an endemic malady affecting crop fields. It is usually detected through periodic scouting by farm workers. Drones, with their swift flight and accurate 3D images, will just replace farm scouts. Manual techniques of soil mapping and depiction of topography with eroded spots may soon become obsolete. It is said that sheet and rill erosion cause loss of soil nutrients immensely and rapidly if not checked. Drones should actually be utilized to alert the farmers about the damage right at the early stages. Perhaps, drones could be utilized to detect loss of surface soil that occurs in the sandy areas due to wind erosion (see Sterk et al., 1996, 1998; Bielders et al., 2002a, 2002b; Krishna, 2008, 2015). Drones could play a major role in locations prone to sand storms and massive erosion, for example, in the Sahelian zone. Detection of loss of seedlings due to storms could be a task accomplished better by drones. As a consequence, replanting seed hills could become easier.

There are other soil maladies such as salinity, acidity, and Al and Mn toxicity. Sensors on drones with ability for imaging crop vigour, leaf chlorophyll, and foliage/canopy can easily lead us to spots that are affected by such maladies. Drone's imagery and ground reality data about soil pH, salinity, or Al and Mn content are usually superimposed with crop's vigour and growth data. This step will reveal great details about the detrimental effect of maladies on crops. The spatial distribution and intensity of maladies could be detected accurately using drones. Satellite data may be less accurate because of problems related to low resolution. There are suggestions by UNEP-GEAS (2013) that small drones could keep a regular watch on crop fields. It is helpful to detect disasters and soil maladies that affect crop belts. For example, floods, drought, soil erosion, gulley formation, and loss of soil fertility are detected easily by drones. Hence, drones are destined to become common in detecting maladies that afflict agricultural soils. Incidentally, drones have also been adopted to apply gypsum granules aerially to soils afflicted with low uncongenial pH range (<5.5). In fact, a report by Dekay (2014) states that tech-savvy farmers are also replacing satellite-guided procedures. They are opting for more accurate close-up shots and data accrued using low-flying drones.

We may have to standardize a few procedures and develop computer software and decision-support systems relevant to routine adoption of drones, particularly during the conduct of agricultural experiments. Drones seem to be easier to handle and collect data about soil fertility experiments. Drones can procure data about large areas of experimental field at one go. Crops exposed to different soil fertility factors, like N inputs, or different N:P:K ratios could be assessed. Drones are cheaper to adopt, and they reduce cost on scouting and collection of data from experimental plots. Drones are less costly compared to skilled research technicians. The digital data that drones collect are easier to store, retrieve, and analyse using different statistical packages. We can also obtain digital maps. Drone imagery allows farmers to compare different soil fertility treatments and their effects simultaneously on a computer screen and select the best option. In future, advantages from drones will surely make them more common in all Agricultural Experimental Stations (field locations).

Agricultural consultancy services that adopt drone technology are proliferating in the North American and European plains, as well as in Far East cropping belts. Since drones are not costly instruments, they may also spread rapidly into other agrarian regions. Drone technology-based soil fertility assessment, suggestions on fertilizer supply and timing may become common. Perhaps, we have to urgently conduct authentic field experiments that compare traditional agronomic methods using farm labour, ground-based techniques and drone technology. *Farmers have to be notified about the great advantages that drones bestow.*

10.3.1.1 DRONES IN PRODUCTION AGRONOMY

Drones may find greater acceptance in farms. It is attributable to their ability to guide agronomic procedures in the crop fields. A few of the agronomic tasks that involve drones are really very difficult for human scouts and skilled technicians to accomplish. Drones scout and offer visual and thermal images of large stretches of fields. Such images are picked from vantage points above the crop canopy in a matter of minutes. This aspect has been utilized shrewdly by several agricultural consultancy agencies. Agricultural drones could become very useful to develop ploughing schedules, seeding plans, deciding about locations of each crop species/genotypes in a field, selecting appropriate planting density, monitoring seeding progress as planters move in fields, and so forth. Drones are also used to monitor ground vehicles that spray plant protection chemicals, apply fertilizers, inter-culture crop rows, and conduct earthing up. Drones may keep a vigil on centre-pivot sprinklers. Drones also monitor combine harvesters, their speed, and collection of grains into transporters and the general work flow. During the crop season, drone-aided thermal imagery helps in collecting data about crop water index. Such data helps in deciding irrigation intervals and quantity. This aspect may attract farmers and agricultural researchers alike to purchase and adopt drones. Drones with their sensors (R, G, and B) have the ability to determine leaf chlorophyll status. Indirectly, it relates to the plant-N status and biomass accumulation rates. Hence, agricultural agencies can advise farmers on basal and in-season split-N supply. We may have to realize that globally, fertilizer-N and water are among the most crucial inputs. They are highly relevant to forage/grain productivity. Since drones can guide farmers about fertilizer-N schedules, they may soon become indispensable gadgets in the hands of farmers. Drones can even distribute liquid formulation of N as foliar sprays. This is to satisfy in-season N needs of crops. Recommendations developed about fertilizer-N are generally accurate. They are not confounded by soil processes. Foliar sprays reach the plant directly without impedance by soil-related physicochemical processes. In fact, fertilizer-N formulation quantities required are exceedingly small if foliar sprays are adopted. Therefore, drones could bring about large savings in fertilizer consumption and costs incurred on them.

Agriculture drones may be adopted frequently during phenotyping of field and plantation crops and also to develop crop surface model. This aspect, if conducted using farm workers and skilled technicians, is simply too costly, tedious, and time consuming. Drones are excellent bets to accomplish collection of data about a series of plant phenotypic traits. Knowledge about crop phenomics is crucial to agronomists. They have to the time various fertilizer and irrigation application events accordingly. Crop breeders need phenomics data and crop surface models to assess the performance (grain yield) of genotypes (Dreiling, 2012; Geipel et al., 2014; see Plate 4.6, 4.7 of Chapter 4).

Drones are already in use, although preliminarily, to assess maize, wheat, barley, soybean, and brassica genotypes for performance in multi-location trials conducted in European and African regions (Case, 2013; Mortimer, 2013; Taylor, 2015). Drones save on the expenditure to be incurred on scouting and collecting detailed data about all the several thousand genotypes. Usually, a large set of genotypes are evaluated for traits such as plant height, leaf number, leaf chlorophyll content, tiller number, panicle number, length, grain number, and final weight. We have to note that unlike data derived from grids, management blocks, or satellite images, the data from drones do not require excessive extrapolations. They offer in situ data, which can be directly calculated and adopted. Extrapolations could add to inaccuracies while drawing inferences. Lumpkin (2012) suggests that drones are non-destructive while assessing the performance of those large collections of genotypes of cereal crops. They are amenable for repeated flights to verify the data. Plant breeders and seed companies may fly them and compare the performance of cultivars side-by-side, on a computer screen. Researchers can follow the growth pattern of genotypes daily and then make decisions. *Thus, drones are destined to be very common instruments during crop breeding efforts in experimental stations.*

Drones and precision farming procedures are said be a good fit (Carlson, 2015; Patas, 2014; Pauly, 2014). Drone technology is effectively utilized to image the topography, soils, and soil fertility using visual imagery. Drones are also used to assess soil water distribution in the surface horizon. Drone-derived data on drought/flood prone regions, disease/pest maps, and grain yield maps from combine harvester are over-layered and then suitable management blocks are formed. Drones with facility to spray and attach with variable-rate nozzles could be used to spray liquid fertilizer formulations, pesticides and herbicides.

Drones may get opted more frequently by farmers adopting precision techniques. The net advantage from drone-aided scouting, mapping, management block formation and variable-rate techniques could be high. There could be a reduction in the use of fertilizers, pesticides and other plant protection chemicals. Drones definitely reduce requirement of farm labour, so obviously, profits are expected to increase. Reports emanating from French plains suggest that adoption of drones and variable-rate fertilizer allocation meant a net profit of 64 € if Brassica was cultivated. It was 100 € if wheat was grown (Economist, 2014). The number of farmers preferring private agencies to analyse soil fertility using drone's imagery is increasing at a rapid rate. In the French plains, it increased from 2000 farmers in 2014 to 5000 in 2015.

10.3.1.2 DRONES IN GRAIN YIELD FORECASTING

Drones are used to assess crop growth. Usually, we utilize vegetation indices, crop vigour index and phenomics data such as plant height, leaf number, LAI, tillers, panicle number and grain number. Together, such data is utilized to develop models and regression equations. It allows us to compute and to arrive at the most probable grain yield. Yield forecasting is essential to farmers and commercial farms. They judge and decide on input levels, yield goals, and profits, accordingly. Yield forecasting, if done using the traditional farm scouts and manual methods, involves high costs. Drones provide the minimum essential data on phenomics and yield factors. However, it is done quickly and at low cost. Such digital data could be effectively utilized in yield forecasting. It is believed that yield forecasting services offered by agricultural consultancy agencies will be sought frequently by farmers. Hence, drones will find regular use in farms throughout the crop season. Comparisons of traditional methods versus drone technology and computer (software)-aided yield forecasting is required. It helps to convince farmers to switch to drone-aided methods.

We may note that drone technology could also be standardized to directly acquire data about tillers, panicles, their size, seed set, seed-fill, panicle, and grain maturity indices. The sensors on drones have to effectively discriminate between senescing leaves and panicles/grains. Ground data about grain moisture percent is required. Using such data, we can arrive at actual grain yield possible at a quick pace. Perhaps, we can even do it in minutes, using appropriate computer software. The spectral signatures of dry senescing leaves, panicles and mature grains have to be known. They have to be deciphered accurately within the images procured by drones. Yield forecasting, again, depends on crop species and its tillering habit, panicle formation, and grain maturity characteristics. Actually, regression equations connecting NDVI, LAI, and grain yield could be developed and utilized for several crop species.

10.3.1.3 AGRICULTURAL DRONES, AGRARIAN REGIONS AND CROP SPECIES

At present, drones are still being tried and tested in farms. There are innumerable start-up drone companies. They are on the verge of flooding a range of drone models into different agrarian zones. Thus, drones sooner may become a frequently adopted 'farm gadget.' Meersman (2015) states that agriculture could be the biggest user of drones in the near future. Major food generating belts in the North American plains and European plains are among earliest to adopt and reap benefits of drone technology. Drones are also getting popular in wheat, rice, and soybean belts of China and Far East. There are reports about drone usage in Asia, Africa, and Australia (see Chapter 4). Drones are amenable for mixed-farming conditions too, particularly, to monitor dairy cattle and other farm animals. Each season, these agrarian regions cultivate millions of hectares of cereals, legumes, oilseeds, and plantations. They have to be regularly monitored, applied with fertilizers, and sprayed with plant protection chemicals. It is a very large area that has to be scouted by human scouts (farm workers). Drones seem to gradually replace them and spread into farm belts. Drones are destined to make scouting and decision-making easy and economically efficient. Drones are versatile and farmers worldwide may adopt them to study, investigate, and obtain relevant data about a wide range of crops (see Chapter 4). In less than a decade from now, each and every farm may own and utilize a drone. Majority of the farmers may own more than one or few models. Perhaps, specialized drone models to suit each intended farm activity will be available. Swarms of drones too could be adopted, but mainly by large-scale grain-producing companies. *No doubt, a drone-aided agricultural revolution is on the rise.*

10.3.2 DRONES TO REVOLUTIONIZE WATER MANAGEMENT AND IRRIGATION

Agricultural drones could become the most useful gadgets that gather data about water resources, soil moisture (surface soil), in-season irrigation needs, impact of droughts, and floods, if any. Drones fitted with sensors that operate at visual band width, infra-red, and near infrared bands have been experimented, tested, and employed to attend irrigation of crop fields (Turner et al., 2011; Tsouvaltsidis et al., 2015; Mac Arthur et al., 2014). In the near future, about 5–10 years, drones could become a routinely used farm equipment conducting surveys to improve irrigation efficiency. Drones are used to monitor irrigation equipment, because they are economically efficient. It is economically useful to monitor irrigation equipment such as center-pivot sprinklers. Farm scouts may find it difficult to move swiftly and identify clogged nozzles or erratic movement of travelling sprinklers. Drones are also used to keep a vigil on water resources such as dams (see Plate 5.1, Chapter 5), lakes, and small ponds. Drones detect fluctuations in water level and storage. Drones are also used to alert farmers about water usage trends from dams.

Drones are useful in surveying crop fields. They procure images of soils, their thermal properties, and topography (3D). This data helps in laying irrigation channels or pipes. During this process, agricultural drones replace human farm scouts and land surveyors. Goli (2014) has shown that drones could obtain thermal imagery of top and surface layers of soil. Therefore, drones could provide data about soil moisture distribution and its variation. In future, drones may be adopted routinely to measure crop water stress index (i.e. canopy temperature minus ambient air temperature) using thermal sensors. Drones offer a great advantage to farmers by detecting crop's water status. Farmers can calculate irrigation requirements. Over all, drones could save a great deal in terms of farm worker and labor requirements during water management (Innova, 2009; Bellvert et al.; 2013).

Drones have already found applications in Agricultural Experimental Stations as a regular work force. They collect data about crops exposed to different quantities of irrigation. In some regions, researchers have adopted them to monitor crop genotypes and their ability to tolerate water deficits, droughts, and floods. They are screening, identifying, and selecting drought-tolerant genotypes using drone technology. In this case, drone's sensors are used to collect data on NDVI, phenomics, including plant height, leaf number, LAI, chlorophyll content, panicle initiation and growth, seed set, seed number, and senescence (see Berni et al., 2009a, 2009b). By adopting drones, crop researchers could screen a large number of genotypes, collect pertinent data, store them in digital form and utilize in computer decision-support systems. Drones are efficient by many folds compared to skilled farm technicians. Hence, drones are destined to become common, during experimental evaluation of crop genotypes for drought stress tolerance. Again, there is an immediate need to experimentally evaluate drone versus farm worker-managed irrigation of crop fields. We have to document various advantages that accrue, due to the usage of drones during irrigation of crop fields.

10.4 DRONES AND CROP PROTECTION

Globally, annual sale and usage of plant protection chemicals is high. It may amount to several billion US dollars. Drones have a potential to alter this to a certain extent. Drones could play a major role in the surveillance and upkeep of crop fields. Drones have been already tried and tested in few cases. Drones help to suppress and destroy weeds by offering spectral imagery. Specific spectral signatures of weeds and crop species are collected aerially. Weeds are mapped accurately and herbicides are applied. Drones are destined to be an important factor in pest and disease control in the future years. They could reduce usage of harmful chemicals (herbicides, insecticides, fungicides, and bactericides). They also reduce cost incurred on farm workers needed to spray the crops. Timeliness, rapidity, and ease with which drones spray plant protection chemicals adds to profits.

Weeds, volunteers from previous season, and rogue vegetation are endemic to all agrarian belts. The diversity of weeds that infest fields of major food crops and plantations is indeed immense. It is not easy to make generalizations regarding weed species, their rapid growth, high proliferation rates, and the extent of damage they cause to main food crop. Weeds often outgrow and suppress food crop species. They compete for soil nutrients, water, and photosynthetic radiation. So far, techniques adopted to cure fields from weeds have ranged widely. They are physical eradication using farm scouts and labor, application of pre-emergent herbicides that suppress weed seed germination and in-season sprays at early stages of weed development. Aerial photography and sprays using manned aircrafts are possible. However, they are not adopted in any measure, since it is costly. Satellite techniques were tested and found good, only to identify and forecast weed spread in very large areas, such as a county/district or an agrarian region. Often, by the time satellite image picks weeds, a good stretch of crop belt would have already experienced devastation by weeds. This is because resolution of satellite imagery is low. Sometimes, even large patches of weed infestation may escape accurate detection.

Most recent method touted for use in crop fields is the drone technology. Drones could be adopted to detect weeds of wide ranging botanical species. Drones can be applied to scout for weed species and their diversity in a field. Drone imagery can show weedy patches. Farmers can then get an estimate about the intensity of weed growth. They can map them accurately using spectral reflectance properties and digital methods. We have still many aspects that need to be studied, sorted, and developed regarding weed identification. We have to first build a data bank of spectral properties of weed species/genotypes. Currently, we know spectral signatures of very few weed species. Drone usage to study weed infestation is still in early stages. We have to collect the spectral signatures of each and every weed species endemic to a particular locality. Computer software to identify weeds accurately, rapidly, and detect them on a field map, are required.

At this juncture, it is highly pertinent to evaluate drones and note their efficiency to control weeds in crop fields. We ought to realize that sensors on drones can help us identify weeds and their species with an accuracy ranging 80–90%. Of course, drones use specific spectral signatures. Drones are rapid, accurate, and economically efficient in collecting digital data and mapping weeds in crop fields. Computer decision-support systems and software that identify and suggest appropriate herbicidal sprays (formulations), their timing, and quantity have yet to be developed. Current reports suggest that drones and computer software at ground stations could easily detect and prescribe herbicides to control weeds that infest wheat fields (Torres-Sanchez et al., 2013; Lopez-Granados, 2011). They could easily trace wild oats, canary grass, and rye grass and provide accurate imagery. Several broad-leaved weeds could be detected accurately in a sunflower field. Later, accurate sprays were made using drone technology (see Chapter 6). Agricultural drones are efficient. They are small aerial robots that can accomplish herbicide sprays swiftly and accurately. They avoid exposure of farm workers to drifts and droplets of harmful chemicals. The savings on herbicides due to drone-based techniques has ranged from 60–90% (Timmerman et al., 2013). However, we may have to conduct a series of location specific and multilocation trials that compare drone techniques with traditional methods. The advantages of drones over traditional methods have to be depicted clearly. Particularly, the reduction in quantity of pesticides, number of farm workers, reduction in accumulation of herbicides in soil and their seepage, and economic gains have to be documented. Such facts will induce adoption of drones during weed control procedures. Drone production companies state that weed control using drone technology may become very common by 2025 (Young et al., 2014; Jenkins and Vasigh, 2013). Application of herbicides prior to seeding or during crop season is perhaps most easily accomplished using drones. They can be adopted irrespective of availability of precision techniques (variable-rate methods). These autonomous flying machines accomplish the task swiftly, accurately, and from a very close range. Drones with their ability for rapid and repeated application of herbicide sprays should be a boon to large farms that cultivate herbicide-tolerant soybean or maize. They can conduct aerial sprays and dispense 8–10 L of herbicides in a matter of few minutes, safely and accurately. Drone technology, it seems, is gaining in popularity with rice farmers in Japan and Far East nations (see Chapters 6 and 7). For the future, spectral signature of all prolific weed species could be collected and stored as digital data in the servers. This helps farm agencies to consult and use spectral signatures of weeds to detect and quantify.

Drones are built to play a major role in the global surveillance of crops for diseases/pests and plant protection initiative. Plant protection specialists from Kansas State University, Manhattan, USA, University of Queensland, Australia, and New Zealand Agricultural Department have proposed a consortium. They aim at use of drones and sensor-based technology on a large scale. Drones could surveil crop diseases world-wide and further the cause of World Plant Protection Initiative (Kansas State University, 2015). Drones have been already tested and used routinely in the rice land of Japan to spray plant protection chemicals. They are making rapid incursions into cereal belts of North American and European plains. Plant protection measures are of utmost necessity in these regions. For example, drones with sensors are able to detect aphid/mite attacks and streak virus spread in the wheat belt of Central plains of United States of America (Rush, 2014; Rush et al., 2008). Drone's sensors with ability to discriminate healthy and disease/pest attacked crop canopies are most useful. A computer-based data bank that stores spectral signatures of disease/pest attacked crops is essential. Right now, we have such data for very few diseases/pests (see Table 7.1 of Chapter 7). It is indeed negligible, hence, there is a need for a global initiative to collect spectral signatures of healthy and disease/pest attacked crops. Perhaps, agricultural agencies and experimental research stations in each county/district could concentrate locally and collect such spectral data. Mathematical models and appropriate regression equations will help farmers to assess disease/pest damage and then forecast grain/fruit yield loss in a crop field. We must be able to fix threshold levels of diseases/pest intensity using drone imagery. Drones, in future, may play a vital role in alerting and warning farmers about disease/pest attack and their spread to and from neighboring regions. Drones are better choices when variable-rate application of pesticides and other plant protection chemicals are adopted. Drones help farmers to focus only on spots afflicted with disease/pests. Adoption of drone technology, therefore, perceptibly reduces use of plant protection chemicals. Use of plant protection chemicals could reduce by a whopping 30–90% compared to traditional methods of blanket sprays (see Chapter 7). Drones could find greater acceptance in plantations. Each tree or a location could be analysed, using spectral data and imagery. The initiative to detect and control citrus greening disease in the orchards of Central Florida is a good example. Drones may beat out other modes of disease/pest scouting. It is mainly because they are rapid, repeatable as many times, and accurate in assessing crop disease/pest. In addition, most importantly, drones cost less than human scouts and are economically highly efficient. Drones offer an overall view of the crop field and disease/pest-attacked locations in shortest time.

Drone usage may introduce economic efficiency in plant protection activities even on a large-scale farm. Drone-based techniques may reduce use of harmful chemicals by several million tonnes. Thus, they delay accumulation of pesticides in soil. Drones could protect farm workers from aerial drifts of harmful chemicals. Therefore, drones have to be deployed to protect crops. We have to note that drones could be used to spray chemicals, irrespective of availability of spectral data. Spectral data showing spatial variations of disease/pest attack is mandatory, only if precision techniques are intended. Drones are efficient, rapid and easy to operate through remote control even when blanket sprays are intended. Clearly, drones have a big role in plant protection irrespective of adoption of precision techniques. Farmers in low-input farming regions too have adopted drones and tested their efficiency in pest/disease control. Drone models that are less expensive are available.

Practically, we have yet to conduct innumerable field trials, experimental evaluations and actual assessment of drones in disease/pest control. The costs and profitability from drone usage has to be compared with the currently established procedures of plant protection. *Thus, there is still to toil with drones, for plant protection specialists, particularly before these aerial robots become common instruments in farms.*

10.5 AGRICULTURAL DRONES AND ECONOMIC ASPECTS

Drones are entering the agricultural world at a time when economic scenarios for farmers change frequently. Fluctuations in productivity due to climate change, disasters and yield depressions are not unusual. Further, for a drone revolution to occur, cost on equipment and safety are deemed prime obstacles (Heatherly, 2015). Drones are garnering investment mainly with a promise of bringing about a transformation of agricultural techniques. Further, they must reduce farm labour needs and improve net crop production. Drones are without doubt versatile. They may turn out to be economically viable with farmers adopting low-input technology or those in high-input intensive farming belts. Farmers earning about 1000 US\$·ac^{-1} may reap an economic advantage of 10% by adopting drone technology. Farmers brake even soon, since payback period is short even for those with 250 ac. Farmers growing cash crops or grain crops make profits if they adopt drones to scout and monitor crops. Savings on scouting itself pays up for costs incurred on drones, say, in 1–2 seasons. It takes 2 US\$·ac^{-1} to scout fields using farm workers. Instead, a drone perhaps accomplishes it in a few minutes of flight at 5–10 cents per acre (Precision Drone, 2013; Precision Farming, 2015a, 2015b; King, 2013; Stutman, 2013; Modern Farmer, 2015). Farmers without doubt need a certain amount of investment initially. This is to purchase the platform, cameras, computer software to transmit images and process them. Drones come in wide range of costs. It is based on tasks they perform, sensors and their sophistication, computer software and purposes for which they are utilized. Drones, especially copters, have facility to carry 8–10 L of pesticide as payload in the tanks. They spread the chemical using digital data as a guide. Such drones may cost 160,000 US\$ (e.g. RMAX by Yamaha Motors Inc.). Most drones available in the market cost in the range of 4000–10,000 US\$. They possess a full complement of accessories to fly them and collect useful data. There are also low-cost drones with ability for just scouting and imaging crop field. They cost 1000–3000 US\$ (see Table 9.1). Homemade drones are also in vogue with small farmers.

Drones will soon find their way into farm lands. They will get used more frequently. As a result, farm workers may eventually lose jobs. Farm workers may have to trace their way out of crop fields. Drone-induced migration of farm workers is a clear possibility. Each drone replaces a sizeable number of farm scouts and workers. Therefore, there is a need to upgrade skills and train as many farm workers as possible in drone technology. Reports suggest that several thousand jobs are being created within the realm of drone technology. For example, a report from North American region states that, in United States of America, over 100,000 jobs could be created due to introduction of drones. It potentially leads to tax revenue of 1.0 billion US\$ by 2025 (Doering, 2015). Several billion US dollars are envisaged to be invested in drone companies that produce the gadgets. The economic turnover due to drone production (instruments) and techniques is forecasted at 112 billion US\$ per year (see Chapter 9). We may note that there are several reports about investment and jobs created due to adoption of drone technology. Forecasts about drone-related economic turnover range widely on the basis of parameters actually considered.

While we standardize drone techniques to become more efficient in farming, we ought to realize that there are rules and regulations for its usage. The regulatory aspects are slightly different based on regions and countries (see Chapter 9).

Discussions in this chapter and previous ones make it clear that drone technology is experiencing a good lift-off into all agrarian belts. A wide range of crop species are amenable for management using drone technology. Drones could easily carve out a niche in farms by scouting and monitoring the activity of semi-autonomous and autonomous ground vehicles (robots). Several aspects of soil and crop management, irrigation, crop protection and harvesting are dealt better by adopting drones. Experimental evaluation of crop genotypes using drones is a clear possibility. Drones seem to have a long way to traverse yet. However, they are destined to make a mark as indispensable farm robots. Drones would definitely revolutionize agricultural crop production systems worldwide. Food generation by major agrarian belts could become less costly and efficient. Total output of grains, forage, fruits and fibre may increase perceptibly. *Drones will no doubt reduce human toil and drudgery in agrarian belts—a boon indeed. Drones have come to stay, fly and perform in agricultural zones, perhaps perpetually. After all, drones are farmers' friends in the agricultural sky.*

KEYWORDS

- **agricultural crop production**
- **soil management**
- **production agronomy**
- **water management and irrigation**
- **grain yield forecasting**
- **drone-aided scouting**
- **agricultural drones**

REFERENCES

Anez Consultancy. Best Source of Information: Precision Agriculture. 2015, pp 1–8. http://anezconsulting.com (accessed June 1, 2016).

Bellvert, Zarco-Tejada, P. J.; Girona, J.; Fereres, E. *Precis. Agric.* DOI 10.1007/s11119-013-9334-5. **2013,** 1–6. (accessed April 10, 2016).

Berni, J. A.; Zarco-Tejada, P. J.; Suarez, L.; Fereres, E. Thermal and Narrowband Multispectral Remote Sensing for Vegetation Monitoring From an Unmanned Aerial Vehicle. *IEEE Xplore.* 2009a, pp 722–738. DOI: 10.1109/tgrs.20082010457. (accessed Jan 20, 2016).

Berni, J. A. J.; Zarco-Tejada, P. J.; Sep.ulcre-Canto, G.; Fereres, E.; Villalobos, F. Mapping Canopy Conductance and CWSI in Olive Orchards Using High Resolution Thermal Remote Sensing. *Remote Sens. Environ.* **2009b,** *113*, 2380–2388.

Bielders, C.; Michels, K.; Rajot, J. On Farm Evaluation of Wind Erosion Control Technologies. 2002a, pp 1–5. ICRISAT.Org CCER.htm (accessed Feb 20, 2016).

Bielders, C.; Rajot, J.; Amadou, M.; Skidmore, E. On Farm Quantification of Wind Erosion Under Traditional Management Practices. 2002b, pp 1–6. ICRISAT.Org CCER.htm (accessed Feb 20, 2016).

Carlson, G.; Precision Agriculture Steering us into Future. Modern Agriculture—A British Columbia's Agricultural Magazine. 2015, pp 1–4. http://modernagriculture.ca/precision-agriculture-steering-us-future/ (accessed April 9, 2016).

Case, P. Rothamsted Unveils Octocopter Crop-monitoring Drone. 2013, pp 1–2. http://www.fwi.co.uk.arable/rothamsted-unveils-octocopter-crop-monitoring-drone.htm (accessed June 25, 2013).

Crop Site, How can Drones Make Farming Profits. 2015, pp 1–3. http://www.thecropsite.com/news/17406/how-can-drones-make-farming-profits/ (accessed June 28, 2015).

Dekay, V. Tech-Savvy Growers Already Replacing Satellite Imagery. 2014, pp 1–6. http://www.producer.com/2014/12/tech-savvy-growers-already-replacing-satellite-imagery (accessed Oct 19, 2015).

Department of Aerospace. *Unmanned Aerial Systems Research Group*; Internal Report. University of Florida: Gainesville: USA, 2014; pp 1–5.

Doering, C. Growing Use of Drones Poised to Transform Agriculture. USA Today. 2015, pp 1–3. http://www.usatoday.com/story/money/business/2014/03/23/drones-agriculture-growth/6665561/ (accessed May 19, 2015).

Draganfly Inc. Innovative UAV Aircraft and Aerial Video Systems: Agriculture and Mapping Package. Draganfly Inc. 2015, pp 1–3. http://draganfly.com/featurette/ag-pack-php?gclid=CLjerpOissYCFQSTjgodczUA3Q.html (accessed June 26, 2015).

Dreiling, L. *New Drone Aircraft to Act as Crop Scout.* Kansas State University: Salina, USA, 2012; pp 1–2. http://www.hpj.com/archives/2012/apr1216/apr16/springPlanting-MACOLDsr.cfm (accessed May 30, 2015).

Economist. The Robot Overhead. The Economist: Technology Quarterly. 2014, pp 1–5. http://www.economist.com/news/technology-quarterly/21635326-after-starting-their (accessed May 24, 2016).

Elmquist, S. Fighter Pilot Turned Farmer to Ply Drones over Crop Land. Stars and Stripes. 2015, pp 1–12. http://www.stripes.com/news/veterans/fighter-pilot-turned-farmer-to-ply-drones-over-cropland. Htm (accessed Dec 27, 2016).

Esfahani, L.; Torres-Rua, A.; Jensen, A.; McKee, M. *Fusion of High Resolution Multi-spectral Imagery for Surface Soil Moisture Estimation Using Learning Machines.* Utah State University: logan, USA, 2014a; p 1. http://www.digitalcommons.usu.edu/do/search/?q=author_iname%3A22Hassan-Esfahani522%20author_fname%22Leila%22& start=0&context=656526 (accessed June 28, 2015).

Esfahani, L.; Torres-Rua, A. M.; Ticlavilca, A.; Jensen, M.; McKee, M. Topsoil Moisture Estimation for Precision Agriculture Using Unmanned Aerial Vehicle Multispectral Imagery. *IEEE International Geoscience and Remote Sensing Symposium-2014.* 2014b, pp 3263–3266. (accessed May 12, 2015).

Esfahani, L.; Torres-Rua, A.; Jensen, A.; McKee, M. Assessment of Surface Soil Moisture Using High Resolution Multi-spectral Imagery and Artificial Neural Network. *Remote Sens* **2015,** *7*, 2627–2646, DOI: 10.3390/rs70302627.

Farm Info News. Drones to have Critical Impact on Precision Agriculture. 2015, pp 14. www.ursulaagriculture.com/.../2015-02-Feedinfo-News-Drones-Critical- (accessed Jan 22, 2016).

Garland, C. Drones May Provide Big Lift to Agriculture when FAA Allows Their Use. Los Angels Times. 2014, pp 1–5. http://www.latimes.com/business/la-fi-drones-agriculture-20140913-story.html#page=1 (accessed June 25, 2015).

Geipel, J.; Link, J.; Claupein, W. Combined Spectral and Spatial Modeling of Corn Yield Based on Aerial Images and Crop Surface Models acquired with an Unmanned Aircraft System. *Remote Sens.* **2014,** *11*, 10335–10355. DOI: 10.3390/rs61110335 (accessed June 23, 2015).

Goli, N. *Researcher Uses Drones to Measure Soil Moisture;* University of Alabama at Huntsville: USA, 2014; p 1. http://www.chargetimes.com/2092/science-and-technology/researcher-uses-drones-to-mesure-soil-moisture.html (accessed January 26, 2016).

Green, R. Unmanned Drones may have Their Greatest Impact on Agriculture. 2013, pp 1–4. http://www.dailybeast.com/articles/2013/03/26/unmanned-drones-may-have-their-greatest-impact-on-agriculture.html#stash.c36uDps.dpuf (accessed May 23, 2015).

Green, R. Drones 'Rapidly Changing' Agriculture. Farming News 10.37.09. 2015, pp 1–4. http://www.farminguk.com/news/Drones-rapidly-changing-agriculture_3729.html (accessed June 29, 2015).

Heatherly, L. Predictions with Agricultural Implications. Mississippi Soybean Producers Board. 2015, pp 1–2. http://msmsoy.org/blog/predictions-with-agricultural-implications/ (accessed May 14, 2016).

Houston, B. Farm Drones Reducing the Cost of Precision Agriculture. Ag Innovators. 2014, pp 1–3. http://www.aginnovators.org.au/news/farm-drones-reduucing-he-cost-of-precision-agricuuture. (accessed Sept 10, 2015).

Innova, A. Estimating Crop Water Needs Using Unmanned Aerial Vehicles. Science Daily. 2009, pp 1–4. https://www.sciencedaily.com/releases/2009/07/090707094702.htm (accessed April 25, 2016).

Jenkins, D.; Vasigh, B. *The Economic Impact of Unmanned Aircraft Systems Integration in the Air Space.* United States Association for Unmanned Vehicle Systems International, Arlington, Virginia, USA, 2013; pp 1–38.

Kaleita, A.; Tian, L.; Hirschi, M. Relationship Between Soil Moisture Content and Soil Surface Reflectance. *Trans. Am. Soc. Agric. Eng.* **2005,** *48*, 1979–1986.

Kansas State University. Project Using Drones to Detect Emerging Pest Insects, Diseases in Crops. Ag Professional. 2015, pp 1–2. http://www.agprofessional.com/news/project-using-drones-detect-emerging-pest-insects-dieases-in-crops (accessed Oct 9, 2015).

King, R. Farmers Experiment with Drones. The CIO Report. 2013, pp 1–5. http://ww.wsh.com/cio/2013/04/18/farmers-experiment-with-drones/ (accessed April 30, 2015).

Koh, L. P.; Wich, S. A. Dawn of Drone Ecology: Low Cost Autonomous Aerial Vehicles for Conservation. *Trop. Conserv. Sci.* **2012,** *5,* 121–132.

Krishna, K. R. *Peanut Agroecosystem: Nutrient Dynamics and Crop Productivity.* Alpha Science International Ltd.: Oxford, England, 2008; pp 75–116.

Krishna, K. R. Savannahs of West Africa: Natural Resources, Environment and Crop Production. In: *Agricultural Prairies: Natural Resources and Crop Productivity.* Apple Academic Press Inc.: Waretown, New Jersey, USA, 2015; pp 246–312.

Krishna, K. R. *Push Button Agriculture: Robotics, Drones, Satellite Aided Soil and Crop Management.* Apple Academic Press Inc.: Waretown, New Jersey, USA, 2016; p 476.

Lamb, K. Indonesia Readies Mass Production of Drones. Voice of America. 2013, pp 1–3. http://www.voanews.com/content/indonesia-readies-mass-production-of-drones (accessed Jan 1, 2016).

Lan, Y.; Thomson, S. J.; Huang, Y.; Hoffmann W. C.; Zhang, H. Current Status and Future Directions of Precision Aerial Application for Site-specific Crop Management in the USA. *Comput. Electron. Agric.* **2010,** *24,* 34–38.

Lopez-Granados, R. Weed Detection for Site-Specific Weed Management Mapping and Real Time Approaches. *Weed Res.* **2011,** *51,* 1–11.

Lumpkin, T. *CGIAR Research Programs on Wheat and Maize: Addressing Global Hunger.* DG's Report; International Centre for Maize and Wheat (CIMMYT): Mexico, 2012; pp 1–8.

Mac Arthur, A.; Robinson, I.; Rossini, M.; Davis, N.; MacDonald, K. A Dual-Field-of-view Spectrometer System for Reflectance and Fluorescence Measurements (Piccolo Doppio) and Correction of Etaloning. *5th International Workshop on Remote Sensing of Vegetation Fluorescence.* Quoted in: Tsouvaltsidis, C., Zaid Salem, N., Bonari, G. Vrekalic, D., Quine, B. 2015 Remote Spectral Imaging Using Low Cost UAV System. *International Archives of the Photogrammetry, Remote Sensing and Spatial Information Sciences.* 2014, XL, pp 25–34.

Matese, A.; Di Gennaro, S. Technology in Precision Viticulture: a State of the Art Review. An Overpass Review. 2015, *7,* pp 69–81. http://dx.doi.org/10.2147/IJWR.S69405 (accessed July 7, 2015).

Meersman, T. Agriculture could be Biggest User of Commercial Drones. Department of Aerospace Engineering and Mechanics. University of Minnesota: St Paul, USA, 2015; pp 1–4. http://www.acm.umn.edu/info/spotlight/Agriculture_UAVs.shtml (accessed Oct 19, 2015).

Ministry of Agriculture. Beijing Applies "Helicopter" in Wheat Pest Control. Ministry of Agriculture of the Peoples of China—A Report. 2013, pp 1–3. http://english.agri.gov.cn/news/dqn/201306/t20130605_19767.htm (accessed Aug 10, 2015).

Modern Farmer. Meet the New Drone that could be a Farmer's Best friend. Modern Farmer. 2015, pp 1–4. http://modernfarmer.com/2014/01/precision-hawk.html (accessed June 29, 2015).

Mortimer, G. 'Skywalker': Aeronautical Technology to Improve Maize Yields in Zimbabwe. International Maize and Wheat Centre, Mexico, DIY Drones. 2013, pp 1–6. http;//www. ubedu/web/ub/en/menu_eines/notices/2013/04/006.html (accessed Feb 10, 2016).

NASA. NASA Flies Dragon Eye Unmanned Aircraft into Volcanic Plume. National Aeronautics and Space Agency, 2013. http://climate.nasa.gov/news/891 (accessed March 19, 2013).

Pacello, R. Drones Help Farmers Minimize Water, Fertilizer Use. 2015, pp 1–8. http://www.delawareonline.com/story/news/local/2015/06/21/drones-help-farmers-minimize-water-fertilizer-use/29094095/ (accessed July 2, 2016).

Patas, M. Drones for Farms a Challenge, but Popular Topic at Precision Ag Meet. Agweek. 2014, pp 1–2. http://www.agweek.com/event/article/article/id/22532 (3, 2014 accessed Dec).

Pauly, K. Applying Conventional Vegetation Vigor Indices to UAS-derived Ortho-mosaics: Issues and Considerations. *Proceedings of 12th International Conference on Precision Agriculture*, Sacramento, California, USA; 2014, p 44.

Precision Drone. Drones for Agricultural Crop Surveillance. 2013, pp 1–4. http://www.precisiondrone.com/agriculture (accessed March 20, 2014).

Precision Farming. Dealer Drone Owners must Register Equipment with FAA-Starting Today. 2015a, pp 1–3. http://www.precisionfarming dealer.com/1881.htm (accessed Dec 22, 2015).

Precision Farming. Dealer FAA to Test Drones that Fly Beyond Pilot's Line of Sight. 2015b, pp 1–5. http://www.precisionfarming dealer.com/articles/1426-faa-to-test-drones-that-fly-beyond-pilot's-line-of-sight (accessed May 11, 2015).

Quattrochi, D.; Luvall, J. Thermal Infrared Remote Sensing for Analysis of Landscape Ecological Processes: Methods and Applications. *Landscape Ecol.* **1999**, *14,* 577–598.

RMAX. RMAX Specifications. Yamaha Motor Company, Japan. 2015; pp 1–4. http://www.max.yamaha-motor.drone.au/specification (accessed Sept 8, 2015).

Rowsey, G. Tennessee Researchers Investigate the Potential for Drones in Agriculture. University of Tennessee Institute of Agriculture: USA, 2014; pp 1–5. http://ag.tennnessee.edu/pages/NR-2014-07-MilanUASs.aspx (accessed June 9, 2015).

Rush, C. Scientist Uses Helicopter Drone to Detect Wheat Disease Progression. Iowa Farmer Today. 2014, pp 1–5. http://www.iowafarmertoday.com/news/crop/scientist-uses-helicopter-drone-to-detect-wheat-disease-progression/article_9d38c4f8-78a5-11e3-bb2963f4.hhtml (accessed Oct 14, 2015).

Rush, C. M.; Workneh, F.; Price, F. Application of Remote Sensing Technologies for Study of Wheat Steak Mosaic Virus. *Proceedings of 9th International Congress of Plant Pathology: Healthy and Safe Food for Everybody.* Torino, Italy Plant Pathology, 2008, 90S, 60–61.

SenseFly. Drones for Agriculture. SenseFly—A Parrot Inc. Chessaeux-Lousanne, Switzerland, 2015, pp 1–5. http://www.sensfly.com/applications/agricuture.html (accessed May 23, 2015).

SenseFly Inc. eBee Sensefly: The Professional Mapping Drone. SenseFly Inc. A Parrot Company, Chesseaux-Lousanne, Switzerland. 2016, pp 1–5. https://www.sensefly.com/drone/ebee.html (accessed Jan 29, 2016).

Spence, K. North Grumman wants to Sell Unmanned Drones to Farmers. The Motley Fool. 2013, pp 1–3. http://www.fool.com/investing/general/2013/12/08/

northrup-grumman-wants-to-sell-unmanned-drones-to-farmers.htm (accessed May 24, 2015).

Sterk, G.; Herman, L.; Bationo, A. Wind Blown Nutrient Transport and Soil Productivity Changes in Southwest Niger. *Land Degrad. Dev.* **1996,** *7,* 325–335.

Sterk, G.; Stroosnijder, L.; Raats, P. A. C. Wind Erosion Process and Control Techniques in Sahelian Zone of Niger. 1998, pp 1–14. http://www.ksu.edu/symposium/proceedings/sterk.pdf (accessed May 25, 2014).

Stutman, J. Agricultural Drone Investing: Unmanned Aircraft Industry Ready to Soar. 2013, pp 1–4. http://www.techinvestingdaily.com/articles/agricultural-drone-investing/229&title/ (accessed May 25, 2015).

Tara de Landgrafit. Flying with Drone Technology in Agriculture. Rural. 2014, pp 1–2. http://www.abc.net.au/news/2014-07-04/wach-drones/5569770 (accessed Jan, 25, 2016).

Taylor, D. L. Salinas-based Drone Company Wows Tech Summit Visitors. 2015, pp 1–3. http://www.thecalifornian.com/story/news/2015/03/26/salinas-based-done-company-wows-tech-summit-visitors.htm (accessed May 29, 2015).

Timmerman, C.; Gerhards, R.; Kuhbauch, W. The Economic Impact of Site-Specific Weed Control. *Precis. Agric.* **2003,** *4,* 349–360

Torres-Sanchez, J.; Lopez-Granados, F.; De Castro, A. L.; Pena-Barragan, J. M. Configuration and Specification of an Unmanned Aerial Vehicle (UAV) for Early Season Site-Specific Weed Management. *PloS One* **2013,** *8*(e58210), 1–3. http://www.ncbi.nlm.nih.gov/pmc/articles/PMC3590160 (accessed June 27, 2015).

Tsouvaltsidis, C.; Zaid Salem, N.; Bonari, G.; Vrekalic, D.; Quine, B. Remote Spectral Imaging Using Low Cost UAV System. *International Archives of the Photogrammetry, Remote Sensing and Spatial Information Sciences* 2015, XL, 25–34.

Turner, D.; Lucieer, A.; Watson, C. Development of an Unmanned Aerial Vehicle (UAV) for Hyper-resolution Vineyard Mapping Based on Visible, Multispectral, and Thermal Imagery. *Proceedings of 34th International Symposium on Remote Sensing of Environment,* 2011; pp 342–347.

UNEP-GEAS. An Eye in the Sky-Eco-drones. United Nations Environment Program-Global Environmental Alert Service. 2013, pp 1–5. http://www.unep.org/gaes (accessed Feb 25, 2016).

Unmanned Vehicle University, Drone UAV Pilot Training: Manage the Unmanned Arial Systems. UXV University. 2015, pp 1–12. http://www.uxvuniversity.com (accessed June 25, 2015)

Yamaha. RMAX-History. 2014, pp 1–4. http://www.rmax.yamaha-motor.com.all/history (accessed Sept 20, 2015).

Yintong Aviation Supplies Ltd. Agricultural Crop protection UAV. 2014, pp 1–3. http://China-yintong.com/article/en/productshow.asp?sortid=7&id=57 (accessed July 13, 2014).

Young, S. L.; George, M.; Wayne, W. Future Directions for Automated Weed Management in Precision Agriculture. West Central Research and Extension Centre, North Platte, paper 79. University of Nebraska: Lincoln, Nebraska, USA 2014, pp 1–19. http://digitalcommons.unl.edu/westcentrexst/79 (accessed Dec 10, 2015).

INDEX

Printed in the United States
by Baker & Taylor Publisher Services